Practical Electronics Handbook

Practical Electronics Handbook

Second Edition

IAN R. SINCLAIR

Heinemann Newnes

Heinemann Newnes
An imprint of Heinemann Professional Publishing Ltd
Halley Court, Jordan Hill, Oxford OX2 8EJ

OXFORD LONDON MELBOURNE AUCKLAND SINGAPORE
IBADAN NAIROBI GABORONE KINGSTON

First published by Newnes Technical Books 1980
Reprinted 1982, 1983 (with revisions)
First published by Heinemann Professional Publishing Ltd 1987
Second edition 1988
Reprinted 1988, 1990

British Library Cataloguing in Publication Data

Sinclair, Ian R. (Ian Robertson), 1932–
Practical electronics handbook. – 2nd ed.
1. Electronic equipment. Circuits
I. Title
621.3815'3

ISBN 0 434 91845 8

Printed and bound in Great Britain by
Redwood Press Limited, Melksham, Wiltshire

Contents

Contents

Preface

Databooks often tend to be simply collections of information, with little or nothing in the way of explanation, and in many cases with so much information that the user has difficulty in selecting what he needs. This book has been designed to include within a reasonable space most of the information which is useful in electronics together with brief explanations which are intended to serve as reminders rather than instruction. The book is not, of course, intended as a form of beginner's guide to the whole of electronics, but the beginner will find here much of interest, as well as a compact reminder of electronic principles and circuits. The constructor of electronic circuits and the service engineer should both find the data in this book of considerable assistance, and the professional design engineer will also find that the items collected here are of frequent use, and would normally only be available in collected form in much larger volumes.

I hope therefore, that this book will become a useful companion to anyone with an interest in electronics, and that the information in the book will be as useful to the reader as it has been to me.

Ian R. Sinclair

Introduction

Mathematical Conventions

Quantities greater than 100 or less than 0.01 are usually expressed in the *standard form:* $A \times 10^n$ where A is a number less than 10, and n is a whole number. A positive value of n means that the number is greater than unity, a negative value of n means that the number is less than unity. To convert a number into standard form, shift the decimal point until a number between 1 and 10 is obtained and count the number of places which have been shifted, which will be the value of n. If the decimal point has had to be shifted to the left n is positive, if the decimal point had to be shifted to the right n is negative. For example:

$$1200 \text{ is } 1.2 \times 10^3, \text{ but } 0.0012 \text{ is } 1.2 \times 10^{-3}$$

To convert numbers back from standard form, shift the decimal point n figures to the right if n is positive and to the left if n is negative. Example:

$$5.6 \times 10^{-4} = 0.000\,56; \quad 6.8 \times 10^5 = 680\,000$$

Numbers in standard form can be entered into a scientific calculator by using the key marked Exp or EE (for details, see the manufacturer's instructions).

Numbers in standard form can be used for insertion in formulae, but component values are more conveniently written using the prefixes shown in *Table 0.1.* Prefixes are chosen so that values can be written without using small fractions or large numbers.

Throughout this book, equations have been given in as many forms as are normally needed, so that the reader should not have to transpose equations. For example, Ohm's law is given in all three forms, $V = RI$, $R = V/I$, and $I = V/R$. The units which must be used with formulae are also shown, and must be adhered to. For example, the equation: $X = 1/(2\pi fC)$ is used to find the reactance of a capacitor in ohms, using C in farads and f in hertz. If the equation is to be used with values given

Table 0.1

Prefix	*Abbreviation*	*Power of Ten*	*Decimal multiplier*
Giga	G	10^9	1 000 000 000
Mega	M	10^6	1 000 000
kilo	k	10^3	1 000
milli	m	10^{-3}	1/1 000
micro	μ	10^{-6}	1/1 000 000
nano	n	10^{-9}	1/1 000 000 000
pico	p	10^{-12}	1/1 000 000 000 000

Note that 1 000 pF = 1 nF; 1 000 nF = 1 μF and so on

Examples: 1 kΩ = 1 000 Ω (sometimes written as 1K0, see *Table 1.3*)
1 nF = 0.001 μF, 1 000 pF or 10^{-9} F
4.5 MHz = 4 500 kHz = 4.5×10^6 Hz

in μF and kHz, then values such as 0.1 μF and 15 kHz are entered as 0.1×10^{-6} and 15×10^3. Alternatively, the equation can be written as $X = 1/(2\pi fC)$ MΩ with f in kHz and C in nF.

In all equations, multiplication may be indicated by use of a dot (A.B) or by close printing ($2\pi fC$). Where brackets ( ) are used in an equation, the quantities within the brackets should be worked out first. For example:

$$2(3+5) \text{ is } 2 \times 8 = 16 \quad \text{and} \quad 2 + (3 \times 5) \text{ is } 2 + 15 = 17$$

Transposing, or changing the subject of an equation, is simple provided the essential rule is remembered: an equation is not altered by carrying out identical operations on both sides.

Example: $Y = \dfrac{5aX + b}{C}$ is an equation

If this has to be transposed so that it reads as a formula to find X, then the procedure is to change both sides so that X is left isolated.
The steps are as follows:

(a) multiply both sides by C result: $CY = 5aX + b$

(b) subtract b from both sides result: $CY - b = 5aX$

(c) divide both sides by $5a$ result: $\dfrac{CY - b}{5a} = X$

Now the equation reads $X = \dfrac{CY - b}{5a}$ which is the desired transposition.

Chapter 1

Passive Components

Resistors

Resistance, measured in ohms (Ω), is defined as the ratio of voltage (in volts) across a length of material to current (in amperes) through the material. When a graph is drawn of voltage across the material plotted

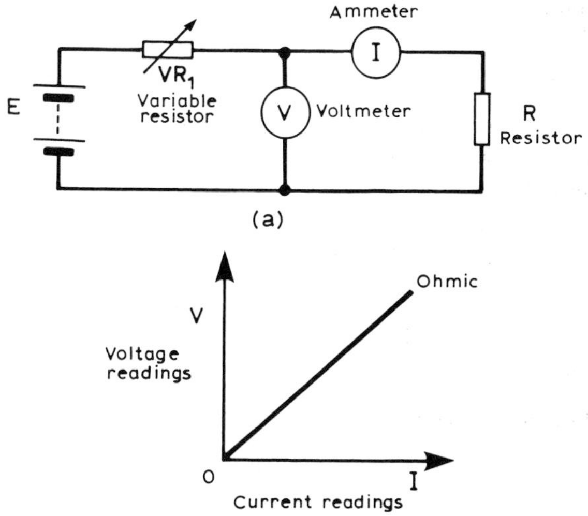

Ammeter
I
E
VR1
Variable resistor
V
Voltmeter
R
Resistor
(a)
Ohmic
V
Voltage readings
O
I
Current readings
(b)

Figure 1.1. (a) A circuit for checking the behaviour of a resistor. (b) The shape of the graph of voltage plotted against current for an ohmic resistor, using the circuit in (a).

against current through the material, the value of resistance is represented by the *slope* of the graph. For a material which is kept at a constant temperature, a straight-line graph indicates that the material is *ohmic*, obeying Ohm's law (*Figure 1.1*). Non-ohmic behaviour is represented on such a graph by curved lines or lines which do not pass through the point, called the origin, which represents zero voltage and

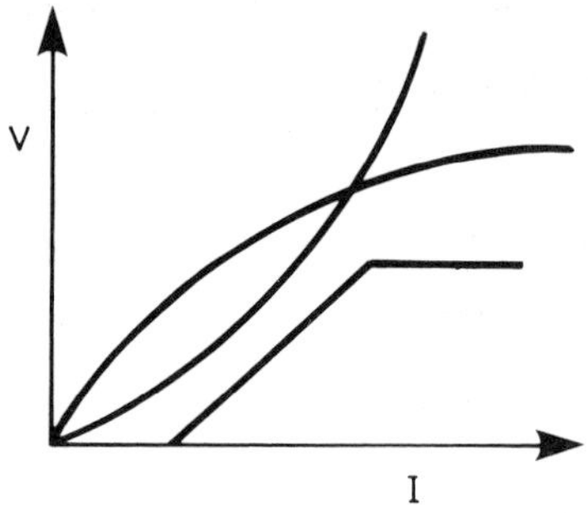
V
I

Figure 1.2. Three types of non-ohmic behaviour indicated by graph curves

zero current. Non-ohmic behaviour can be caused by temperature changes (light bulbs, thermistors), by voltage-generating effects (thermocouples), or by conductivity being affected by voltage (diodes), as in *Figure 1.2*.

Resistance values on components are either colour coded, as noted in *Table 1.2*, or have values printed on using the convention of BS 1852: 1970 (*Table 1.3*).

Resistivity

The resistance of any sample of material is determined by its dimensions and by the value of resistivity of the material. Wire drawn from a single reel will have a resistance value depending on the length cut; for example, a 3 m length will have three times the resistance of a 1 m length. When equal lengths of wire of the same material are compared, the resistance multiplied by the square of the diameter is the same for each. For example, if a given length of a sample wire has a resistance of 12 ohms and its diameter is 0.3 mm, then the length of wire of diameter 0.4 mm, of the same material, will have resistance R such that

$$R \times 0.4^2 = 12 \times 0.3^2 \text{ so that } R = \frac{12 \times 0.3^2}{0.4^2} = \frac{12 \times 0.09}{0.16} = 6.75 \text{ ohms}$$

Resistivity measures the effect which different materials contribute to the resistance of a wire. The resistivity of the material can be found by measuring the resistance R of a sample, multiplying by the area of cross-section, and then dividing by the length of the sample. As a formula, this is

$$\rho = \frac{R.A}{L}$$

where ρ (Greek rho) is resistivity, R is resistance, A is area of cross-section and L is length. When R is in ohms, A in square metres, and L

Table 1.2. RESISTOR COLOUR CODE

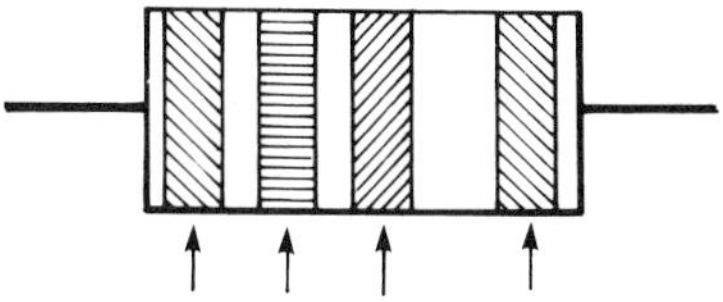
1st
2nd
3rd
Tolerance
(if used)

Band Coding
First Band – first figure of resistance value (first *significant* figure)
Second Band – second figure of resistance value (second *significant* figure)
Third Band – number of zeros following second figure (multiplier)
Tolerance Band – percentage tolerance of value (5% or 10%). No tolerance band is used if the resistor has 20% tolerance.

Code Colours (also used for capacitor values)

Figure	*Colour*	
0	black	
1	brown	
2	red	
3	orange	
4	yellow	
5	green	
6	blue	
7	violet	
8	grey	
9	white	
0.01	silver	used as multipliers (3rd band) only
0.1	gold	used as multipliers (3rd band) only
Tolerance		
10%	silver	
5%	gold	

Table 1.3. RESISTANCE VALUE CODING

Resistance values on components and in component lists are often coded according to BS 1852. In this scheme no decimal points are used, and a value in ohms is indicated by R, kilohms by K (not k), and megohms by M. The letter R, K or M is used in place of the decimal point, with a zero in the leading position if the value is less than 1 ohm.

Examples 1K5 = 1.5k or 1500 ohms; 2M2 = 2.2MΩ; 0R5 = 0.5 ohms

in metres, ρ is in ohm-metres. Since most wires are of circular cross-section, $A = \pi r^2$ (or $1/4\pi d^2$ where d is diameter).

Because the resistivities of materials are well known, this formula is much more useful in the form

$$R = \frac{\rho L}{A}$$

with ρ in ohm-metres, L in metres, A in metres^2, to find the resistance of a given length of wire of known area. This formula can be rewritten as

$$R = 1.27 \times 10^{-3} \frac{\rho L}{d^2}$$

with ρ in nano-ohm metres, as in *Table 1.4*, L in metres, and d (diameter) in millimetres. *Table 1.4* shows values of resistivities, in nano-ohm

Table 1.4. VALUES OF RESISTIVITY AND CONDUCTIVITY

The values of resistivity are in nano-ohm metres. The values of conductivity are in megasiemens per metres.

	Metal	*Resistivity*	*Conductivity*
	Aluminium	27.7	37
	Copper	17	58
	Gold	23	43
	Iron	105	9.5
Pure	Nickel	78	12.8
elements	Platinum	106	9.4
	Silver	16	62.5
	Tin	115	8.7
	Tungsten	55	18.2
	Zinc	62	16
	Carbon-steel (average)	180	5.6
	Brass	60	16.7
	Constantan	450	2.2
	Invar	100	10
	Manganin	430	2.3
Alloys	Nichrome	1105	0.9
	Nickel-silver	272	3.7
	Monel metal	473	2.1
	Kovar	483	2.0
	Phosphor-bronze	93	10.7
	18/8 stainless steel	897.6	1.11

metres, for various metals, both elements and alloys. The calculation of resistance by either formula follows the pattern of the example below.

Example A: Find the resistance of 6.5 m of wire, diameter 0.6 mm, if the resistivity value is 430 nano-ohm metres.

Using $R = \frac{\rho L}{A}$, $\rho = 430 \times 10^{-9}$, $L = 6.5$ m, $A = \pi \frac{(0.6 \times 10^{-3})^2}{4}$

(remembering that 1 mm = 10^{-3} m).

$$R = \frac{430 \times 10^{-9} \times 6.5}{2.82 \times 10^{-7}} = 9.88 \text{ ohms, about 10 ohms.}$$

Using $R = 1.27 \times 10^{-3} \frac{L}{d^2}$

$$R = \frac{1.27 \times 10^{-3} \times 430 \times 6.5}{0.36} = 9.86 \text{ ohms, nearly 10 ohms.}$$

To find the length of wire needed for a given resistance value, the formula is transposed to

$L = \frac{RA}{\rho}$, using R in ohms, A in metres^2, and ρ in ohm-metres to obtain L in metres. An alternative formula is

$L = 785.4 \times \frac{Rd^2}{\rho}$ with R in ohms, d in millimetres, ρ in nano-ohm metres.

To find the diameter of wire needed for a resistance R and length L metres, using ρ in nano-ohm metres, the formula is

$$d = 3.57 \times 10^{-2} \sqrt{\frac{\rho L}{R}} \text{ in millimetres}$$

For some purposes, conductivity is used in place of resistivity. The conductivity, symbol σ (Greek sigma), is defined as $\frac{1}{\text{resistivity}}$, so that $\rho = 1/\sigma$. The unit of conductivity is the siemens per metres, S/m. The resistivity formulae, using basic units, can be rearranged in terms of conductivity as follows

$$R = \frac{L}{\sigma A}, \qquad L = RA\sigma.$$

Conductivity values are also shown in *Table 1.4.*

Resistor Construction

The materials used for resistor construction are generally metal alloys, pure metal or metal oxide films, or carbon. Wirewound resistors use metal alloy wire wound onto ceramic formers. The winding must have a low inductance, so the wire is wound in the fashion shown in *Figure 1.3*, with each half of the winding wound in the opposite direction. Wirewound resistors are used when very low values of resistance are needed, or when very precise values must be specified. Large resistance values, in the region of 20 kΩ to 100 kΩ need such fine-gauge wire that failure can occur due to corrosion in humid conditions, so that high-value wirewound resistors should not be used for marine or tropical applications unless the wire can be protected satisfactorily.

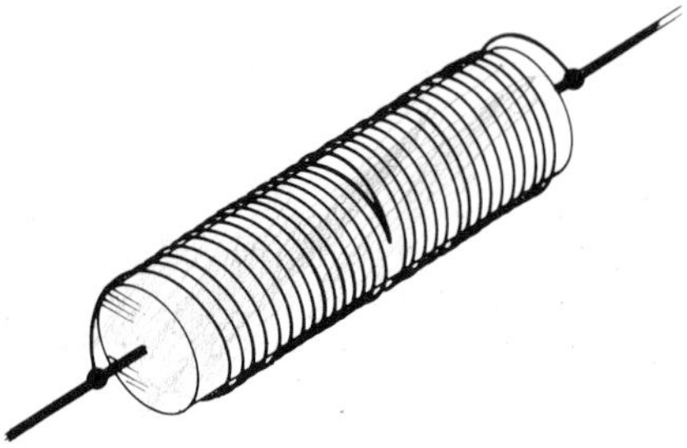

Figure 1.3. Non-inductive winding of a wirewound resistor. The two halves of the total length of wire are wound in opposite directions so that their magnetic fields oppose each other.

The majority of fixed resistors still use carbon composition, a mixture of graphite and clay whose resistivity value depends on the proportions of these materials used in the mixture. Because the resistivity value of such a mixture can be very high, greater resistance values can be obtained without the need for physically large components. Resistance value tolerances (see later) are high, however, because of the greater difficulty in controlling the resistivity of the mixture and the final dimensions of the carbon composition rod after heat treatment.

Metal film, carbon film and metal oxide film resistors are more recent types which are made by evaporating metals (in a vacuum) or tin oxide films (in air) onto ceramic rods. The resistance value is controlled (1) by controlling the thickness of the film; (2) by cutting a spiral pattern into the film after formation. These resistors are considerably cheaper to make than wirewound types, and can be made to closer tolerance values than carbon composition types.

Variable resistors and potentiometers can be made using all the methods that are employed for fixed resistors. The component is termed a potentiometer when connections are made to both ends as well as to the sliding connection; a variable resistor when only one end connection and the sliding connection is used. By convention, both are wired so that the quantity being controlled is *increased* by clockwise rotation of the shaft as viewed by the operator.

Any mass-production process aimed at producing one dimension will inevitably produce a range of values whose maximum tolerance can be specified. The tolerance is the maximum difference between any actual value and the target value, usually expressed as a percentage. For example a 10 kΩ 20% resistor may have a value of

$$10\,000 + \left(\frac{20}{100} \times 10\,000\right) = 12\ \text{k}\Omega \text{ or } 10\,000 - \left(\frac{20}{100} \times 10\,000\right) = 8\ \text{k}\Omega.$$

Tolerance series of preferred values (shown in *Table 1.5*), are ranges of target values chosen so that no component can be rejected. The mathematical basis of these preferred value figures is the sixth root of ten ($\sqrt[6]{10}$) for the E6 20% series (there are 6 steps of value between 1 and 6.8) and the twelfth root of 10 ($\sqrt[12]{10}$) for the E12 10% series. The figures produced by this series are rounded off. For example,

$$\sqrt[6]{10} = 1.46, \quad (\sqrt[6]{10})^2 = 2.15, \quad (\sqrt[6]{10})^3 = 3.16, \quad (\sqrt[6]{10})^4 = 4.64, \quad (\sqrt[6]{10})^5 = 6.8.$$

These figures are rounded to the familiar 1.5, 2.2, 3.3, 4.7, 6.8 used in the 20% series, and similar rounding is used for the 10% and 5% series.

A simpler view of the tolerance series is that, taking the 20% series as an example, 20% up on any value will overlap in value with 20% down on the next value. For example

4.7 + 20% = 5.64, and 6.8 − 20% = 5.44, allowing an overlap.

Table 1.5. PREFERRED VALUES TOLERANCE SERIES

E6 series (20%)	*E12 series* (10%)	*E24 series* (5%)
1.0	1.0	1.0
		1.1
	1.2	1.2
		1.3
1.5	1.5	1.5
		1.6
	1.8	1.8
		2.0
2.2	2.2	2.2
		2.4
	2.7	2.7
		3.0
3.3	3.3	3.3
		3.6
	3.9	3.9
		4.3
4.7	4.7	4.7
		5.1
	5.6	5.6
		6.2
6.8	6.8	6.8
		7.5
	8.2	8.2
		9.1

The numbers then repeat, but each multiplied by ten, up to 91Ω, then multiplied by 100 up to 910Ω and so on.

After manufacture resistors are graded with the 1%, 5% and 10% tolerance values removed. The remaining resistors are sold as 20% tolerance. Because of this, it is pointless to sort through a bag of 20% 6K8 resistors hoping to find one which will be of exactly 6K8 – such values will have been removed in the first sorting process by the manufacturer. Electronic circuits are designed to make use of wide tolerance components as far as possible. When close tolerance components are specified, it will be for a good reason, and 20% tolerance components cannot be substituted for 10% or 5% types.

Resistor characteristics

Important characteristics of resistor types include resistance range, usable temperature range, stability, noise level and temperature coefficient. Wirewound resistors are available in values ranging from fractions of an ohm (usually 0.22 Ω) up to about 10 kΩ; carbon compositions from about 2.2 Ω to 10 MΩ, with film resistors available in ranges which are typically 1 Ω to 1 MΩ. Typical usable temperature ranges are –40° to +105°C for composition, –55°C to +150°C for metal oxide. Wirewound resistors can be obtained which will operate at higher temperatures (up to 300°C) depending on construction and value of resistance. The stability of value means the maximum change of value which can occur during shelf life, on soldering, or in adverse conditions, particularly operation at high temperature in damp conditions. Composition resistors have poorest stability, with typical shelf life change of 5%, soldering change of 2% and 'damp heat' change of 15% in addition to normal tolerance. Metal oxide resistors can have shelf life changes of 0.1%, soldering change of 0.15%, 'damp heat' changes of 1%, typically. The noise level of a resistor is specified in terms of microvolts (μV) of noise signal generated for each volt of d.c. across the resistor, and range from 0.1 μV/V for metal oxide to a minimum of 2 μV/V for composition (increasing with resistance value for composition resistors). The formula generally used for noise level of carbon composition resistors is $2 + \log_{10} \frac{R}{1\,000}$ μV/V. For example, a 680 kΩ resistor would have a noise level of $2 + \log_{10} \frac{680\,000}{1\,000} = 2 + \log_{10} 680$ μV/V = 4.8 μV/V.

The temperature coefficient of resistance measures the change of resistance value as temperature changes. The basic formula is

$$R_\theta = R_o\,(1 + a\,\theta)$$

where R_θ = resistance at temperature θ°C, R_o = resistance at 0°C, a = temperature coefficient, θ = temperature in °C.

The value of temperature coefficient is usually quoted in parts per million per °C (ppm/°C), and this has to be converted to a fraction (by dividing by one million) to use in the above formula.

Example: What is the value of a 6.8 kΩ resistor at 95°C, if the temperature coefficient is +1 200 ppm/°C?

Solution: Converting +1 200 ppm/°C to standard form,

$$= \frac{+1\,200}{1\,000\,000} = 1.2 \times 10^{-3} \qquad (0.0012)$$

Using the formula: $R_o = 6.8\ (1 + 1.2 \times 10^{-3} \times 95)$

(The multiplication of $1.2 \times 10^{-3} \times 95$ must be carried out before adding 1).

$= 6.8\ (1 + 0.114)$

$= 7.57\ \text{k}\Omega$

Note that if the resistance at some other temperature ϕ is given (as distinct from the resistance at 0°C) then the formula becomes

$$R_o = R\phi \left(\frac{1 + \alpha\theta}{1 + \alpha\phi}\right).$$ The 1s cannot be cancelled.

For example: if a resistor has a value of 10 Ω at 20°C, its resistance at 80°C can be found. If the value of a is 1.5×10^{-3}, then

$$R_o = 10 \times \frac{1 + 80 \times 1.5 \times 10^{-3}}{1 + 20 \times 1.5 \times 10^{-3}} = 10 \times \frac{1 + .12}{1 + .03} = 10 \times \frac{1.12}{1.03} = 10.87\Omega$$

Temperature coefficients may be positive, meaning that the resistance will increase as the temperature rises, or negative meaning that the resistance will decrease as the temperature rises. Carbon composition resistors have temperature coefficients which vary from +1200 ppm/°C, metal oxide types have the lowest temperature coefficient values of ±250 ppm/°C.

Dissipation and temperature rise

The dissipation rating, measured in watts (W), for a resistor indicates how much power can be converted to heat without damage to the resistor. The rating is closely linked to the physical size of the resistor, so that ¼ W resistors are much smaller than 1 W resistors of the same resistance value. These ratings assume 'normal' surrounding (ambient) temperatures, and for high temperature use, derating must be applied according to the manufacturer's specification. For example, a ½ W component may have to be used in place of a ¼ W component when the ambient temperature is 70°C.

Figure 1.4 shows the graph of temperature rise plotted against dissipated power for average ½ W and 1 W composition resistors. Note that these figures are of temperature rise *above* the ambient level. If such a temperature rise takes the resistor temperature above the maximum temperature permitted in its type, a higher wattage resistor must be used.

The dissipation in watts is given by $W = VI$ with V the voltage across a conductor in volts and I the current through the conductor in amps.

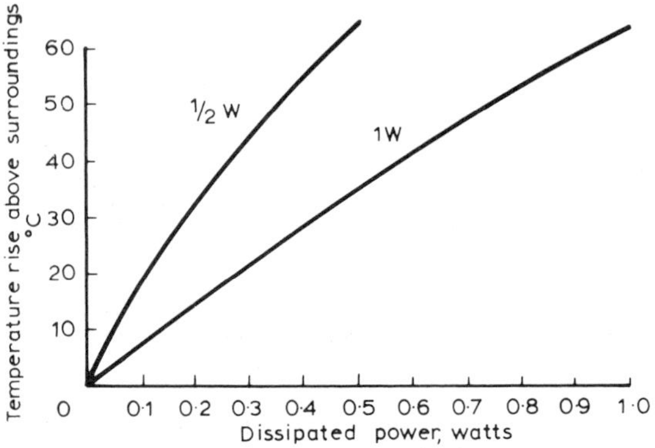

Temperature rise above surroundings °C
60
50
40
30
20
10
0
½ W
1W
0·1
0·2
0·3
0·4
0·5
0·6
0·7
0·8
0·9
1·0
Dissipated power, watts

Figure 1.4. Temperature rise and power dissipation for typical carbon resistors. The temperature scale is in °C above surrounding (ambient) temperature. For example, in a room at 20°C, a ½W resistor dissipating 0.1 W will be at a temperature of 40°C

When current is measured in mA, then VI gives power dissipation in milliwatts. This expression for dissipated power can be combined with Ohm's law when the resistance of R of the conductor is constant, giving

$$W = V^2/R \text{ or } I^2R$$

The result will be in watts for V in volts and R in ohms or I in amps and R in ohms. When R is in kΩ, V^2/R gives W in milliwatts; when I is in mA and R in kΩ, W is also in milliwatts.

Note that *power* is energy transformed (from one form to another) per second. The unit of energy is the joule; the number of joules of energy dissipated is found by multiplying the power in watts by the time in seconds for which the power has been dissipated.

Variables and laws

The *law* of a variable resistor or potentiometer must be specified in addition to the quantities specified for any fixed resistor. The potentiometer law describes the way in which resistance between the slider and one contact varies as the slider is rotated; the law is illustrated by plotting a graph of resistance against shaft rotation angle (*Figure 1.5*). A linear law potentiometer (*Figure 1.5a*) produces a straight line graph, hence the name linear. Logarithmic (log) law potentiometers are

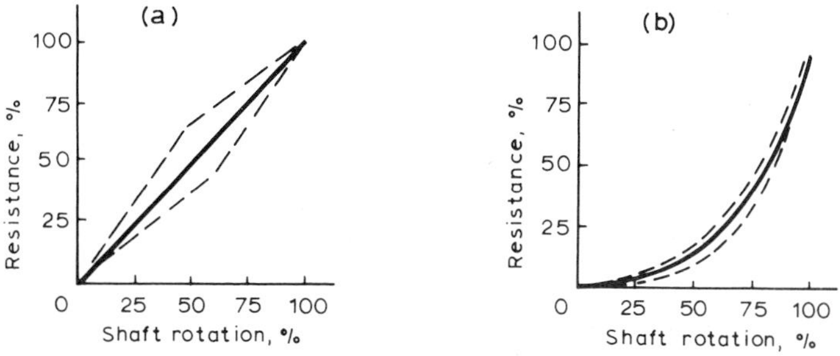
(a)
100
75
50
25
Resistance, %
0
25 50 75 100
Shaft rotation, %
(b)
100
75
50
25
Resistance, %
0
25 50 75 100
Shaft rotation, %

Figure 1.5. Potentiometer laws: (a) linear, (b) logarithmic. In the USA the word 'taper' is used in place of 'law', and 'audio' in place of 'log'. Broken lines show tolerance limits

extensively used as volume controls and have the graph shape shown in *Figure 5.1b.* Less common laws are anti-log and B-law; specialised potentiometers with sine or cosine laws are available.

Resistors in circuit

Resistors in circuit obey Ohm's law and Kirchhoff's laws. Ohm's law is written in its three forms as

$$V = RI; \qquad R = VI; \qquad I = V/R$$

where V is voltage, R is resistance, I is current.

The units of these quantities are as shown in *Table 1.6*. These equations can be applied even to materials *which do not obey Ohm's law*, if the value of R is known. Materials which do not obey Ohm's law do not have a *constant* value of resistance, but the relationships given above hold good. The usefulness of the equations is greatest when the resistance values of resistors are constant.

Table 1.6. OHM'S LAW AND UNITS

Forms of the law: $V = RI$; $R = V/I$; $I = V/R$

Units of V	*Units of R*	*Units of I*
Volts, V	Ohms, Ω	Amps, A
Volts, V	Kilohms, kΩ	Milliamps, mA
Volts, V	Megohms, MΩ	Microamps, μA
Kilovolts, kV	Kilohms, kΩ	Amps, A
Kilovolts, kV	Megohms, MΩ	Milliamps, mA
Millivolts, mV	Ohms, Ω	Milliamps, mA
Millivolts, mV	Kilohms, kΩ	Microamps, μA

Kirchhoff's laws relate to the conservation of voltage and current. In a circuit, the voltage across each series component can be added to find the total voltage. Similarly the total current entering a junction must equal the total current leaving the junction. These laws are illustrated in *Figure 1.6.*

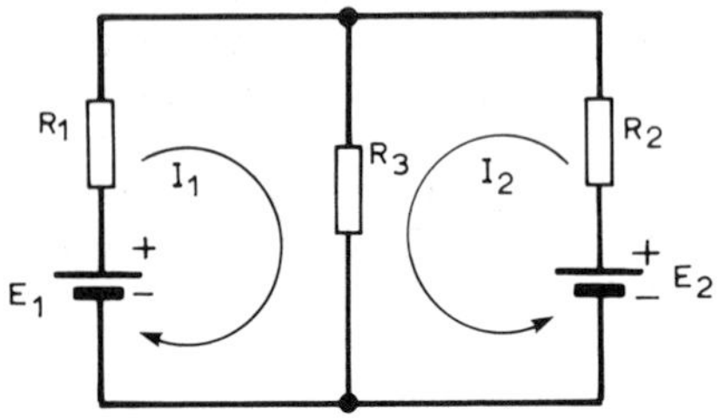
R1
I1
R3
I2
R2
+
E1
–
+
E2
–

Current law: Current in $R_3 = I_1 + I_2$

Voltage law: $E_1 = R_1 I_1 + R_3(I_1 + I_2)$

$E_2 = R_2 I_2 + R_3(I_1 + I_2)$

Figure 1.6. Kirchhoff's laws. The current law states that the total current leaving a circuit junction equals the total current into the junction – no current is 'lost'. The voltage law states that the driving voltage (or e.m.f.) in a circuit equals the sum of voltage drops (IR) around the circuit

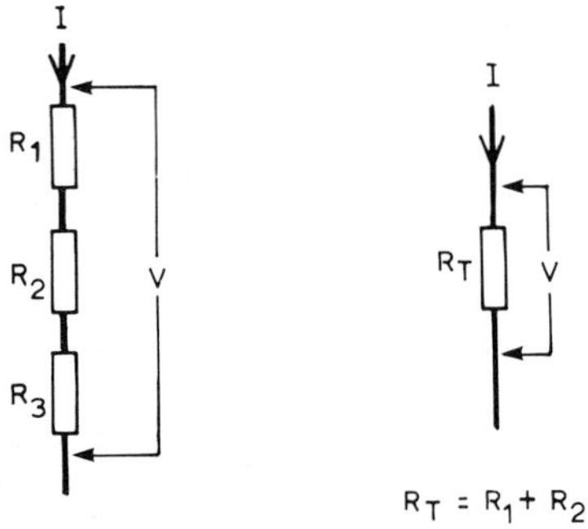
I
R1
R2
V
R3
I
RT
V

$R_T = R_1 + R_2 + R_3$

(a)

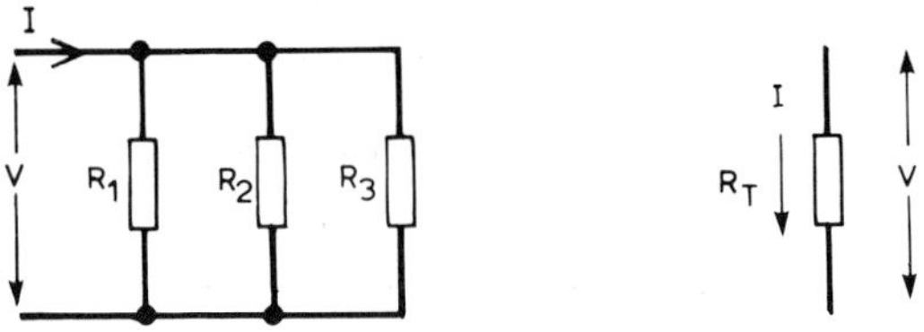
I
V
R1
R2
R3
I
RT
V

$$\frac{1}{R_T} = \frac{1}{R_1} + \frac{1}{R_2} + \frac{1}{R_3}$$

For two resistors in parallel $R_T = \dfrac{R_1 R_2}{R_1 + R_2}$

(b)

Figure 1.7. Resistors in series (a) and in parallel (b)

Figure 1.7 shows the rules for finding the total resistance of resistors in series or in parallel. When a combination of series and parallel connections is used, the total resistance of each series or parallel group must be found first before finding the grand total.

The superposition theorem is of great value in finding the voltages and currents in a circuit with two or more sources of voltage. *Figure 1.8* shows an example of the theorem in use. One supply is selected and the circuit redrawn to show the other supply or supplies short circuited. The voltage and current caused by the first supply can then be calculated, using Ohm's law and the rules for combining parallel resistors. Each supply is treated in this way in turn, and finally the currents and voltages caused by each supply are added.

SUPERPOSITION PRINCIPLE

In any linear network, the voltage at any point is the sum of the voltages caused by each generator in the circuit. To find the voltage caused by a generator, replace all other generators in the circuit by their internal resistances, and use Ohm's law.
A linear network means an arrangement of resistors and generators, with the resistors obeying Ohm's law and the generators having a constant voltage output and constant internal resistances.

Example: In the network shown, find the voltage across the 2.2kΩ resistor.

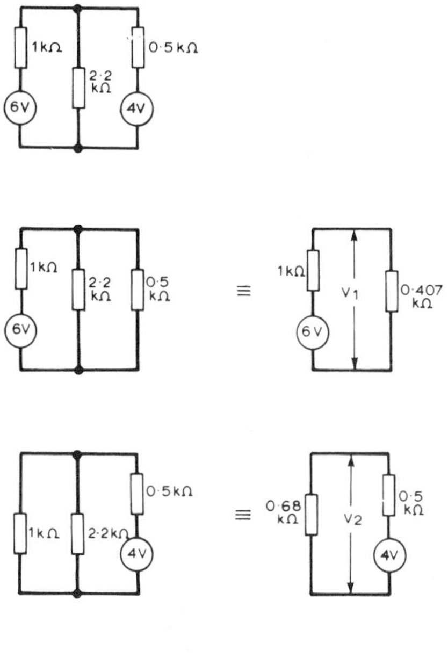
1kΩ
0·5kΩ
2·2 kΩ
6V
4V
1kΩ
2·2 kΩ
0·5 kΩ
6V
≡
1kΩ
V1
0·407 kΩ
6V
0·5kΩ
1kΩ
2·2kΩ
4V
≡
0·68 kΩ
V2
0·5 kΩ
4V

In this network, there are two generators and three resistors. The generators might be batteries, oscillators, or other signal sources.

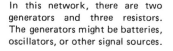

To find the voltage caused by the 6V generator, replace the 4V generator by its internal resistance of 0.5kΩ. Using Ohm's law, and the potential divider equation, V_1 = 1.736 V

To find the voltage caused by the 4V generator, the 6V generator is replaced by its 1kΩ internal resistance. In this case, V_2 = 2.315 V

Now the total voltage in the original circuit across the 2.2kΩ resistor is simply the sum of these, 4.051 V.

Figure 1.8. Using the superposition theorem. This is a simple method of finding the voltage across a resistor in a circuit where more than one source is present.

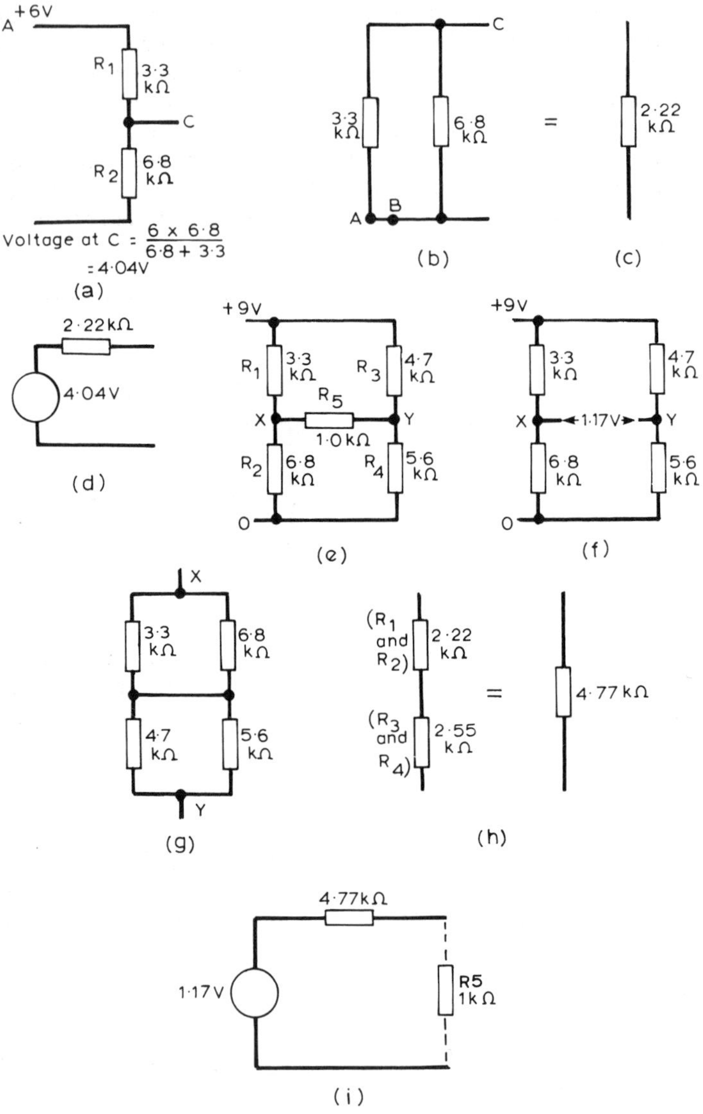
+6V
A
R1
3·3 kΩ
C
R2
6·8 kΩ
Voltage at C = 6 x 6·8 / 6·8 + 3·3
= 4·04V
(a)
C
3·3 kΩ
6·8 kΩ
A
B
(b)
=
2·22 kΩ
(c)
2·22kΩ
4·04V
(d)
+9V
R1
3·3 kΩ
R3
4·7 kΩ
R5
X
Y
1·0 kΩ
R2
6·8 kΩ
R4
5·6 kΩ
0
(e)
+9V
3·3 kΩ
4·7 kΩ
X
1·17V
Y
6·8 kΩ
5·6 kΩ
0
(f)
X
3·3 kΩ
6·8 kΩ
4·7 kΩ
5·6 kΩ
Y
(g)
(R1 and R2)
2·22 kΩ
=
4·77 kΩ
(R3 and R4)
2·55 kΩ
(h)
4·77kΩ
1·17V
R5 1kΩ
(i)

Figure 1.9. Using Thévenin's theorem. The potential divider (a) has an output voltage, with no load, of 4.04 V. It is equivalent to a 4.04 V source whose internal resistance is found by imagining the voltage supply short-circuited (b and c), so that the equivalent is as shown in (d). This makes it easy to find the output voltage when a current is being drawn. Similarly the bridge circuit (e) will have an open-circuit voltage, with R5 removed, of 1.17 V across X and Y (f), and the internal resistance between these points is found by imagining the supply short circuited (g). The combination of resistors in (g) is resolved (h) to give the single equivalent (i)

Thévenin's theorem is, after Ohm's law, one of the most useful electrical circuit laws. The theorem states that any linear network (such as resistors and batteries) can be replaced by an equivalent circuit consisting of a voltage source with a resistance in series. The size of the equivalent voltage is found by taking the open circuit voltage between two points in the network, and the series resistance by calculating the resistance between the same two points in the network, assuming that the voltage source is short circuited. Examples of the use of Thévenin's theorem are shown in *Figure 1.9.*

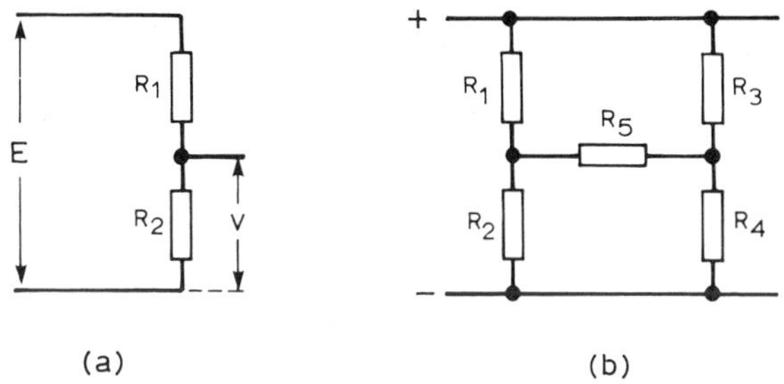
E
R1
R2
V
(a)
+
−
R1
R2
R3
R4
R5
(b)

Figure 1.10. Potential divider (a) and bridge (b) circuits

Figure 1.10 shows two important networks, the potential divider and the bridge. When no current is taken from the potential divider, its output voltage is

$$V = \frac{R_2 E}{R_1 + R_2}$$

as shown, but when current is being drawn (as when a transistor is biased by this circuit) the equivalent circuit (using Thévenin's theorem) is more useful. The bridge circuit is said to be balanced when there is no voltage across R_5 (which is often a galvanometer or microammeter). In this condition,

$$\frac{R_1}{R_2} = \frac{R_3}{R_4}$$

If the bridge is *not* balanced, the equivalent circuit derived from Thévenin's theorem is, once again, more useful.

Thermistors

Thermistors are resistors made from materials which have large values of temperature coefficients. Both p.t.c. and n.t.c. types are produced for applications ranging from measurement to transient current suppression. Miniature thermistors in glass tubes are used for temperature measurement, using a bridge circuit (*Figure 1.11*), for timing, or for stabilising the amplitude of sine wave oscillators (see Chapter 3). Such thermistors are self heating if the current through them is allowed to exceed the limits laid down by the manufacturers, so that the current flowing in a bridge measuring circuit must be carefully limited. Larger thermistor types, with lower values of cold resistance (at 20°C) are

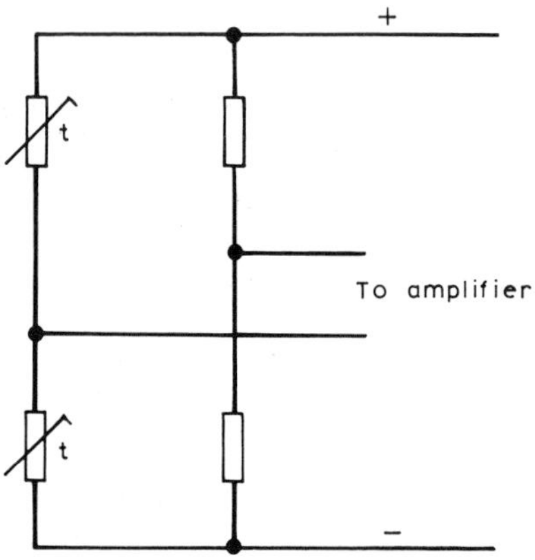
+
t
t
To amplifier
−

Figure 1.11. Thermistor bridge for temperature measurement

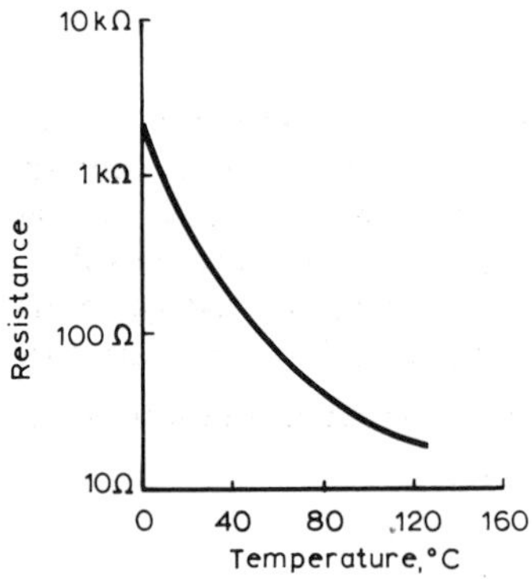
10 kΩ
1 kΩ
100 Ω
10Ω
Resistance
0
40
80
120
160
Temperature, °C

Figure 1.12. Graph of resistance plotted against temperature for typical thermistor

used for current regulation, such as circuits for degaussing colour TV tubes or controlling the surge current into light bulbs or valve heaters. The form of graphs of resistance plotted against temperature is that of *Figure 1.12* and the formula for finding the resistance at any temperature is shown in *Figure 1.13* with examples.

$$R_{\theta 1} = R_{\theta 2} \cdot e^{B\left(\frac{1}{\theta_1} - \frac{1}{\theta_2}\right)}$$

$R_{\theta 1}$ – resistance at temperature θ_1 (kelvins)

$R_{\theta 2}$ – resistance at temperature θ_2 (kelvins)

B – thermistor constant

Note: kelvin temperature = °C + 273

Calculator procedure:

Enter value of known temperature θ_1 then [1/x], [–], enter value of θ_2, [1/x], [=], [×], enter value of B [=] [e^x] [×] enter value of $R_{\theta 2}$ [=] read answer

Example: A thermistor has a resistance of 47 kΩ at 20°C. what is its resistance at 100°C if its B value is 3900?

Equation is $R = 47 \times e^{3900\left(\frac{1}{373} - \frac{1}{293}\right)}$

$= 2 \cdot 7\,\text{k}\Omega$

Figure 1.13. Finding thermistor resistance at any temperature, knowing the thermistor constant, B, and the resistance at a given temperature

Capacitors

Two conductors which are not connected and are separated by an insulator constitute a capacitor. When a cell is connected to such an arrangement, current flows *momentarily*, transferring charge (in the form of electrons) from one conducting plate (the + plate) to the other. When a quantity of charge Q has been transferred, the voltage across the plates equals the voltage V across the battery. For a given arrangement of conductors, the ratio $\frac{Q}{V}$ is a constant, and is called capacitance. The relationship can be written in three forms

$$Q = CV \qquad C = \frac{Q}{V} \qquad V = \frac{Q}{C}$$

The parallel-plate capacitor is the simplest practical arrangement, and its capacitance value is relatively easy to calculate. For a pair of parallel plates of equal area A, separation d, the capacitance is given by

$$C = \frac{\epsilon_r \epsilon_0 A}{d}$$

The quantity ϵ_o is called the permittivity of free space and has the constant value of 8.84×10^{-12} farads per metre.

Air has approximately this value of permittivity also, but other insulating materials have values of permittivity which are higher by the factor ϵ_r, which is different from each material. Values of this quantity, the relative permittivity (formerly called dielectric constant) are shown in *Table 1.7*.

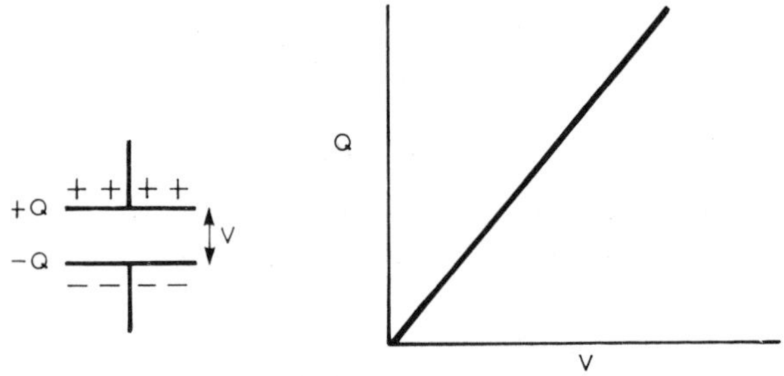
+ + + +
+Q
−Q
– – – –
V
Q
V

Figure 1.14. Basic principles of the capacitor. The relationship Q/V, shown in the graph, is defined as the capacitance, C

The formula can be recast, using units of cm^2 for area and mm for spacing, as

$$C = \frac{0.88 \times \epsilon_r \times A}{d} \text{ pF}$$

For small plate sizes, these units are more practical, but some allowance must be made for stray capacitance between any conductor and the metal surrounding it. Even a completely isolated conductor will have some capacitance.

Table 1.7. VALUE OF ϵ_o = 8.84 pF PER METRE

Material	*Relative permittivity value*
Aluminium oxide	8.8
Araldite resin	3.7
Bakelite	4.6
Barium titanate	600–1200 (varies with voltage)
Magnesium silicate	5.6
Nylon	3.1
Polystyrene	2.5
Polythene	2.3
PTFE	2.1
Porcelain	5.0
Quartz	3.8
Soda glass	6.5
Titanium dioxide	100

Example 1: Find the capacitance of two parallel plates 2 cm × 1.5 m, spaced by a 0.2 mm layer of material of relative permittivity 15.

Solution: Using $C = \dfrac{\epsilon_r \epsilon_o A}{d}$ $\epsilon_r = 15$; $\epsilon_o = 8.84 \times 10^{-12}$ F/m; $A = 0.02 \times 1.5\ \text{m}^2$; $d = 0.2 \times 10^{-3}$ m

$$C = \frac{15 \times 8.84 \times 10^{-12} \times 0.02 \times 1.5}{0.2 \times 10^{-3}} = 1.989 \times 10^{-8}\ \text{F}$$

about 2×10^{-8} F or 0.02 μF

Example 2: Find the capacitance of two parallel plates 2 cm × 1 cm spaced 0.1 mm apart by a material with relative permittivity 8.

Solution: Using $C = \dfrac{0.88 \times \epsilon_r \times A}{d}$ with $A = 2 \times 1 = 2\ \text{cm}^2$, $d = 0.1$ mm

$$C = \frac{0.88 \times 8 \times 2}{0.1}\ \text{pF} = 140.8 \text{ or } 141\ \text{pF}$$

Construction

Small value capacitors can be made using thin plates of insulating material metallised on each side to form the conductors. Thin plates can be stacked and interconnected (*Figure 1.15c*), to form larger

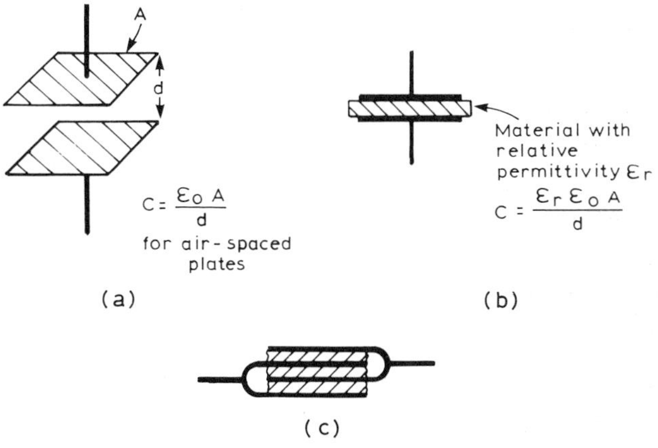
A
d
C = εo A / d
for air-spaced plates
(a)
Material with relative permittivity εr
C = εr εo A / d
(b)
(c)

Figure 1.15. The parallel-plate capacitor. The amount of capacitance is determined by the plate area, plate separation, and the relative permittivity of the material between the plates

capacitance values up to 1000 pF or more. Silver mica types are used when high stability of value is important as in oscillators; and ceramic for less important uses (such as decoupling). Ceramic tubular capacitors make use of small tubes silvered inside and outside.

Rolled capacitors use as dielectric strips of paper, polyethylene, polyester, polycarbonate or other flexible insulators which are metallised (by vacuum coating) on each side and then rolled up (*Figure 1.16*),

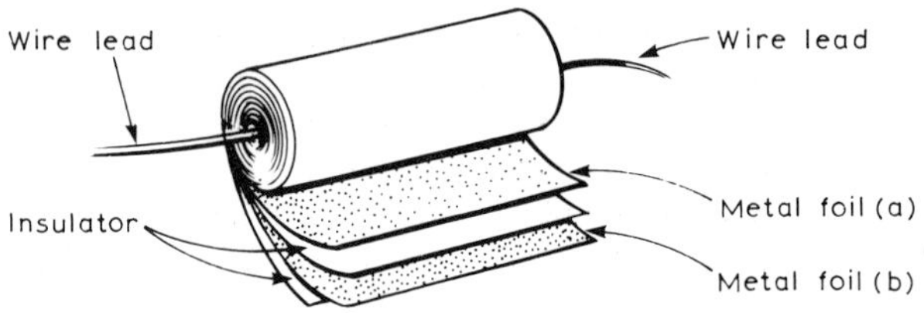
Wire lead
Wire lead
Insulator
Metal foil (a)
Metal foil (b)

Figure 1.16. The rolled construction used for capacitors which make use of sheet dielectrics such as paper, polyester, polystyrene or polycarbonate

with another insulating strip to prevent the metallising on one side shorting against the metallising on the other side. Using this construction, quite large capacitance values can be achieved in small volume; up to a few μF is common.

Electrolytic capacitors are used when very large capacitance values are needed. One 'plate' is of aluminium in contact with an aluminium perborate solution in the form of a jelly or paste. The insulator is a film of aluminium oxide which forms on the positive 'plate' when a voltage, called the forming voltage, is applied during manufacture. Because the layer of oxide can be very thin (of only a few molecules thick) and the surface area of the aluminium can be made very large by roughening the surfaces, very large capacitance values can be achieved. The disadvantage of aluminium electrolytics include leakage current which is high compared to that of other capacitor types, polarisation (the + and – markings must be observed) and comparatively low voltage operation (less important when only transistor circuits are used). Incorrect polarisation can break down the oxide layer and if large currents can flow (as is the case in a power supply reservoir capacitor) the capacitor can explode, showering its surroundings with corrosive jelly.

Tantalum electrolytes can be used unpolarised (but not necessarily reverse polarised) and have much lower leakage currents that aluminium types, making them more suitable for some applications.

Capacitor characteristics

The same series of preferred values (20%) as is used for resistor values is also applied to values of capacitance (other than electrolytics) though older components will generally be marked with values such as 0.02 μF which are for all practical purposes equivalent to the preferred value 0.022 μF. Some manufacturers mark the values in pF only, using the rather confusing k to indicate thousands of pF (*Table 1.8*). Colour coded values are always in pF.

Electrolytic capacitors are always subject to very large tolerance values, of the order of −50% +100%, so that the actual capacitance value may range from half of the printed value to double that value. The insulation resistance between the plates is often so low that capacitance meters are unable to make accurate measurements. Capacitor values marked in circuit diagrams can use the BS 1852 method (6n8, 2μ2, etc) but are often marked in μF and pF. Quite commonly, fractions refer to μF and whole numbers to pF unless marked otherwise so that values of 0.02, 27, 1000, 0.05 mean 0.02 μF, 27 pF, 1000 pF and 0.05 μF respectively.

For all capacitors, the working voltage rating (abbreviated VW) must be carefully observed. Above this voltage, sparking between the

Table 1.8. CAPACITOR COLOUR CODING

Most capacitors are marked with values in μF or pF. The letter k is sometimes used in place of nF, i.e. 10 k = 10 nF = 0.01 μF. Colour coding is sometimes used:

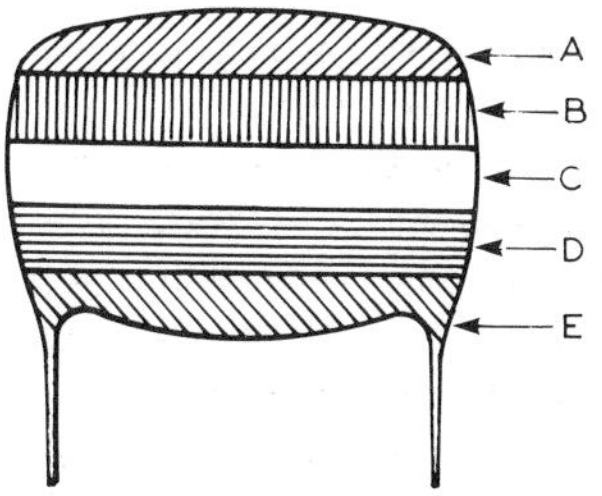
A
B
C
D
E

Bands A, B, C are used for coding values in pF in the same way as for resistors — remember that 1000 pF = 1 nF = 0.001 μF
Band D — Black = 20% White = 10%
Band E — Red — 250 V d.c. working
Yellow — 400 V d.c. working

Colour code for small block capacitors (mainly polyester)

Tantalum electolytic capacitors are also sometimes colour coded, but with values in μF rather than pF

Band	*1*	*2*	*3*	*4*
Black	—	0	×1	10 V
Brown	1	1	×10	
Red	2	2	×100	
Orange	3	3	—	
Yellow	4	4	—	6.3 V
Green	5	5	—	16 V
Blue	6	6	—	20 V
Violet	7	7	—	
Grey	8	8	×0.01	25 V
White	9	9	×0.1	3V
Pink				35 V

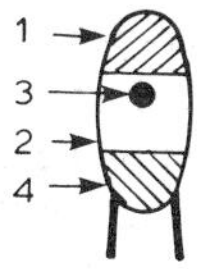
1
3
2
4

conductors can break down the insulation, causing leakage current and eventual destruction of the capacitor. The maximum voltage that can be used is much lower at high temperatures than at low temperatures.

Values as low as 3 V may be found in high capacitance electrolytics; and values as high as 20 kV for ceramic capacitors intended for transmitters. The common voltage ranges used for capacitors in transistor circuits are shown in *Table 1.9.*

Table 1.9. CAPACITORS – COMMON WORKING VOLTAGES

10 V	16 V	20 V	25 V	35 V	40 V
63 V	100 V	160 V	250 V	400 V	1000 V

Changes of temperature and of applied d.c. voltage affect the value of capacitors because of changes in the dielectric. Both p.t.c. and n.t.c. types can be obtained, and the two are often mixed to ensure minimal capacitance change in, for example, oscillator circuits. Paper and polyester capacitors have, typically, positive temperature coefficients of around 200 ppm/°C, but silver micas have much lower (positive) temperature coefficients. Aluminium electrolytics have large positive temperature coefficients, with a considerable increase in leakage current as temperature increases. In addition, electrolytics cannot be operated below about –20°C, as the electrolyte paste freezes. The normal working range for other types is –40°C to +125°C, though derating may be needed at the higher temperature. Voltage ratings are generally for 70°C working temperature.

A few types of capacitors, notably 'High-K' ceramics, change value as the applied voltage is changed. Such capacitors are unsuitable for use in tuning circuits and should be used only for non-critical decoupling and coupling applications.

Variable capacitors can make use of variation of overlapping area or of spacing between plates. Air dielectric is used for the larger types (360 pF or 500 pF), but miniature variables make use of mica or plastic sheets between the plates. Compression trimmers are manufactured mainly in the smaller values, up to 50 pF. In use, the moving plates are always earthed, if possible, to avoid changes of capacitance (due to stray capacitance) when the control shaft is touched.

Energy and charge storage

The amount of charge stored by a capacitor is given by

$$Q = CV$$

When C is in μF and V in volts, Q is in microcoulombs (μC).

Example: How much charge is stored by a 0.1 μF capacitor charged to 50 V?

Solution: Using $Q = CV$ with C in μF, V in volts

$Q = 0.1 \times 50 = 5\ \mu$C.

When charged capacitors are connected to each other (but isolated from a power supply), the total charge is constant, equal to the sum of all charges on the capacitors. If the voltages are not equal, energy will be lost (as electromagnetic radiation) when the capacitors are connected. The amount of energy, in joules, stored by a charged capacitor is most conveniently given by $W = \frac{1}{2}CV^2$.

Other equivalent expressions are $\frac{1}{2}\frac{Q^2}{C}$ or $\frac{1}{2}QV$.

Example: How much energy is stored by a 5 μF capacitor charged to 150 V?

Solution: Using $\frac{1}{2}CV^2$, $C = 5 \times 10^{-6}$, $V = 150$

$W = \frac{1}{2} \times 5 \times 10^{-6} \times (150)^2$

$= 0.056$ J

This is used in connection with the use of capacitors to fire flash bulbs or in capacitor discharge car ignition systems.

In circuits, the laws concerning the series and parallel connections of capacitors are the *inverse* of those for resistors:

For capacitors in parallel: $C_{\text{total}} = C_1 + C_2 + C_3 + \ldots$

For capacitors in series: $\dfrac{1}{C_{\text{total}}} = \dfrac{1}{C_1}\;\;\dfrac{1}{C_2}\;\;\dfrac{1}{C_3}$

Time Constants

The charging and discharging of a capacitor is never instant. When a sudden step of voltage is applied to one plate of a capacitor, the other plate will step by the same amount. If a resistor is present, connecting the second plate to another voltage level, the capacitor will then charge to this other voltage level. The time needed for this change is about four time constants, as shown by *Figure 1.17*. The quantity, time constant T, is measured by $R \times C$ where R is the resistance of the charge/discharge resistor and C is the capacitance. For C in farads and R in ohms, T is in seconds. For the more practical units of μF and with resistance in kΩ, T is in milliseconds (ms) or for C in nF and resistance in kΩ, T is in microseconds (μs).

Example: In the circuit of *Figure 1.18*, how long does the voltage at the output take to die away?

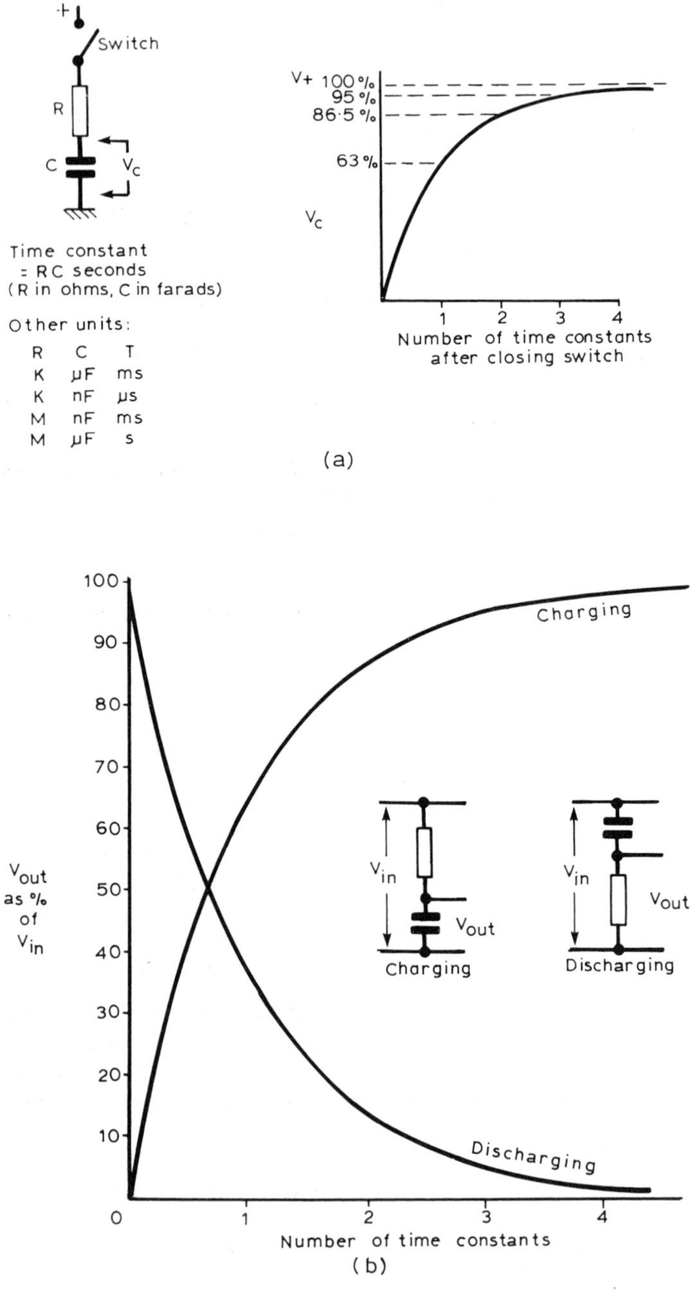

+
Switch
R
C
V_c
Time constant
= RC seconds
(R in ohms, C in farads)
Other units:
R C T
K μF ms
K nF μs
M nF ms
M μF s
V+ 100%
95%
86·5%
63%
V_c
1 2 3 4
Number of time constants
after closing switch
(a)
100
90
80
70
60
50
40
30
20
10
0
V_{out} as % of V_{in}
Charging
Discharging
V_{in}
V_{out}
Charging
V_{in}
V_{out}
Discharging
1 2 3 4
Number of time constants
(b)

Figure 1.17. Capacitor charging and discharging. (a) Principles of charging, (b) Universal charge/discharge curves

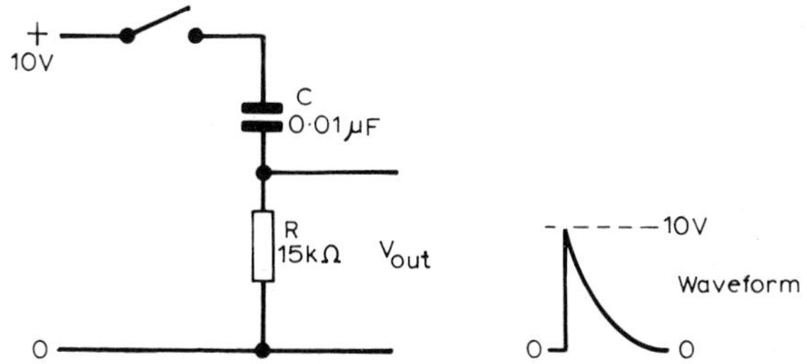
+
10V
C
0·01μF
R
15kΩ
Vout
10V
Waveform
0
0
0

Figure 1.18. Time constant: differentiating circuit

Solution: With C = 0.01 μF = 10 nF and R = 15 kΩ, T = 150 μs.

Four time constants will be 4 × 150 μs = 600 μs, so that we can take it that the output voltage has reached zero after 600 μs.

Example: In the circuit of *Figure 1.19*, how long does the capacitor take to charge to 10 V?

Solution: With C = 0.22 μF = 220 nF and R = 6.8 kΩ, T = 6.8 × 220 = 1496 μs

Four time constants will be 4 × 1496 = 5984 μs or 5.98 ms, approx. 6 ms to charge.

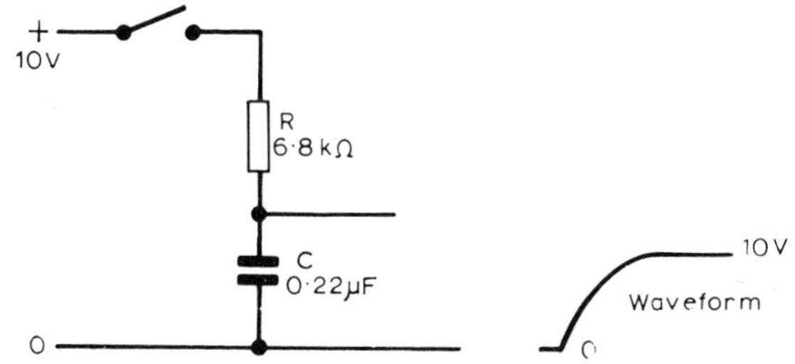
+
10V
R
6·8 kΩ
C
0·22μF
10V
Waveform
0
0

Figure 1.19. Time constant: integrating circuit

Note that a time of four time constants is taken for charging or discharging, because in practice charging or discharging is virtually complete by this time. The shape of the graph, however, indicates that charge is still being moved even several hundred time constants later.

Reactance

The reactance of a capacitor for a sine wave signal is given by

$$X_C = \frac{1}{2\,\pi\, fC} \qquad (2\,\pi = 6.28)$$

where C is capacitance in farads, f is frequency in hertz.

Reactance is measured in ohms, and is the ratio

$$\frac{\tilde{V}}{\tilde{I}}$$

where $\tilde{V}$ is a.c. voltage across the capacitor and $\tilde{I}$ is the a.c. current in the circuit containing the capacitor.

Unlike resistance, reactance is not a constant but varies inversely with frequency (*Figure 1.20*). In addition, the current sine wave is ¼ cycle (90°) ahead of the voltage sine wave across the capacitor plates. For phase and amplitude graphs of *CR* circuits, see later this chapter.

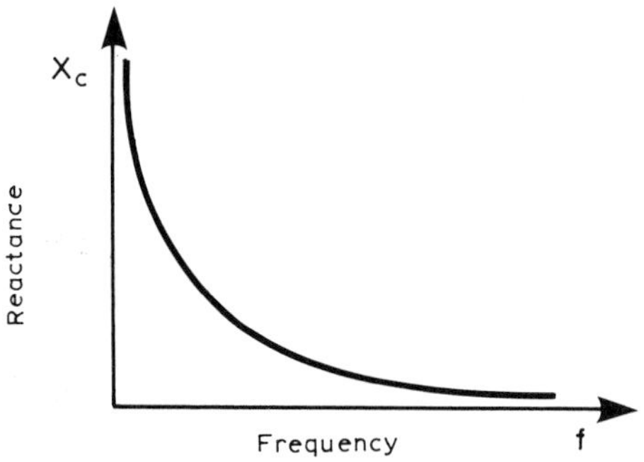
Xc
Reactance
Frequency
f

(a)

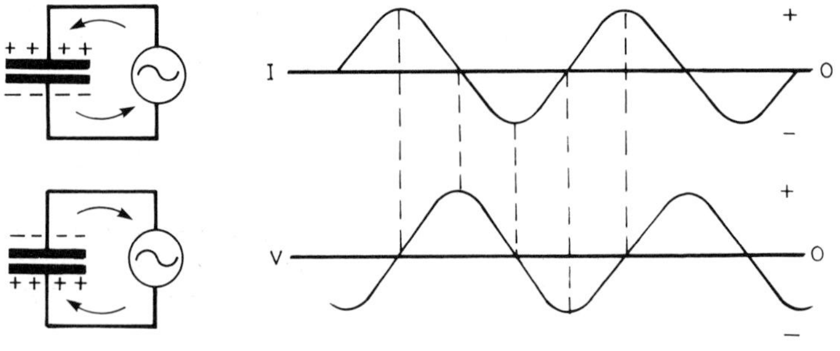
+ + + +
– – – –
I
V
+
O
–
+
O
–

(b)

Figure 1.20. Capacitive reactance to a.c. signals. (a) Graph showing how capacitive reactance varies with frequency of signal. (b) Phase shift. As the capacitor charges and discharges, current flows alternately in each direction. The maximum current flow occurs when the capacitor is completely uncharged (zero voltage), and the maximum voltage occurs when the capacitor is completely charged (zero current). The graph of current is therefore ¼ cycle (90°) ahead of the graph of voltage

Inductors

An inductor is a component whose action depends on the magnetic field which exists around any conductor when a current flows through that conductor. When the strength of such a magnetic field (or magnetic flux) changes, a voltage is induced between the ends of the conductor. This voltage is termed an induced e.m.f., using the old term e.m.f. (electromotive force) to mean a voltage which has not been produced by current flowing through a resistor.

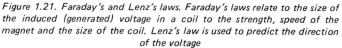

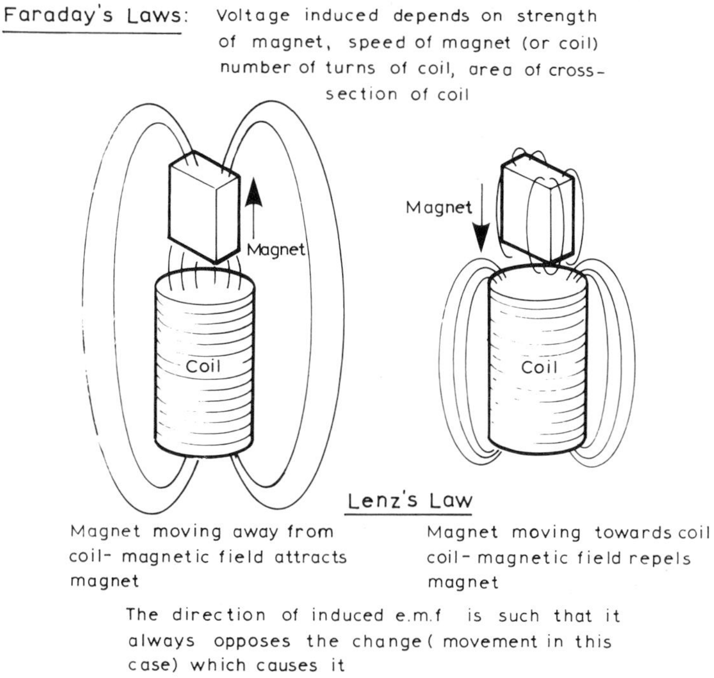
Faraday's Laws:
Voltage induced depends on strength of magnet, speed of magnet (or coil) number of turns of coil, area of cross-section of coil
Magnet
Coil
Magnet
Coil
Lenz's Law
Magnet moving away from coil- magnetic field attracts magnet
Magnet moving towards coil coil- magnetic field repels magnet
The direction of induced e.m.f is such that it always opposes the change (movement in this case) which causes it

Figure 1.21. Faraday's and Lenz's laws. Faraday's laws relate to the size of the induced (generated) voltage in a coil to the strength, speed of the magnet and the size of the coil. Lenz's law is used to predict the direction of the voltage

If we confine out attention to static devices (coils and transformers rather than electric motors), then the change of magnetic field or flux can only be due to a change of current through one conductor. The induced e.m.f. is in such a direction that it opposes this change of current, and the faster the rate of change of current, the greater the opposing e.m.f. Because of its direction, the induced e.m.f. is called a *back e.m.f.* The laws governing these effects are Faraday's law and Lenz's law, summarised in *Figure 1.21.*

The size of the back e.m.f. can be calculated from the rate of change of current through the conductor and the details of construction of the conductor, straight wire or coil, number of turns of coil, use of a core etc. These constructional factors are lumped together as one quantity called inductance, symbol L. By definition,

$$E = L\frac{dI}{dt}$$

where E is the back e.m.f., L is inductance and $\frac{dI}{dt}$ is rate of change of current. The symbol 'd' in this context means 'change of' the quantity written after 'd'. If E is measured in volts and $\frac{dI}{dt}$ in amperes per second, then L is in henries (H).

Example: What back e.m.f. is developed when a current of 3A is reduced to zero in $\frac{1}{50}$ s (20ms) through a 0.5 H coil?

Solution: The amount of back e.m.f. is found from

$$E = L\frac{dI}{dt}$$

$$= 0.5 \times \frac{3}{20 \times 10^{-3}}$$

$$= 75 \text{ V}.$$

Note that this 75 V back e.m.f. will exist only for as long as the current is changing (20 ms), and may be *much* greater than the voltage drop across the coil when a steady current is flowing.

The rate of change of current is seldom uniform, so the back e.m.f. is usually a pulse waveform, whose maximum value must be found by measurement.

The existence of inductance in a circuit causes a reduction in the rate at which current can increase or decrease in the circuit. For a coil with inductance L and resistance R the time constant T is L/R seconds (L in henries, R in ohms). *Figure 1.22* shows how the current at a time t after switch-on varies in an inductive circuit — once again we take the time of four time constants to represent the end of the process.

The large e.m.f. which is generated when current is suddenly switched off in an inductive circuit can have destructive effects, causing sparking at contacts or breakdown of transistor junctions. *Figure 1.23* shows the commonly used methods of protecting switch contacts and transistor junctions from these switching transients.

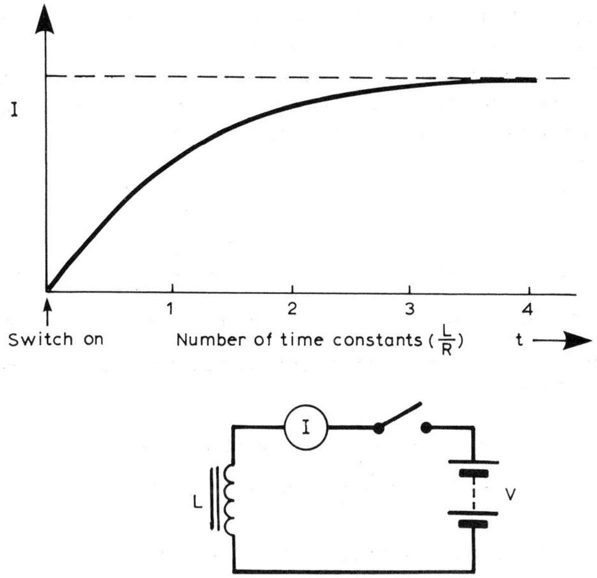
I
1
2
3
4
Switch on
Number of time constants (L/R)
t
I
L
V

Figure 1.22. The growth of current in an inductive circuit

The changing magnetic field around one coil of wire will also affect windings nearby. The two windings are then said to have mutual inductance, symbol M. By definition

$$M = \frac{\text{back e.m.f. induced in second winding}}{\text{rate of change of current in first winding}}$$

using the same units as before, so that the unit of M is the henry.

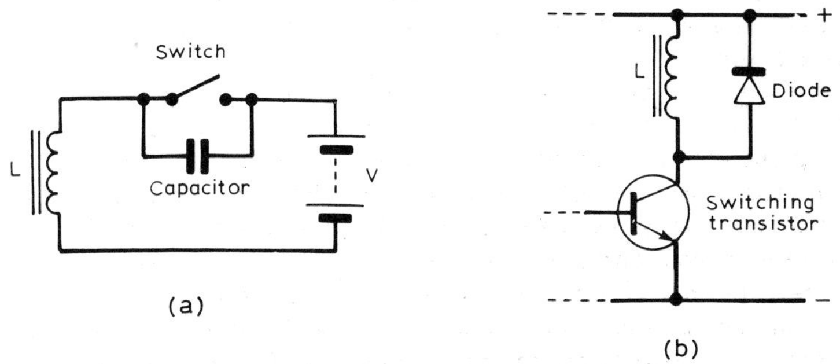
Switch
L
Capacitor
V
(a)
+
L
Diode
Switching transistor
−
(b)

Figure 1.23. Protection against voltage surges in inductive circuits. (a) Using a capacitor across switch contacts, (b) using a diode across the inductor

Inductance calculations

Of all electronics calculations, those of inductance are the least precise. When an air-cored coil is used, the changing magnetic field does not affect all turns equally. Using a magnetic core makes the shape of the magnetic field more predictable, but makes its size less predictable. In addition, the permeability of the core changes considerably if d.c. flows in windings. Any equations for inductance are therefore very approximate, and should be used only as a starting point in the construction of an inductor. *Table 1.10* shows a formula for the number of

Table 1.10. APPROXIMATE INDUCTANCE FOR AIR-CORED SINGLE-LAYER COIL (A SOLENOID)

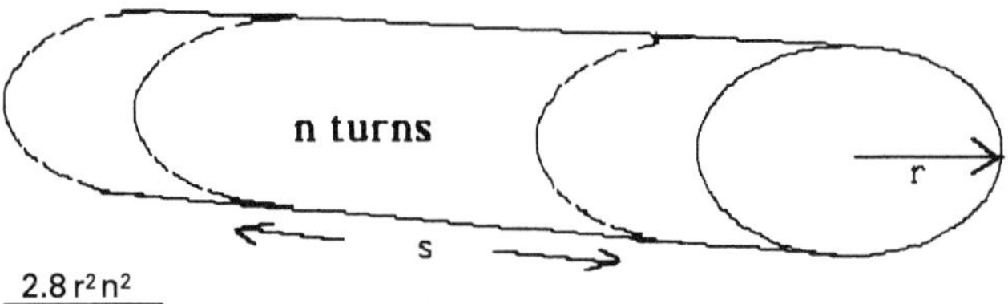
n turns
r
s

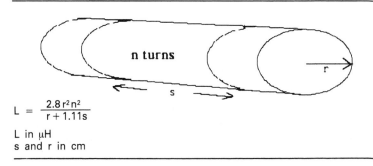

$$L = \frac{2.8\,r^2 n^2}{r + 1.11s}$$

L in μH
s and r in cm

turns of a single-layer close wound coil (solenoid) to achieve a given inductance. This approximate formula gives reasonable results for air-cored coils of values commonly used for radio circuit tuning. The addition of a core (generally of ferrite material) will cause an increase in inductance which could be by a factor as great as the relative permeability

Table 1.11. RELATIVE PERMEABILITY VALUES

$$\text{Relative permeability, } \mu_r = \frac{\text{Inductance of coil with core}}{\text{Inductance of coil without core}}$$

Alternatively, inductance value with core = μ_r X inductance value without core.

Material	*Relative permeability maximum value*
Silicon-iron	7 000
Cobalt-iron	10 000
Permalloy 45	23 000
Permalloy 65	600 000
Mumetal	100 000
Supermalloy	1 000 000
Dustcores	10 to 100
Ferrites	100 to 2 000

(*Table 1.11*) of the ferrite. The multiplying effect is seldom as large as the value of relative permeability, because the ferrite does not normally enclose the coil. Manufacturers of ferrite cores which enclose coils

Example:

Inductance of a 120 turn coil is measured as 840μH

How many turns need to be removed to give 500μH ?

Since: $L \propto n^2$ (L- inductance, n- number of turns)

Then: $\frac{L_1}{L_2} = \frac{n_1^2}{n_2^2}$ and $\frac{840}{500} = \frac{120^2}{n_2^2}$

$\therefore \; n_2^2 = \frac{120^2 \times 500}{840} = 8571$

$\therefore \; n = \sqrt{8571} = 93$ approx

Figure 1.24. Adjusting inductors to different values

provide winding data appropriate for each type and size of core. *Figure 1.24* shows how inductors can be adjusted for a different inductance value, using the principle that inductance is proportional to the square of the number of turns.

Inductive reactance

The reactance of an inductor for a sine wave signal of frequency *f* hertz is $2\pi fL$, where L is inductance in henries. The reactance is the ratio $\tilde{V}/\tilde{I}$

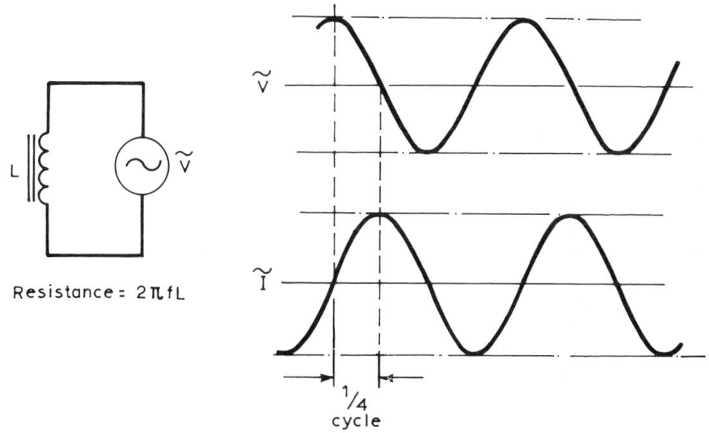
L
Ṽ
Resistance = 2πfL
Ṽ
Ĩ
1/4
cycle

Figure 1.25. Reactance and phase shift of a perfect (zero resistance) inductor

($\tilde{V}$ is signal voltage, $\tilde{I}$ is signal current) and is measured in ohms. For a coil whose reactance is much greater than its resistance, the voltage sine wave is 90° (¼ cycle) ahead of the current sine wave (*Figure 1.25*).

Untuned transformers

An untuned transformer consists of two windings, primary and secondary, neither or which is tuned by a capacitor, on a common core. For low frequency use, a massive core made from laminations (thin sections) of transformer steel alloy (such as silicon iron) must be used. Transformers which are used only for higher audio frequencies can make use of considerably smaller cores. At radio frequencies, the losses caused by transformer steels make such materials unacceptable, and ferrite materials are used as cores. For the highest frequencies, no core material is suitable and only self supporting air-cored coils (or pieces of straight wire or strip) can be used. In addition, high frequency currents flow mainly along the outer surfaces of conductors. This has two practical consequences – tubular conductors are as efficient as solid conductors (but use much less metal and can be water cooled) and silver plating can greatly decrease the effective resistance of a conductor.

For an untuned transformer with 100% coupling, the ratio of voltages $\tilde{V}_s/\tilde{V}_p$ is equal to the ratio of winding turns N_s/N_p, where $\tilde{V}$ refers to a.c. voltage, N to number of turns and s, p to secondary, primary

Table 1.12. REACTIVE CIRCUIT RESPONSE

Circuit	$\frac{V_{out}}{V_{in}}$	*Phase angle*
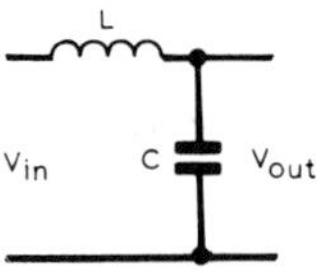L C Vin Vout	$\frac{1}{1 - f^2/f_o^2}$ $\approx -f_o^2/f^2$	0° when $f < f_o$ 180° when $f > f_o$
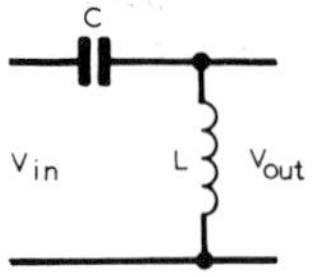C L Vin Vout	$\frac{1}{1 - f_o^2/f^2}$ $\approx -f^2/f_o^2$	0° when $f > f_o$ 180° when $f < f_o$

Notes: f_o is frequency of resonance $= \frac{0.16}{\sqrt{(LC)}}$
f is frequency at which response is to be found
$>$ greater than
$<$ less than
$\approx$ approximately equal to

Circuits:

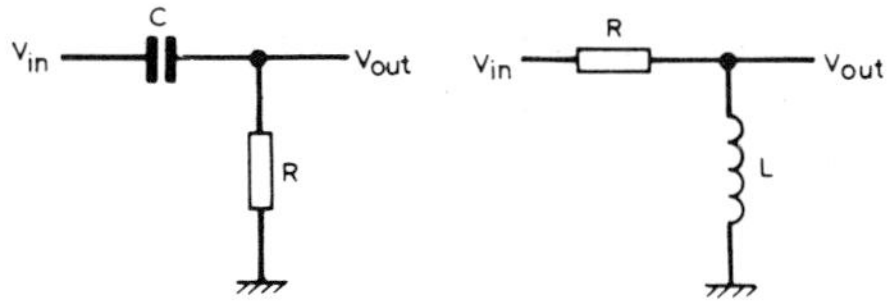
C
Vin
Vout
R
R
Vin
Vout
L

$$G = \frac{Vout}{Vin}$$

ϕ = phase angle

T = CR or L/R

ω = 2π X frequency

ωT	$\phi°$	G	ωT	$\phi°$	G
0.1	84.3	0.099	0.9	48.0	0.67
0.2	78.7	0.196	1.0	45.0	0.707
0.3	73.3	0.287	1.5	33.7	0.83
0.4	68.2	0.37	2.0	26.6	0.9
0.5	63.4	0.45	3.0	18.4	0.95
0.6	59.0	0.51	4.0	14	0.97
0.7	55.0	0.57	5.0	11.3	0.98
0.8	51.34	0.62			

Circuits:

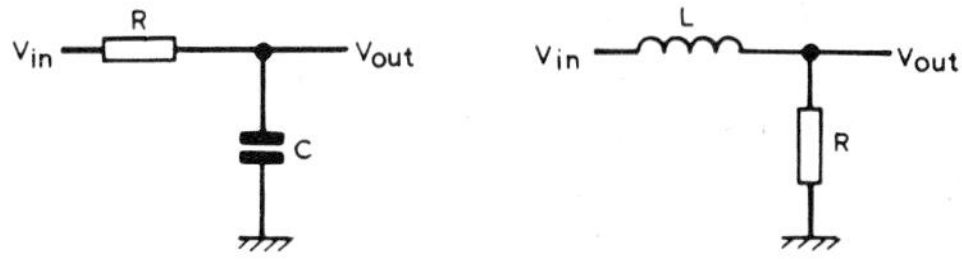
R
Vin
Vout
C
L
Vin
Vout
R

ωT	$\phi°$	G	ωT	$\phi°$	G
0.1	−5.7	0.99	0.9	−41.8	0.74
0.2	−11.3	0.98	1.0	−45	0.707
0.3	−16.7	0.96	1.5	−56	0.55
0.4	−21.8	0.93	2.0	−63	0.45
0.5	−26.5	0.89	3.0	−72	0.32
0.6	−31	0.85	4.0	−76	0.24
0.7	−35	0.82	5.0	−79	0.2
0.8	−38.6	0.78			

Figure 1.26. Amplitude/phase tables for LR and CR circuits. The quantity T is the time constant (CR or L/R), and ω is equal to $2\pi f$

respectively. When an untuned transformer is used to transfer power between circuits of different impedance Z_p, Z_s, then the best match (maximum power transfer) condition is

$$\frac{N_s}{N_p} = \sqrt{\frac{Z_s}{Z_p}}$$

LCR circuits

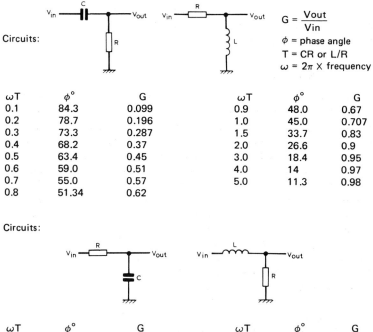

The action of *CR* and *LR* circuits upon a sine wave signal is to change both the amplitude and the phase of the signal. Universal amplitude/phase tables can be prepared, using the time constant *T* of the *CR* and

LR circuit and the frequency f of the sine wave. These tables are shown, with examples, in *Figure 1.26*.

When a reactance (L or C) is in circuit with a resistance R, the general formula for the total impedance (Z) are as shown in *Table 1.12*. Impedance is defined as $Z = \tilde{V}/\tilde{I}$, but the phase angle between V and I will not be 90°, and will be 0° only when resonance (see later) exists.

The combination of inductance and capacitance produces a tuned circuit which may be series (*Figure 1.27a*) or parallel (*Figure 1.27b*). Each type of tuned (or resonant) circuit has a frequency of resonance, symbol, f_o, at which the circuit behaves like a resistance so that there is no phase shift between voltage and current. At other frequencies, the circuit may behave like an inductor or like a capacitor. Below the frequency of resonance, the parallel circuit behaves like an inductor, the series circuit behaves like a capacitor. Above the frequency of resonance the parallel circuit behaves like a capacitor, the series circuit like an inductor. At resonance, the parallel circuit behaves like a large value resistor and the series circuit like a small value resistor.

The series resonant circuit can provide voltage amplification at the resonant frequency when the circuit of *Figure 1.28* is used. The amount of voltage amplification is given by $\frac{2\pi fL}{R}$ or $\frac{1}{2\pi fCR}$ at the frequency of

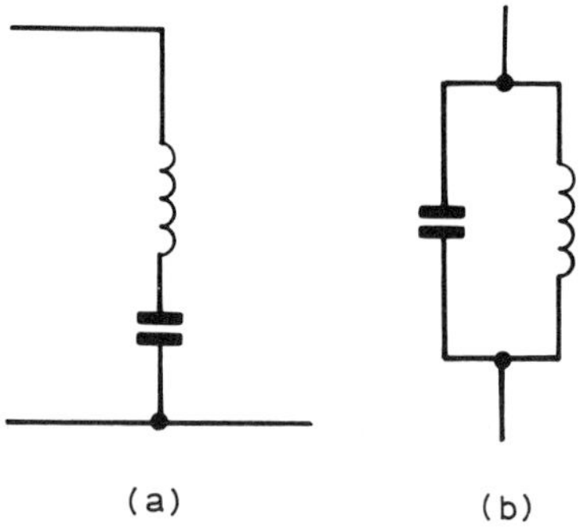
(a)
(b)

Figure 1.27. Tuned circuits (a) series, (b) parallel

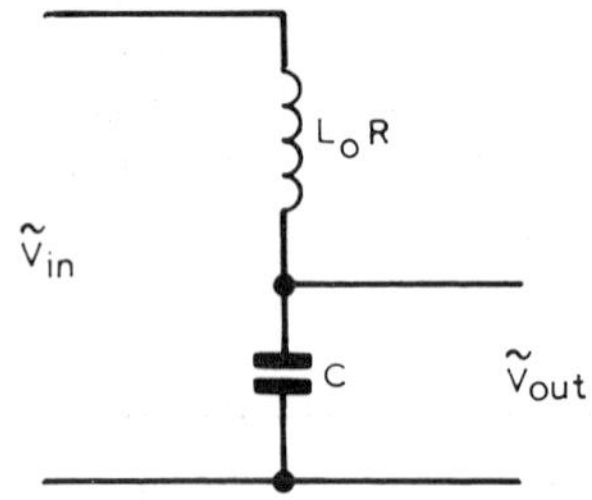
L_o R
Ṽin
C
Ṽout

Figure 1.28. Voltage amplification of a tuned series circuit. The amplification is of the resonant frequency only, and can occur only if the signal source is of comparatively low impedance

resonance, and this quantity is often termed the circuit magnification factor, Q. There is no *power* amplification, as the voltage step up is achieved by increasing the current through the circuit, assuming that the signal input voltage is constant. *Table 1.12* shows typical phase and amplitude response formulae in universal form for the series resonant circuit. Note that the tuning capacitance may be a stray capacitance.

The parallel resonant circuit is used as a load which is a pure resistor (with no phase shift) only at the resonant frequency, f_0. The size of the equivalent resistance is called dynamic resistance and is calculated by the formula shown in *Figure 1.29*. The effect of adding resistors in

At resonance: $R_d = \frac{L}{CR}$

L = inductance in henries
C = capacitance in farads
R = resistance of coil in ohms

Figure 1.29. Dynamic resistance of a parallel resonant circuit

parallel with such a tuned circuit is shown in *Figure 1.30*, the dynamic resistance at resonance is reduced, but the resistance remains fairly high over a greater range of frequency. Such 'damping' is used to extend the bandwidth of tuned amplifiers. *Table 1.12* shows the amplitude and phase response of a parallel tuned circuit in general form. Once again, the tuning capacitance may be a stray capacitance.

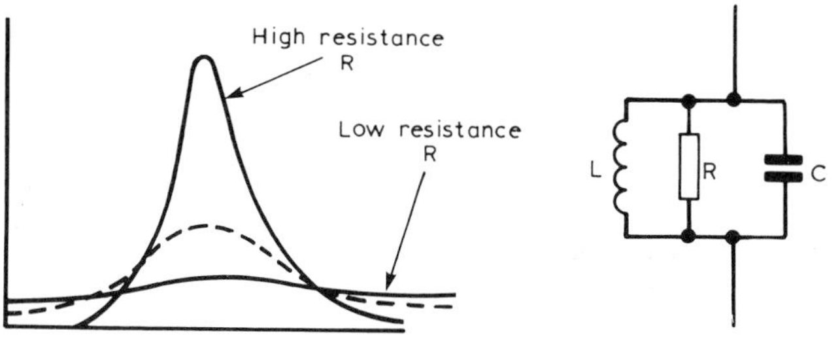
High resistance
R
Low resistance
R
L
R
C

Figure 1.30. The effect of damping resistance on the resonance curve

The impedance of a series circuit is given by the formula shown, with example, in *Table 1.12a.* Note that both amplitude (in ohms) and phase angle are given. The corresponding expression for a parallel circuit in which the only resistance is that of the coil (R) is also shown in *Table 1.13b*. When a damped parallel circuit is used, the resistance of the coil has generally a negligible effect compared to the damping resistor, and the formula of *Table 1.13c* applies.

Coupled tuned circuits

When two tuned circuits are placed so that their coils have some mutual inductance M, the circuits are said to be coupled. The size of the

Table 1.13. IMPEDANCE Z AND PHASE ANGLE ϕ

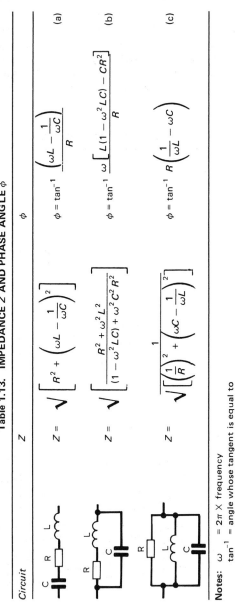

Circuit	Z	ϕ	
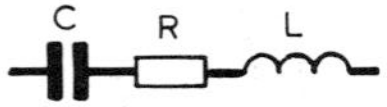	$Z = \sqrt{\left[R^2 + \left(\omega L - \frac{1}{\omega C}\right)^2\right]}$	$\phi = \tan^{-1} \frac{\left(\omega L - \frac{1}{\omega C}\right)}{R}$	(a)
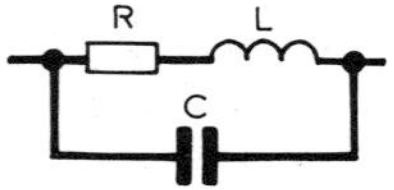	$Z = \sqrt{\left[\frac{R^2 + \omega^2 L^2}{(1 - \omega^2 LC) + \omega^2 C^2 R^2}\right]}$	$\phi = \tan^{-1} \frac{\omega\left[L(1 - \omega^2 LC) - CR^2\right]}{R}$	(b)
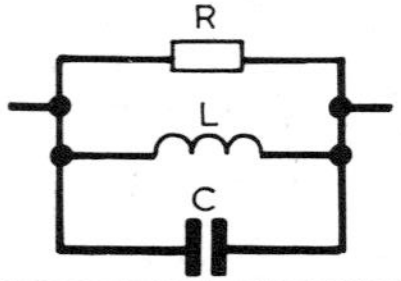	$Z = \frac{1}{\sqrt{\left[\left(\frac{1}{R}\right)^2 + \left(\omega C - \frac{1}{\omega L}\right)^2\right]}}$	$\phi = \tan^{-1} R\left(\frac{1}{\omega L} - \omega C\right)$	(c)

Notes: ω = $2\pi \times$ frequency
$\tan^{-1}$ = angle whose tangent is equal to

mutual inductance is not simple to calculate; one approximate method is shown in *Table 1.14*. When the mutual inductance (M) between the coils is small compared to their self inductances (L_1, L_2) then the coupling is said to be loose, and the response curve shows a sharp peak. When the mutual inductance between the coils is large compared to their self inductances, the coupling is tight (or overcoupled) and the response curve shows twin peaks. For each set of coupled coils there is an optimum coupling at which the peak of the response curve is

Table 1.14. MUTUAL INDUCTANCE

1. From values of coil size S and X (note that these are lengths divided by coil diameter) find K.
2. Knowing inductance values, L_1, L_2, find M.
   $M = k\sqrt{(L_1 L_2)}$

S = spacing/diamater
X = coil winding length/diameter

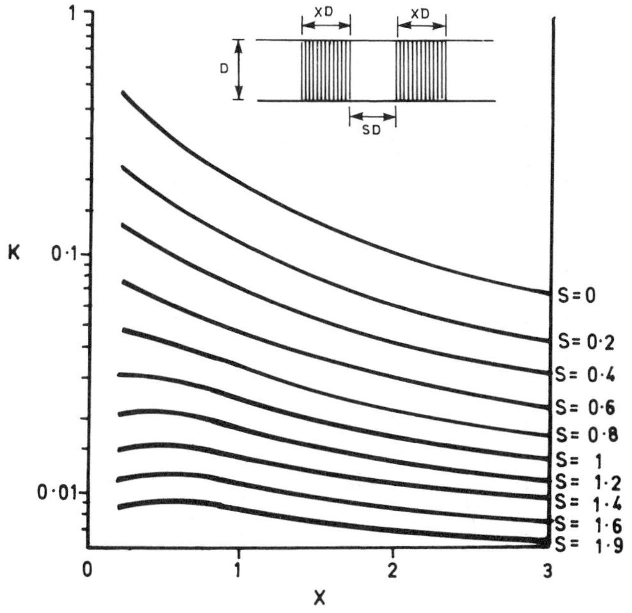
XD
XD
D
SD
1
K
0·1
0·01
0
1
2
3
X
S=0
S= 0·2
S= 0·4
S= 0·6
S= 0·8
S= 1
S= 1·2
S= 1·4
S= 1·6
S= 1·9

flattened and the sides steep. This type of response is an excellent compromise between selectivity and sensitivity.

The coefficient of coupling k is defined by

$$k = \frac{M}{\sqrt{(L_1 L_2)}}$$

(or $\frac{M}{L}$ if both coils have the same value of L) and critical coupling occurs when $k = \frac{1}{Q}$, assuming that both coils have the same Q factor – if they do not, then $Q = \sqrt{(Q_1 Q_2)}$. The size of the coefficient of coupling depends almost entirely on the spacing between the coils and no formulae are available to calculate this quantity directly.

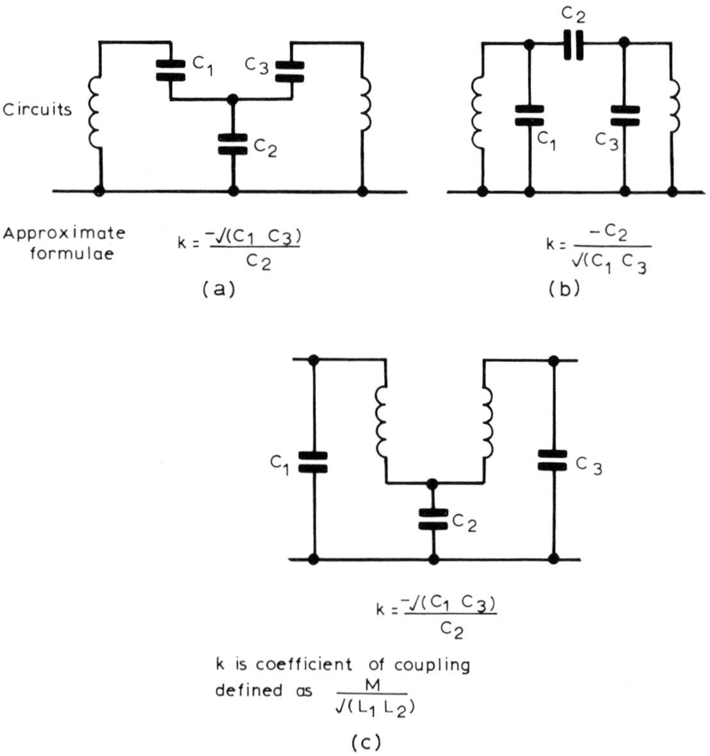
Circuits
C1
C3
C2
C2
C1
C3
Approximate formulae
k = −√(C1 C3) / C2
(a)
k = −C2 / √(C1 C3
(b)
C1
C3
C2
k = −√(C1 C3) / C2
k is coefficient of coupling defined as M / √(L1 L2)
(c)

Figure 1.31. Other methods of circuit coupling, and their design formulae

Other types of coupled circuits, with some design data, are shown in *Figure 1.31.* These make use of a common impedance or reactance for coupling and are not so commonly used.

Quartz crystals

Quartz crystals, cut into thin plates and with electrodes plated onto opposite flat faces, can be used as resonant circuits, with *Q* values ranging from 20 000 to 1 000 000 or more. The equivalent circuit of a crystal is shown in *Figure 1.32*. The crystal by itself acts as a series

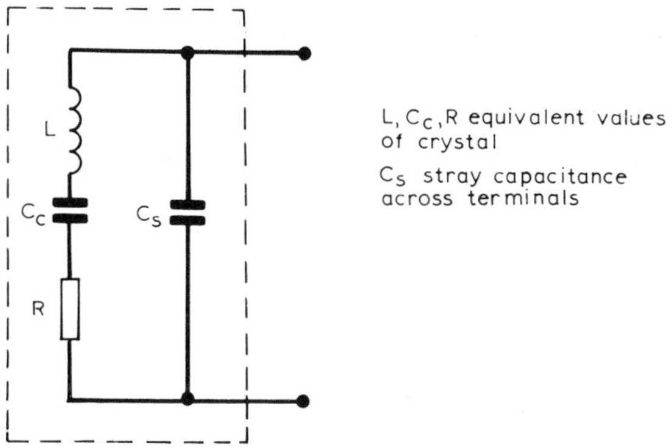
L
C_C
R
C_S
L, C_C, R equivalent values of crystal
C_S stray capacitance across terminals

Figure 1.32. Equivalent circuit of a quartz crystal

resonant circuit with very large inductance, small capacitance and fairly small resistance (a few thousand ohms). The stray capacitance across the crystal will also permit parallel resonance at a frequency slightly higher than that of the series resonance. *Figure 1.33* shows how the

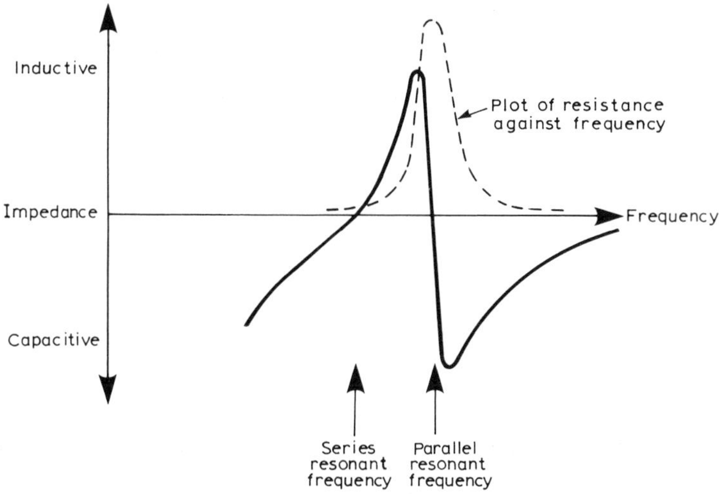
Inductive
Impedance
Capacitive
Plot of resistance against frequency
Frequency
Series resonant frequency
Parallel resonant frequency

Figure 1.33. Variation of reactance and resistance of a crystal near its resonant frequencies

reactance and the resistance of a crystal vary as frequency is varied – the reactance is zero at each resonant frequency, the resistance is maximum at the parallel resonance frequency. Usually the parallel or the series frequency is specified when the crystal is manufactured.

Wave filters

Wave filter circuits are networks containing reactive components (*L*,*C*) which accept or reject frequencies above or below cut-off frequencies (which are calculated from the values of the filter components). Output amplitude and phase vary considerably as the signal frequency approaches a cut-off frequency, and the calculations involved are beyond the scope of this book. Much more predictable response can be obtained, for audio frequencies at least, by using active filters (see Chapter 3).

Measuring R, L, C

Resistance measurements can be made using the multimeter; the scale is non-linear but readings can be precise enough to indicate whether or not the value is within tolerance. Measurements of resistance, capacitance, self and mutual inductance can be carried out using bridge circuits (*Figure 1.34*) which rely on a 'null reading'. This means that potentiometers or switches are adjusted until a meter reading reaches a minimum,

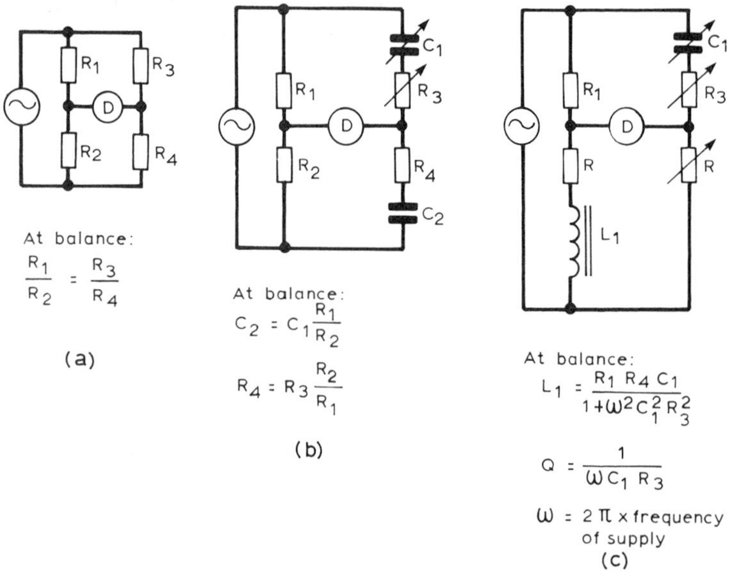
R1
R3
D
R2
R4
At balance:
$\frac{R_1}{R_2} = \frac{R_3}{R_4}$
(a)
C1
R1
R3
D
R2
R4
C2
At balance:
$C_2 = C_1 \frac{R_1}{R_2}$
$R_4 = R_3 \frac{R_2}{R_1}$
(b)
C1
R1
R3
D
R
R
L1
At balance:
$L_1 = \frac{R_1 R_4 C_1}{1 + \omega^2 C_1^2 R_3^2}$
$Q = \frac{1}{\omega C_1 R_3}$
ω = 2π x frequency of supply
(c)

Figure 1.34. Measuring bridge circuits, with balance conditions. (a) Simple Wheatstone resistance bridge, (b) capacitance bridge, (c) inductance bridge

upon which the value of the quantity being measured can be read from the dials. Such bridge circuits use an audio frequency oscillator as a source of bridge voltage.

Direct-reading capacitance meters use a rather difference principle. Referring to *Figure 1.35*, the capacitor C is charged to a known voltage V, and then discharged through the meter, M. The amount of charge

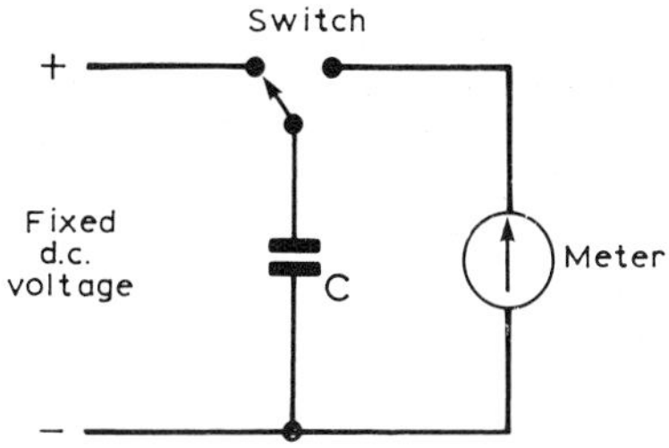
Switch
+
Fixed
d.c.
voltage
C
Meter
–

Figure 1.35. Principle of a direct-reading capacitance meter

passing through the meter on each discharge is CV, so that if the switch is actuated f times per second, the amount of charge flowing per second is fCV. This is the average current I, read by the meter, so that $I = fCV$ or $C = \frac{I}{fV}$. By a suitable choice of switching frequency, charging voltage and meter range, values of capacitance ranging from 10 pF to several μF can be measured, though erratic results are sometimes experienced with electrolytic capacitors. The switching is carried out by transistors, as the switching speeds needed are beyond the range of mechanical switches, even reed switches.

Chapter 2

Active Discrete Components

Diodes

Semiconductor diodes may use two basic types of construction, point contact or junction. Point contact diodes are used for small signal purposes where a low value of capacitance between the terminals is important – their main use nowadays is confined to r.f. demodulation. Junction diodes are obtainable with much greater ranges of voltage and current and are used for most other purposes. Apart from diodes intended for specialised purposes, such as light-emitting diodes, the fabrication materials are silicon or germanium, with germanium used almost exclusively for point contact demodulator diodes.

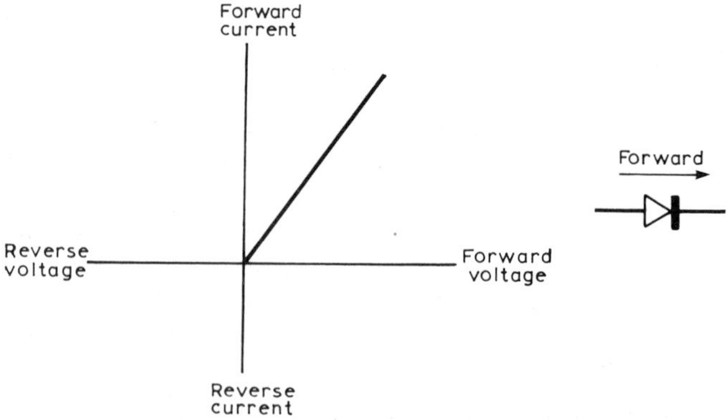
Forward current
Reverse voltage
Forward voltage
Reverse current
Forward

Figure 2.1. Characteristic of an ideal diode.

An ideal diode would conduct in one direction only, and the characteristic (the graph of current plotted against applied voltage) would look as in *Figure 2.1*. Practical diodes have a low, forward resistance (not a *constant* value, however) and a high reverse resistance; and they conduct when the anode voltage is a few hundred millivolts higher than the cathode voltage.

The diode can be destroyed by excessive forward current, which causes high power dissipation at the junction or contact, or by using excessive reverse voltage, causing junction breakdown (see later), allowing it to conduct. Because reverse voltages are much higher than the voltage across a forward conducting diode, breakdown causes excessive current to flow so that once again the junction or contact is destroyed by excessive dissipation. For any diode therefore, the published ratings of peak forward current and peak reverse voltage should not be exceeded at any time, and should not be approached if reliable operation is to be achieved.

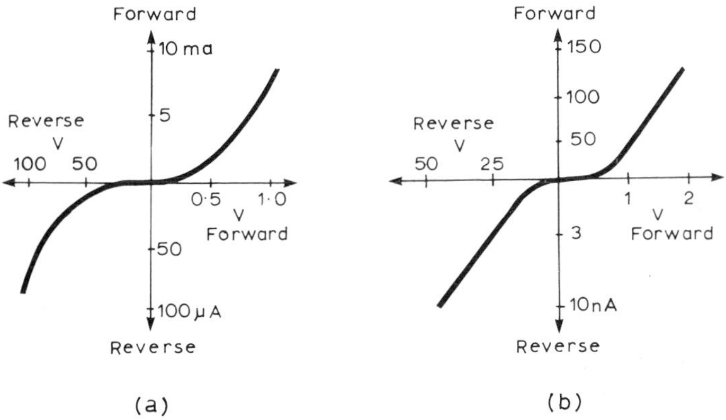
Forward
10 ma
5
Reverse
V
100 50
0·5 1·0
V
Forward
50
100 μA
Reverse
(a)
Forward
150
100
50
Reverse
V
50 25
1 2
V
Forward
3
10nA
Reverse
(b)

Figure 2.2. Characteristics of real diodes. (a) Germanium point diode, (b) silicon junction diode. Note the different scales which have to be used to allow the graphs to be fitted into a reasonable space

Characteristics for a typical point contact germanium diode and a typical small signal silicon junction diode are also shown in *Figure 2.2.* Comparing the two type of diode:

(a) Germanium point contact diodes have lower reverse resistance values, conduct at a lower forward voltage (about 0.2 V) but have higher forward resistance because of their small junction area. They also have rather low peak values of forward current and reverse voltage.

(b) Silicon junction diodes have very high values of reverse resistance, conduct at around 0.55 V forward voltage, can have fairly low forward resistance, and have fairly high peak values of forward current and reverse voltage.

The foward resistance of a diode is not a fixed quantity but is (very approximately) inversely proportional to current. Another approximation, useful for small currents, is that the forward voltage of a silicon diode increases by only 60 mV for a tenfold increase in current.

The effect of temperature change on a silicon diode is to change the forward voltage at any fixed value of current. A change of about 2.5 mV per °C is a typical figure with the voltage reducing as the temperature is raised. The reverse (leakage) current is much more temperature dependent, and a useful rule of thumb is that leakage current doubles for each 10°C rise in temperature.

Zener diodes are used with reverse bias, making use of the breakdown which occurs across a junction when the reverse voltage causes a large electrostatic field across the junction. This breakdown limit occurs at low voltages (below 6 V) when the silicon is very strongly doped, and this type of breakdown is called zener breakdown. For such

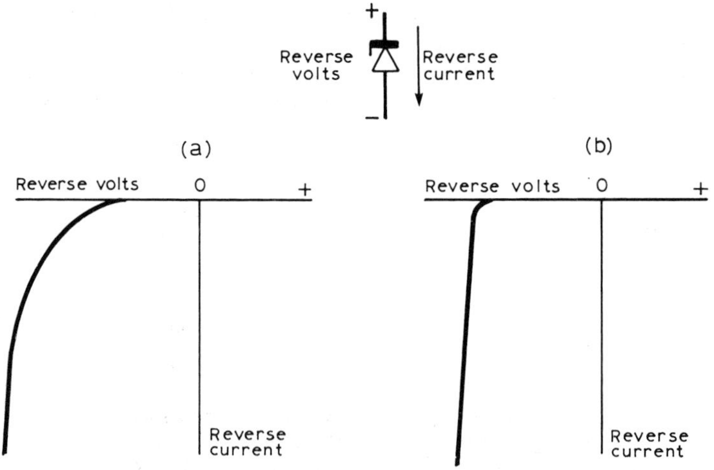
+
Reverse
volts
Reverse
current
–
(a)
Reverse volts
0
+
Reverse
current
(b)
Reverse volts
0
+
Reverse
current

Figure 2.3. Zener diode. The true zener effect causes a 'soft' breakdown (a) at low voltages, the avalanche effect causes a sharper turnover (b)

a true zener diode, the reverse characteristic is as shown in *Figure 2.3a*. As the graph shows, the reverse current does not suddenly increase at the zener voltage, and the voltage across the diode is not truly stabilised unless the current is more than a few milliamps. This type of characteristic is termed a 'soft' characteristic. In addition to this, a true zener diode has a negative temperature coefficient – the voltage across the diode (at a constant current value) decreases as the junction temperature is increased.

Avalanche breakdown occurs in diodes with lower doping levels, at voltages above about 6 V. The name is derived from the avalanche action in which electrons are separated from holes by the electric field across the junction, and these electrons and holes then cause further electron-hole separation by collision. These diodes have 'hard' characteristics (*Figure 2.3b*) with very little current flowing below the avalanche voltage and large currents above the avalanche voltage. In addition, the temperature coefficient of voltage is positive, so that the voltage across the diode *increases* as the junction temperature is raised.

Both types of diodes are known, however, as zener diodes, and these with breakdown voltages between 4 V and 6 V combine both effects. At a breakdown voltage of around 5.6 V, the opposing temperature characteristics balance, so that the breakdown voltage of a 5.6 V diode is practically unaffected by temperature. The stabilisation of the diode is measured by its dynamic resistance, defined as the ratio

$$\frac{\text{Voltage change, } V}{\text{Current change, } I}\text{, units of ohms,}$$

when *V* is the change of voltage across the diode caused by a change of current *I* when the diode is stabilising.

This ratio should be below 50 ohms, and reaches a minimum value of about 4 ohms for a diode with a breakdown voltage of about 8 V.

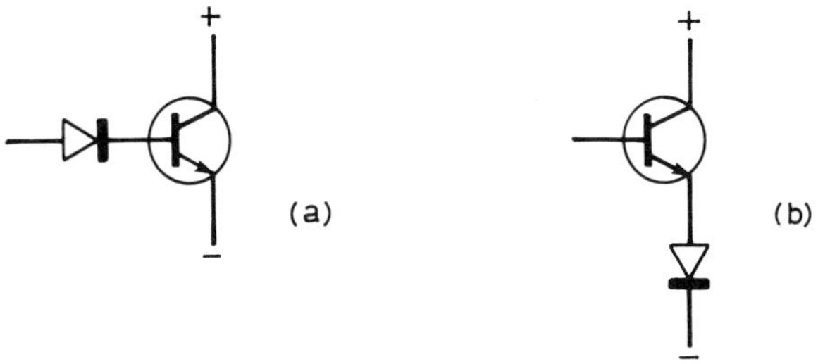
+
−
(a)
+
−
(b)

Figure 2.4. Protecting the base-emitter junction of a transistor against excessive reverse voltage, (a) diode in base circuit, (b) diode in emitter circuit. The base circuit is preferred, since the diode does not have to pass the emitter current of the transistor

Note that the base-emitter junction of many types of silicon transistors will break down by avalanche action at voltages ranging from 7 V to 20 V reverse bias, though this action does not necessarily cause collector current to flow. The base-emitter junction can be protected by a silicon diode (with a high breakdown voltage) wired in series (*Figure 2.4*).

Reference diodes are doped to an extent which makes the breakdown voltage practically constant despite changes in temperature. Voltages of 5 V to 6 V are used, and temperature changes ranging from ± 0.01% per degree to 0.000 5% per degree can be achieved. Reference diodes are used for very precise voltage stabilisation.

Varactor Diodes

All junction diodes have a measurable capacitance between anode and cathode when reverse biased, and this capacitance varies with voltage, being least when high reverse voltage are used. This variation is made use of for varactor diodes, in which the doping is arranged for the maximum possible capacitance variation. A typical variation is of 10 pF at 10 V bias to 35 pF at 1 V reverse bias. Varactor diodes are used for electronic tuning applications; a typical circuit is shown in *Figure 3.26* (Chapter 3).

LEDs

Light emitting diodes use compound semiconductors such as gallium arsenide or indium phosphide. When forward current passes, light is emitted from the junction. The colour of the light depends on the material used for the junction, and the brightness is approximately proportional to forward current. LEDs have higher forward voltages when conducting; 1.6 to 2.2 V as compared to the 0.5 to 0.8 V of a silicon diode. The maximum permitted reverse voltages are low, typically 3 V, so that a silicon diode must be connected across the LED as

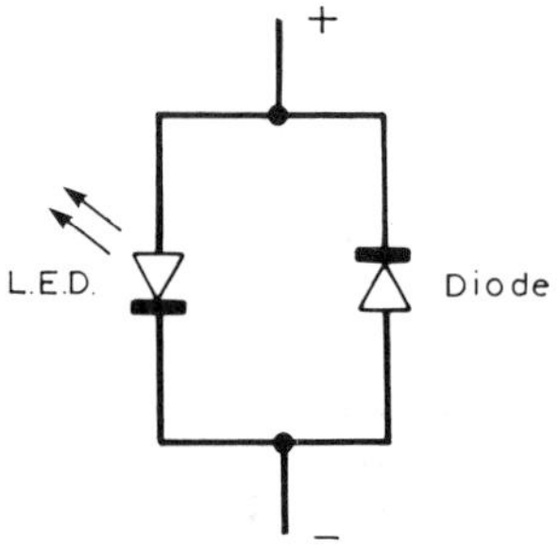
+
L.E.D.
Diode
–

Figure 2.5. Protecting an LED from reverse voltage

shown in *Figure 2.5* if there is any likelihood of reverse voltage (or a.c.) being applied to the diode. A series resistor must always be used to limit the forward current unless pulsed operation is used.

For more specialised applications, microwave diodes of various types can be obtained which emit microwave radiation when forward biased and enclosed in a suitable resonant cavity. Tunnel diodes are diodes with an unstable portion of the characteristic (a reverse slope, indicating 'negative resistance'). When the tunnel diode is biased to the unstable region, oscillations are generated at whatever frequency is permitted by the components connected to the diode (*RC, LC,* cavity, etc.)

Diode circuits

Figure 2.6 shows some application circuits for diodes, with approximate design data where appropriate. Diode types should be selected with reference to the manufacturers data sheets, having decided on the basic reverse voltage and load current quantities required by the circuit.

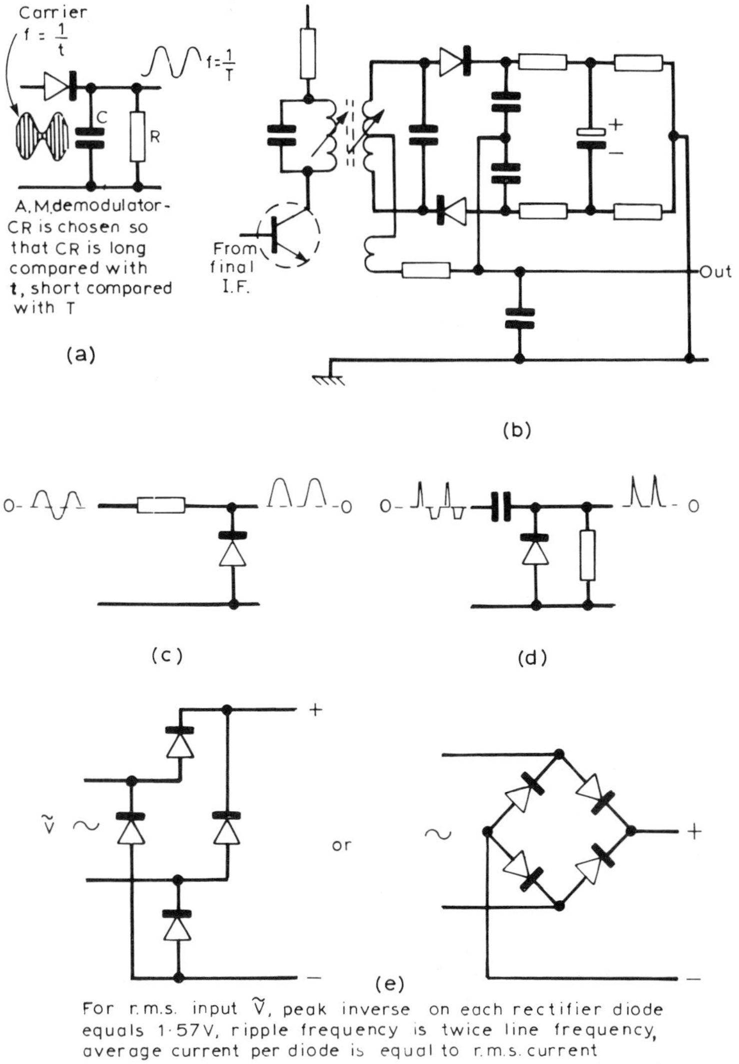
Carrier
f = 1/t
f = 1/T
C
R
A.M. demodulator - CR is chosen so that CR is long compared with t, short compared with T
(a)
From final I.F.
Out
(b)
(c)
(d)
Ṽ
or
(e)
For r.m.s. input Ṽ, peak inverse on each rectifier diode equals 1·57V, ripple frequency is twice line frequency, average current per diode is equal to r.m.s. current

Figure 2.6. Some diode applications: (a) amplitude demodulation, (b) ratio detector for f.m. (c) signal clipping, (d) d.c. restoration, (e) bridge rectification

Transistors

Like signal diodes, transistors may be constructed using either silicon or germanium, but virtually all transistors now use silicon. The design data of this section refer to silicon transistors only.

The working principle of a transistor is that current flows between the collector and emitter terminals only when current is flowing between the base and emitter terminals. The ratio of these currents is called

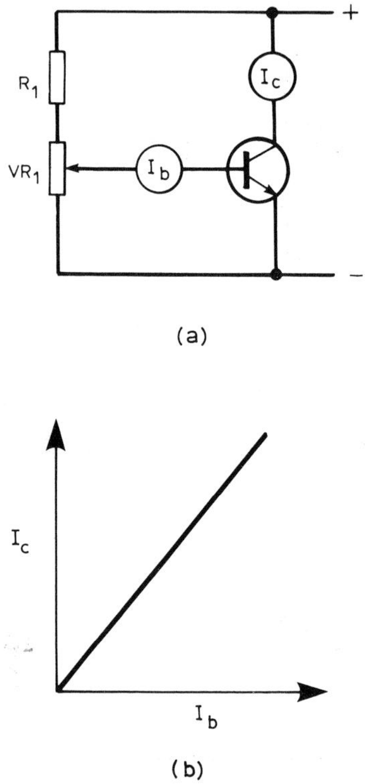

+
R1
VR1
Ic
Ib
–
(a)
Ic
Ib
(b)

Figure 2.7. Forward current transfer ratio (a) measuring circuit, (b) graph. The slope of the graph (I_c/I_b) is equal to the forward current transfer ratio, h_{fe}

forward current transfer ratio, symbol h_{fe}. For the arrangement of *Figure 2.7*, the ratio is

$$h_{fe} = \frac{i_c}{i_b}$$

In databooks, a distinction is made between h_{FE}, for which I_c and I_b are d.c. quantities; and h_{fe}, for which i_c and i_b are a.c. quantities. The two quantities are however generally close enough in value to be interchangeable, and the symbol h_{fe} will be used here to indicate both values. The size of h_{fe} for any transistor can be measured in the circuit of *Figure 2.7a*; a simpler method, used in many transistor testers, is shown in *Figure 2.8*. Values vary from about 25 (power transistors operating at high currents) to over 1000 (some high-frequency amplifier types).

Base current will not flow unless the voltage between the base and the emitter is correct. The precise voltage at which current starts to

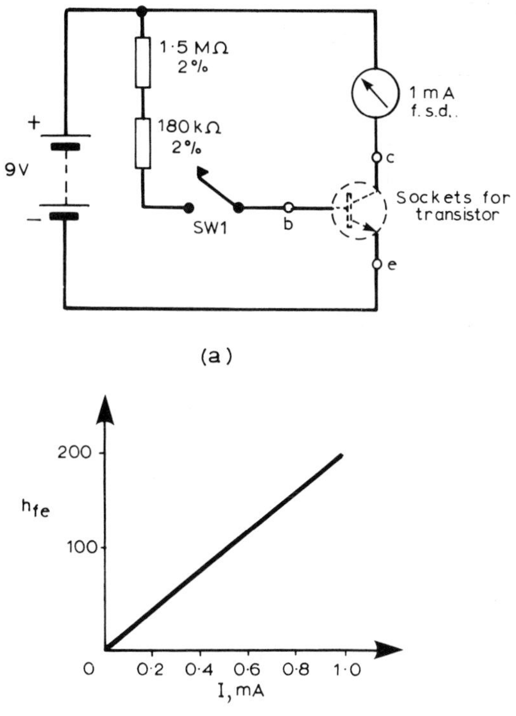
1·5 MΩ
2%
180 kΩ
2%
+
9V
−
1 mA
f. s.d.
c
Sockets for
transistor
SW1
b
e
(a)
200
100
h_{fe}
0
0·2
0·4
0·6
0·8
1·0
I, mA
(b)

Figure 2.8. A simple transistor tester (a) and its calibration graph (b)

become detectable varies from one specimen (even of the same type number) of transistor to another, but for silicon transistors is generally about 0.5 V. The pnp type of transistor will require the emitter to be at a more positive voltage than the base, the npn type will require the base to be more positive than the emitter. When the transistor has the correct d.c. currents flowing (with no signal applied) it is said to be correctly biased. Amplification is carried out by adding a signal voltage to the steady voltage at the input of the transistor. The vast majority of transistor circuits use the base as the input terminal, though a few (common base amplifiers) use the emitter as an input terminal.

Bias for linear amplifiers

A linear amplifier produces at its output a waveform which is a perfect copy, but of greater amplitude, of the waveform at the base. The voltage gain of such an amplifier is defined as

$$G = \frac{V_{out}}{V_{in}}$$

where V indicates a.c. (signal) voltage measurements. If the output waveform is not a perfect copy of the input, then the amplifier exhibits distortion of one sort or another. One type of distortion is non-linear distortion, in which the shape of the waveform is changed by the 'copying' process. Such non-linear distortion is caused by the action of the transistor and can be minimised by careful choice of transistor type (see later) and by correct bias.

A transistor is correctly biased when the desired amount of gain can be obtained with minimum distortion. This is easiest to achieve when the (peak to peak) output signal from the amplifier is much smaller than the supply voltage. This may be achieved with the no-signal voltage at the collector of the transistor at almost any reasonable level, but to allow for unexpected signal overloads, the preferred collector voltage is half-way between supply positive voltage and the voltage of the emitter.

When the value of collector resistor has been chosen, bias is applied by passing current into the base so that the collector voltage drops to the desired values of 0.5 V_{ss} where V_{ss} is supply voltage. For any bias system, the desired base current must be equal to

$$\frac{0.5\ V_{ss}}{R_L \times h_{fe}}$$ mA, with V_{ss} in volts, R_L in kΩ, h_{fe} as a ratio.

Figure 2.9 shows three bias systems, with design data for obtaining a suitable bias voltage. The method of *Figure 2.9a* is the most difficult to use, as a different resistor value must be chosen for each transistor used. It may be necessary to use resistors in series or in parallel to achieve the correct value and the collector voltage will decrease noticeably as the temperature of the transistor increases. The method of *Figure 2.9b* is a considerable improvement over that of *Figure 2.9a*. The bias system may be designed around an 'average' transistor (with an average value of h_{fe} for that type) and can then be used unchanged for other transistors without too serious a change in the collector voltage. In addition the collector voltage changes much less as the temperature changes.

The bias system of *Figure 2.9c* is one which can be used for any transistor provided that the current flowing through the two base bias resistors R_1, R_2 is much greater than the base current drawn by the transistor. Unlike the other two systems, the design formula does not

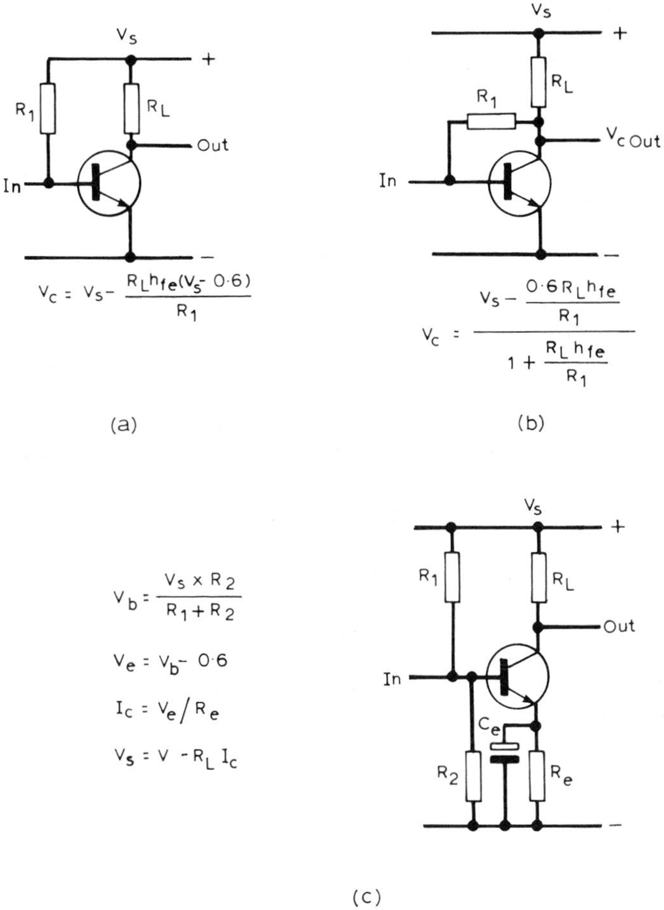
Vs
+
R1
RL
Out
In
–
Vc = Vs – RLhfe(Vs – 0·6) / R1
(a)
Vs
+
RL
R1
Vc Out
In
–
Vc = (Vs – 0·6RLhfe / R1) / (1 + RLhfe / R1)
(b)
Vb = Vs x R2 / (R1 + R2)
Ve = Vb – 0·6
Ic = Ve / Re
Vs = V – RL Ic
Vs
+
R1
RL
Out
In
Ce
R2
Re
–
(c)

Figure 2.9. Transistor bias circuits: (a) simple system, usually unsatisfactory; (b) using negative feedback of bias and signal; (c) potential divider method

require the h_{fe} value of the transistor to be known if the standing current through the transistor is to be only a few milliamps. For power transistors, the quantities that are needed are the V_{be} and I_{be} at the bias current required. This system does not, however, stabilise the collector voltage so effectively against changes caused by changes of temperature.

Transistor parameters and linear amplifier gain

Transistor parameters are measurements which describe the action of the transistor. The name parameter is used to distinguish these quantities from constants. Transistor parameters are not generally constants, they vary from one transistor to another (even of the same type) and from one value of bias current to another. One such parameter, the common-emitter current gain, h_{fe}, has already been described.

Of the parameters for linear amplifiers, G_m is probably the most useful. G_m, called mutual conductance and measured in units of milli-siemens (mS) (equal to milliamps per volt) is defined by

$$G_m = \frac{\tilde{I}_c}{\tilde{V}_{be}}$$

where I_c = a.c. signal current, collector to emitter

where V_{be} = a.c. signal voltage, base to emitter.

The usefulness of G_m as a parameter arises from the fact that the voltage gain of a transistor amplifier for small signals is given by

$G = G_m R_L$ where R_L is the load resistance for signals (if G_m is measured in mS and R_L in kΩ, gain is correctly specified)

Note that R_L will generally be less than the resistance connected between the collector and the positive supply, because this value will be shunted by any other resistors connected through a capacitor to the collector

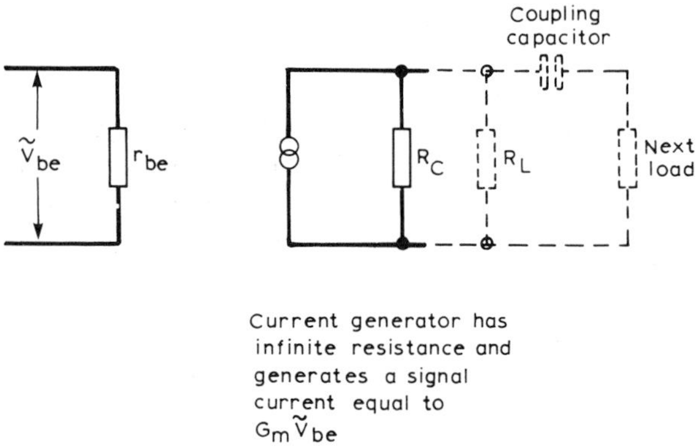
Coupling capacitor
Ṽbe
rbe
RC
RL
Next load
Current generator has infinite resistance and generates a signal current equal to GmṼbe

Figure 2.10. A useful equivalent circuit for the transistor. The signal voltage $\tilde{V}_{be}$ between the base and the emitter causes an output signal current $G_m\tilde{V}_{be}$. This current flows through the parallel resistor R_c (the transistor output resistance), R_L, the load resistor, and any other load resistors in the circuit

(*Figure 2.10*). The input resistance of the next transistor (if used) will also be in parallel with the collector resistor.

A graph of collector current plotted against base-emitter voltage is not a straight line, so that the ratio $\frac{i_c}{V_{be}}$ which is G_m, is not constant. A useful rule of thumb for small bias currents is that G_m = 40 × bias current in milliamps with G_m in mS (mA/V). The shape of the graph is always curved for low currents, but can vary in shape at higher currents. For a few transistors, the i_c, V_{be} graph has a noticeably straight portion, making these transistors particularly suitable for linear amplification applications. It is this straightness of the G_m characteristic which makes some types of power transistor much more desirable (and high priced) for audio output stages than others.

To take advantage of these linear characteristics, of course, the bias must be arranged so that the working point is at the centre of the linear region with no signal input. 'Working point' in this context means the combination of collector volts and base volts which represents a point on the characteristic.

Two other useful parameters for silicon transistors used in the common-emitter circuit are the input and output resistance values. The input resistance is defined as

$$\frac{\text{signal voltage, base input}}{\text{signal current into base}}$$

with no signal at the collector, symbol h_{ie}.

Then output resistance is, using a similar definition

$$\frac{\text{signal voltage at collector}}{\text{signal current at collector}}$$

with no signal at the base, symbol h_{oe}.

The output resistance h_{oe} has about the same range of values, from 10 kΩ to 50 kΩ for a surprisingly large number of transistors, irrespective of operating conditions, provided these are on the flat part of the V_{ce}/I_{ce} characteristic (*Figure 2.11*). An average value of 30 kΩ can usually be taken.

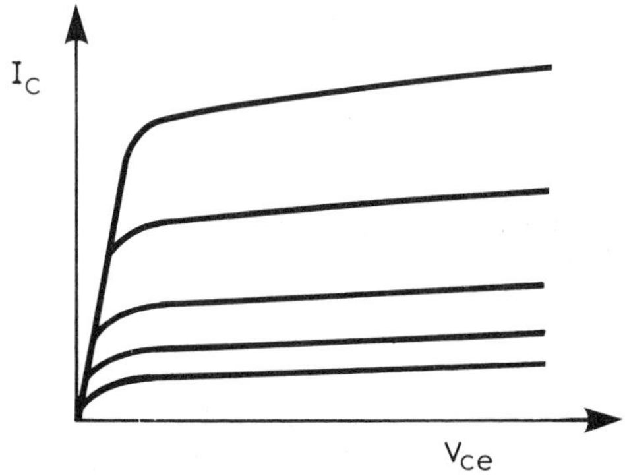
Ic
Vce

Figure 2.11. The I_c/V_{ce} characteristic. The flat portion is the operating part. The small amount of slope indicates that the output resistance, R_c, is high, usually 40 kΩ or more

The input resistance is not a constant, because the input of a transistor is a diode, the base-emitter junction. The value of input resistance h_{ie} is related to steady bias current, and to the other parameters

$$h_{ie} = \frac{h_{fe}}{G_m} \quad \text{and since } G_m = 40I_c,$$

$$= \frac{h_{fe}}{40I_c} \quad \text{where } I_c \text{ is the steady, no-signal, bias collector current.}$$

For example: if a transistor has an h_{fe} value of 120, and is used at a current of 1 mA, then the input resistance (in kΩ) is

$$h_{ie} = \frac{h_{fe}}{40I_c} = \frac{120}{40 \times 1} = 3\text{k}\Omega.$$

Noise

Any working transistor generates electrical noise, and the greater the current flowing through the transistor the greater the noise. For bipolar transistors, the optimum collector current for low noise operation is given approximately (in milliamps) by

$$I_c = \frac{28\sqrt{h_{fe}}}{R_g} \quad \text{where } R_g \text{ is the signal source resistance in ohms.}$$

Low-noise operation is most important for the first stages of audio preamplifiers and for r.f. tuner and early i.f. stages. The noise generated by large value resistors is also significant, so that the resistors used for small signal input stages should be fairly low value, high stability film types, with small currents flowing. Variable resistors must never be used in a low-noise signal stage.

Voltage gain

The voltage gain of a simple single stage voltage amplifier can be found from a simple rule of thumb. If V_{bias} is the steady d.c. voltage *across the collector load resistor*, then the voltage gain is

$$G = 40 \times V_{bias}$$

For a single stage amplifier, the signal is attenuated both at the input and at the output by the resistance of devices connected to the transistor (microphones, tape heads, other amplifying stages). If the resistance of the signal source is R_s and the resistance of the next stage is R_{load}, then the measured gain will be

$$G \times \frac{h_{ie}}{R_s + h_{ie}} \times \frac{R_{load}}{R_{load} + R_c}$$

where R_c is the collector output resistance as shown in *Figure 2.12*, and G is the value of gain given by $40V_{bias}$. This method gives gain values accurately enough for most practical purposes. When precise values of gain are needed, negative feedback circuits (see below) must be used.

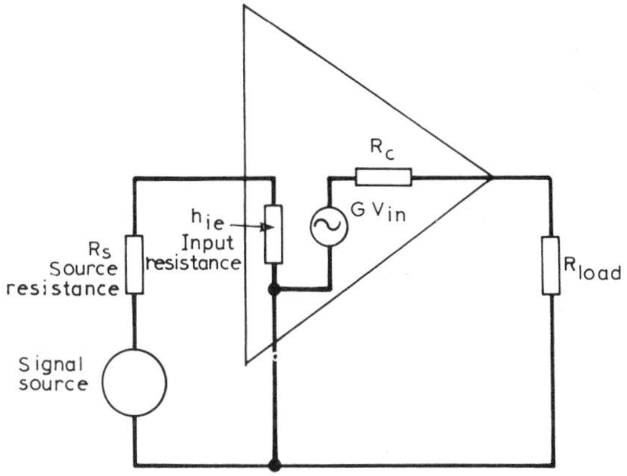

R_c
h_{ie}
Input
resistance
G V_{in}
R_s
Source
resistance
R_{load}
Signal
source

Figure 2.12. The voltage signal equivalent. The voltage gain, G, obtained by the transistor is reduced by the potential divider networks at the input and at the output

For a multi-stage amplifier, the gains of individual stages are multiplied together, and the attenuations caused by the potential dividing actions of R_s and R_{load} are also multiplied together.

FETs

Field effect transistors (FETs) are constructed with no junctions in the main current path between the electrodes, which are called drain and source respectively. The path between these contacts, called the channel, may be p-type or n-type silicon, so that both p-channel and n-channel FETs are found. Control of the current flowing in the

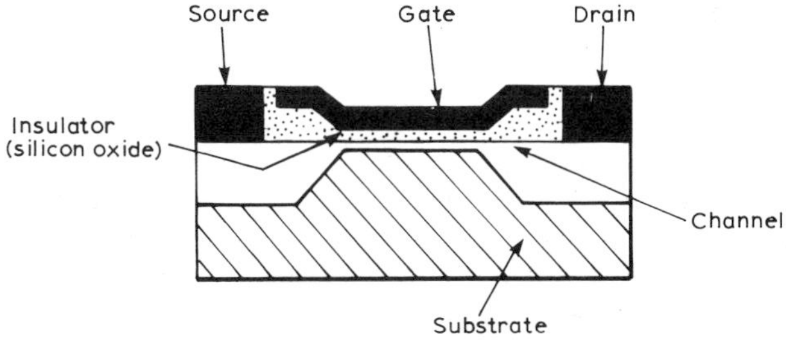

Source
Gate
Drain
Insulator
(silicon oxide)
Channel
Substrate

Figure 2.13. MOSFET structure

channel is achieved by varying the voltage at the third electrode, the gate. In junction FETs, the gate is a contact to a junction formed on the channel and reverse biased in most circuit applications. MOSFETs use a capacitor structure (*Figure 2.13*), so that the gate is completely insulated from the channel. No bias is needed, since the gate is insulated, but care has to be taken to avoid gate breakdown caused by excess voltage. Even electrostatic voltages, generated by contact, can cause damage, so that the gate electrode should be shorted to the source until the MOSFET is wired into circuit. In any circuit application, there must be a resistor connected from gate to source.

By altering the geometrical shape of the FET, power output FETs can be constructed. These are usually known as VFETs, the 'V' meaning 'vertical' and describing the construction, which is arranged so that the drain can be large and easily put into contact with a heat sink. Matched complementary pairs of these VFETs have been used to a considerable extent in hi-fi amplifiers.

The input resistance of either type of FET is very high, and low noise levels can be achieved, even with source resistances as high as $1\text{M}\Omega$.

Negative feedback

Feedback means using a fraction of the output voltage of the amplifier as an input. When the signals at input and output are oppositely phased (mirror-image waveform), then the feedback signal is said to be negative. Negative feedback signals subtract from the input signals to the amplifier, so reducing the overall gain of the amplifier. The effect on gain is as follows:

Let G = Gain of amplifier with no feedback, known as the open loop gain

n = feedback fraction (or loop gain), so that V_{out}/n is fed back.

Then the gain of the amplifier when negative feedback is applied is

$G/(1 + \frac{G}{n})$ = the closed loop gain.

For example, if open loop gain, $G = 100$ and $n = 20$ (so that 1/20 of the output voltage is fed back) then the closed loop gain is $\frac{100}{1 + \frac{100}{20}} = \frac{100}{6} =$ 16.7. A very useful approximation is that if the open loop gain G is very much greater than the loop gain, then the closed loop gain (with the negative feedback connected) is simply equal to n. This is because G/n is large, so that in the expression above it is much greater than unity and the expression is approximately $G/(G/n) = n$.

Negative feedback, in addition to reducing gain, reduces noise signals which originate in the components of the amplifier, and will also

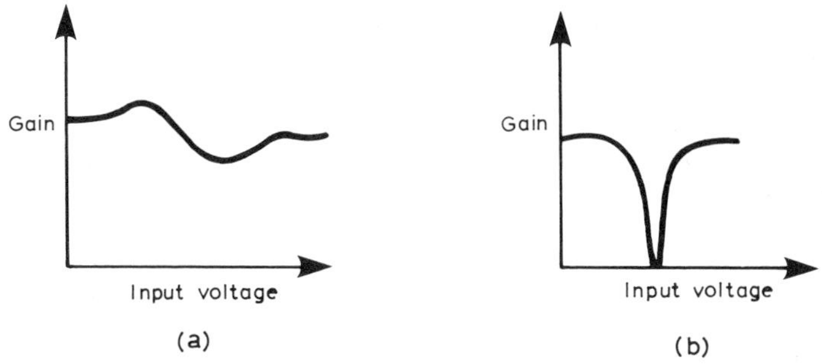
Gain
Input voltage
(a)
Gain
Input voltage
(b)

Figure 2.14. Negative feedback can correct non-linear distortion provided that the gain (before feedback) remains reasonably high (a) over the full range of input voltages. If the gain is zero for any input (b) or is unusually low, then feedback cannot correct the distortion. Cross-over distortion is an example of a fault which causes zero gain.

reduce distortion provided that the distortion does not cause loss of open-loop gain (*Figure 2.14*). Input and output resistances are also affected in the following way. If the feedback signal shunts the input (*Figure 2.15*) (applied to the same terminal), then input resistance is reduced, often to such an extent that the input terminal is practically at earth potential for signals (a *virtual* earth). If the feedback signal is in series with the input signal (*Figure 2.16*), the input resistance of the amplifier is increased. When the feedback network is connected in parallel with the *output* load (*Figure 2.16*) the effect is to reduce output

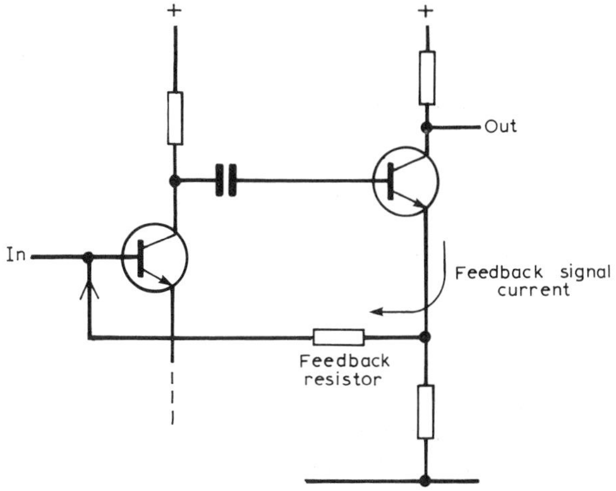
+
+
Out
In
Feedback signal current
Feedback resistor

Figure 2.15. A feedback circuit in which the feedback signal is in shunt with the input signal. At the output, the feedback resistor is connected in series with the output load

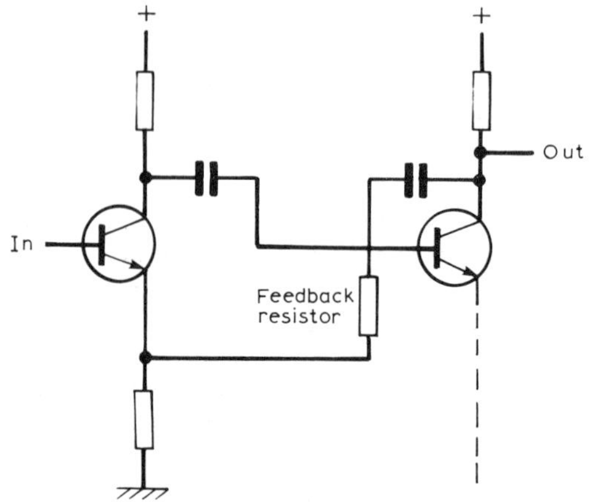
+
+
Out
In
Feedback
resistor

Figure 2.16. A feedback circuit in which the feedback signal is in series with the input signal through the emitter junction of the first transistor. The feedback resistor is also connected in parallel with the output load

resistance, and when the feedback is connected in series (*Figure 2.15*), the effect is to increase output resistance. The effects on output resistance are generally small compared to the effects on input resistance.

Heatsinks

A transistor passing a steady (or average) current *I* and with a steady (or average) voltage *V* between collector and emitter dissipates a power of *VI* watts. This electrical power is converted into heat at the collector-base junction, and unless this heat can be removed the temperature of the junction will rise until the junction fails irreversibly. Heat is removed in two stages, by conduction to the case or other metal work of the transistor, and into heatsinks if fitted, then by convection into the air. The temperature of the junction will stabilise when the rate of removing heat, measured in watts, is exactly equal to the electrical power dissipation; this may, however, happen only when the junction temperature is too high for continuous operation. The power dissipation of a transistor is limited therefore mainly by the rate at which heat can be removed.

For practical purposes, the resistance to heat transfer is measured by the quantity called thermal resistance, θ, whose units are °C/W. The same measuring units are used for convection from heatsinks as for conduction through the transistor, so that all the figures of thermal resistance from the collector-base junction to the air can be added together as for resistor values in series. The temperature difference between the junction and the air around the heatsink is then found by multiplying the total thermal resistance by the number of watts dissipated

$$T^\circ = \theta \times W$$

This latter figure is a temperature *difference,* so that to find the actual junction temperature, the temperature of the air around the heatsink (the ambient temperature) must be added to this figure. An ambient figure of between 30°C (for domestic equipment) and 70°C (for industrial equipment) may be taken. If this procedure ends with a calculated junction temperature higher than the manufacturer's rated values (120°C to 200°C for silicon transistors), then the dissipated power must be reduced, a larger heatsink used, or a water cooled heatsink used. Large power transistors are designed so that the transfer of heat from junction to case is efficient, with a low value of thermal resistance, and the largest thermal resistance in the 'circuit' is that of the heatsink-to-air. Small transistors generally have much higher thermal resistance values, so that the heatsinking is not so effective.

To ensure low thermal resistance, the collector of medium or high-power transistors is connected directly to the case. To prevent unintended short circuits, the heatsink may have to be insulated from other metal work, or the transistor insulated from the heatsink using mica washers. Such washers used with silicone heatsink grease can have thermal resistance values of less than 1°C/W and are available from transistor manufacturers or components specialists. The use of mica washers makes it possible to use a metal chassis as heatsink or to mount several transistors on the same heatsink.

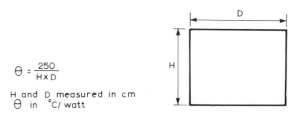

$$\theta = \frac{250}{H \times D}$$

H and D measured in cm
θ in °C/ watt

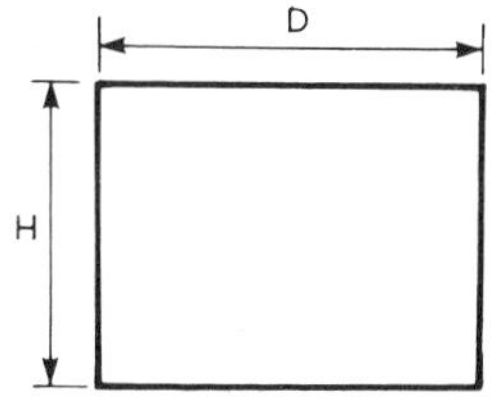
D
H

Figure 2.17. An approximate guide to the thermal resistance of a metal fin

The calculation of thermal resistance for heatsinks is not particularly simple, but *Figure 2.17* shows an approximate formula. Measurement of thermal resistance can be carried out by bolting a 25 W wirewound resistor of the metal-cased type to the heatsink. A 2.2 Ω value is suitable and will dissipate 4 W at 3 V and 16.4 W at 6 V. The temperature of the heatsink is measured when conditions have stabilised (no variation of temperature in one minute), and the electrical power figure is divided by the difference between heatsink temperature and ambient (air) temperature. This method is not precise, but gives values which are suitable for practical work.

Switching circuits

A linear amplifier circuit creates an 'enlarged' copy of a waveform. A pulse (logic) switching circuit charges rapidly from one value of voltage or current to another for a small change of voltage or current at the input. The output waveform need not be similar in shape to the input waveform, but the changes of voltage or current at the output should taken place with only a small time delay (a microsecond or less) after the changes at the input.

The bipolar (or junction) transistor has a good switching action because of its large G_m figure. A useful rule-of-thumb is that the collector current of a transistor will be increased tenfold by an extra 60 mV at the base, provided that the transistor is conducting before the extra base voltage is added, and is not saturated by the extra current. Current switches can thus be easily achieved, and a stage of current amplification can be added if larger current swings or smaller voltage swings are needed (*Figure 2.18*). A voltage switching stage must

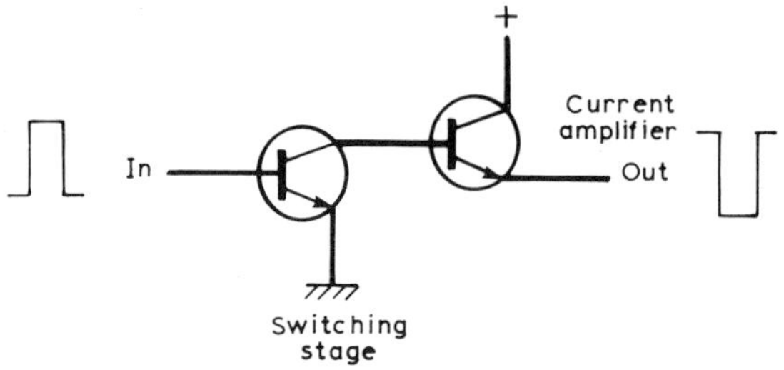
+
Current
amplifier
In
Out
Switching
stage

Figure 2.18. Adding a current-amplifying stage to a simple switching transistor

use some form of load to convert the current changes at the collector into voltage changes. If this load is a resistor, the switch-on of the transistor may be faster than the switch-off. Stray capacitances between the collector and earth are discharged rapidly by the current through the collector at switch-on, but must be charged through the load resistor when the transistor is switched off (*Figure 2.19*). If the rise time of the wave does not need to be short, this can be overcome by using a comparatively low value resistor (1 kΩ or less). An alternative method is shown in *Figure 2.20* using series connected transistors switching in either direction.

For fast switching, the stored charge of transistors can cause problems. During the time when the transistor is conducting, the emitter is injecting charges into the base region. These charges cannot disappear instantly when the base bias is reversed, so that the transistor will conduct momentarily in the reverse direction. As a result, the

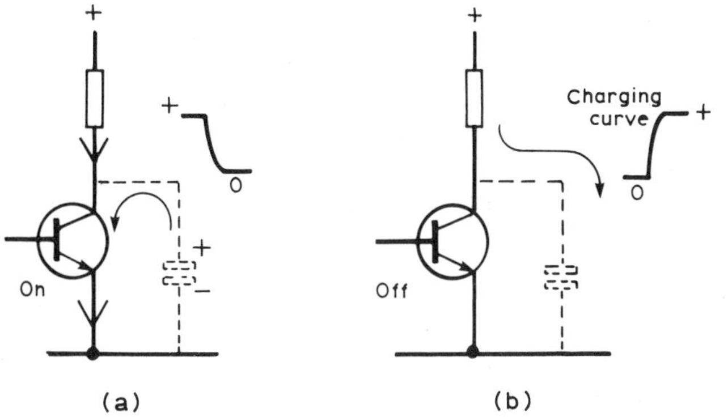
+
+
0
+
–
On
(a)
+
Charging
curve
+
0
Off
(b)

Figure 2.19. Charging and discharging stray capacitances. When the transistor conducts (a) the stray capacitance is rapidly discharged, and the voltage drop at the collector is sharp. When the transistor cuts off (b), the stray capacitance is recharged through the load resistor, causing a slower voltage rise

circuit of *Figure 2.20* can suffer from excessive dissipation at high switching speeds, since for short intervals, both transistors will be conducting. Manufacturers of switching transistors at one time quoted figures of stored charge, Q in units of picocoulombs (pC) or nano-coulombs (nC), but nowadays, generally quote the more useful turn-on

Figure 2.20. Using a two-transistor output circuit so that the switching is equally rapid in both directions. This type of output stage is used in TTL digital i.c.s

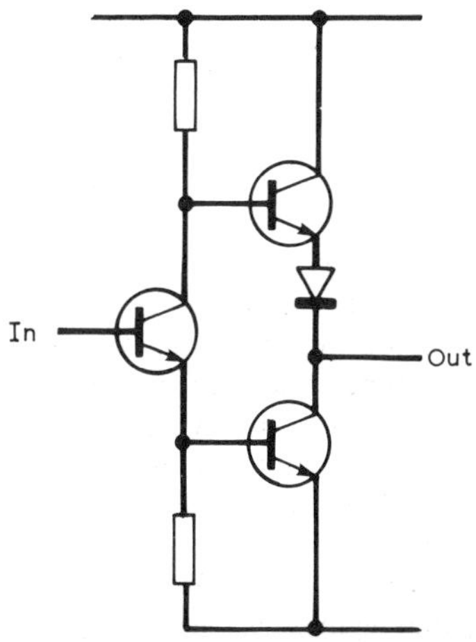
In
Out

and turn-off times, in nanoseconds (ns) under specified conditions. If only stored charge figures are given, an approximate value for turn-off time can be obtained from the equation

$$t = \frac{Q}{I}$$

where t is turn-off time in nanoseconds (10^{-9}s), Q is stored charge in pC and I is the current in mA which is to be switched off. Transistor switch-off times are improved by reverse-biasing the base, but some care has to be taken not to exceed the reverse voltage limits, since the base-emitter junction will break down at moderate voltages.

A considerable improvement in switch-off times is also obtained if the transistor is not allowed to saturate during its switch-on period; this has to be done by clamping the base voltage and is not easy because of the considerable variation of switch-on voltage between one transistor and another. The fastest switching times are achieved by 'current switching' circuits, in which the transistor is never saturated nor cut-off.

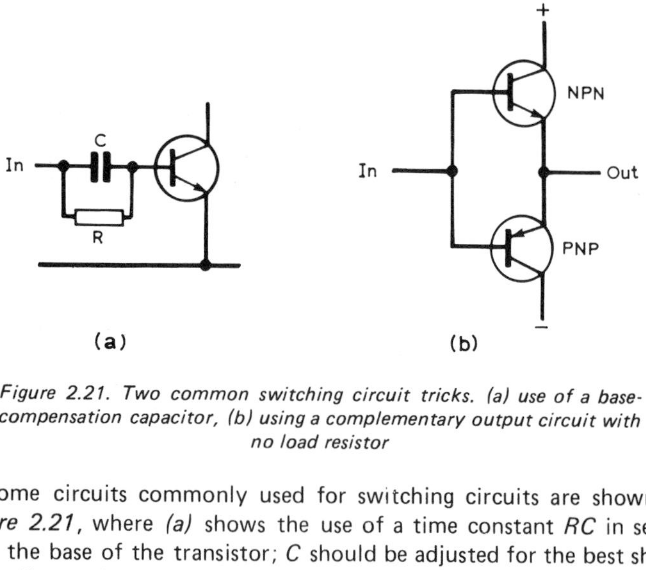
C
In
R
(a)
+
NPN
In
Out
PNP
−
(b)

Figure 2.21. Two common switching circuit tricks. (a) use of a base-compensation capacitor, (b) using a complementary output circuit with no load resistor

Some circuits commonly used for switching circuits are shown in *Figure 2.21*, where *(a)* shows the use of a time constant *RC* in series with the base of the transistor; *C* should be adjusted for the best shape of leading and trailing edges. *Figure 2.21b* shows the familiar double-emitter-follower circuit which uses transistors both to charge and to discharge stray capacitance.

Gating of analogue signals is an action similar to that of pulse (logic) switching, but the switch may be a series component rather than a shunt component, with the added reduction that it should not distort the analogue signal while in the ON state. Diodes, bipolar transistors, and FETs can all be used in such gating circuits. *Figure 2.22* shows a

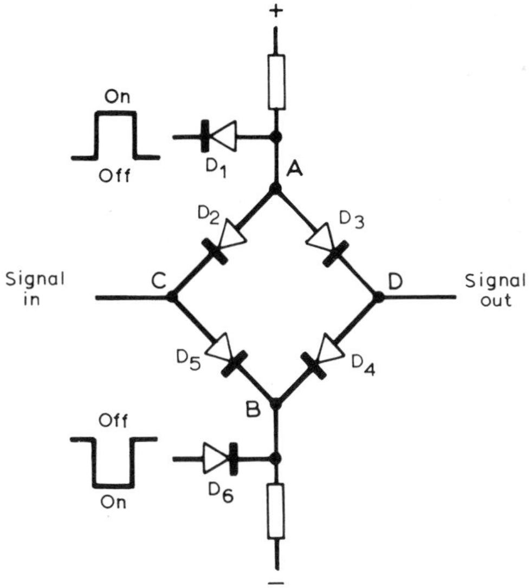
+
On
Off
D_1
A
D_2
D_3
Signal in
C
D
Signal out
D_5
D_4
B
Off
On
D_6
–

Figure 2.22. A diode-bridge gate circuit. In the off state, diodes D_1 and D_6 conduct, so that no current flows through D_2, D_3, D_4, D_5. When the gate is switched on by symmetrical pulses, D_1 and D_6 shut off, allowing the other diodes to conduct, so that the input signal can reach the output. Using symmetrical switching signals ensures that very little of the switching waveform appears in the output signal

diode bridge gate. When current flows through the diodes, assuming that there is a large resistance between point A and earth and between point B and earth, then there is a low resistance path C – D for signals in either direction. If the diodes are well matched, the d.c. level at D should be identical to that at C. Such a voltage difference is called an offset and is undesirable. When current ceases to flow, the diodes

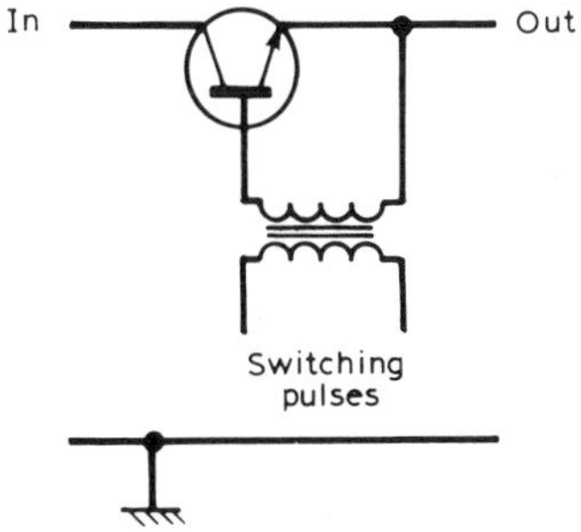
In
Out
Switching pulses

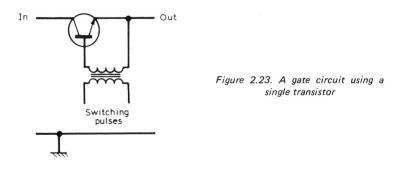

Figure 2.23. A gate circuit using a single transistor

become non-conducting, the gate closes. Offset voltages may be around 10 mV (comparatively high), but very high operating speeds are possible when switching diodes are used.

Bipolar transistors can be used as gates, with the offset voltage between collector and emitter being lowest (less than 2 nV) for a saturated transistor when the normal collector and emitter terminals are *reversed*. As a series switch, however, the transistor suffers from the requirement of a switching pulse applied between base and emitter, so that a transformer must be used to apply the switching voltage if a single transistor is to be used (*Figure 2.23*). A very common gating circuit is the long-tailed pair of *Figure 2.24*, which is, however, useful

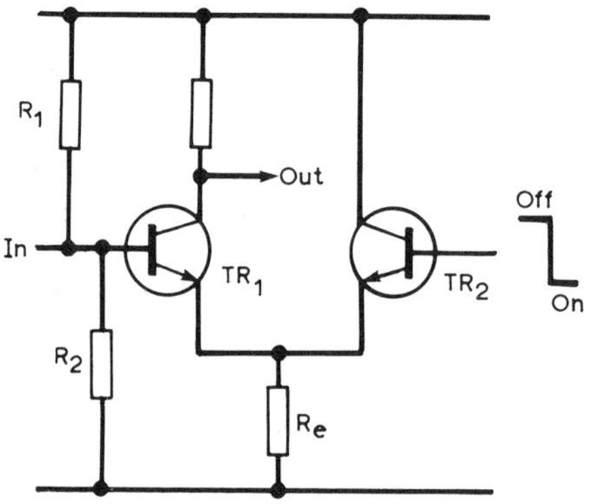
R1
Out
In
TR1
TR2
Off
On
R2
Re

Figure 2.24. The long-tailed pair gate. When Tr_2 is switched off Tr_1 is normally biased by R_1, R_2, and acts as an inverting amplifier. When Tr_2 is switched on, with its base voltage several volts higher than the normal bias voltage of Tr_1, then Tr_1 is biased off

only when the offset voltages (v_{be}) and the voltage change caused by switching are unimportant.

FETs may be used as shunt or series switches (*Figure 2.25*). The shunt switch is considerably easier to drive because the source terminal is earthed. Offset voltages of less than 10 μV are obtainable, and practically no drive current to the gate is needed. The disadvantage of the FET is that its resistance when switched ON is much higher (up to 1 kΩ) than that of a bipolar transistor.

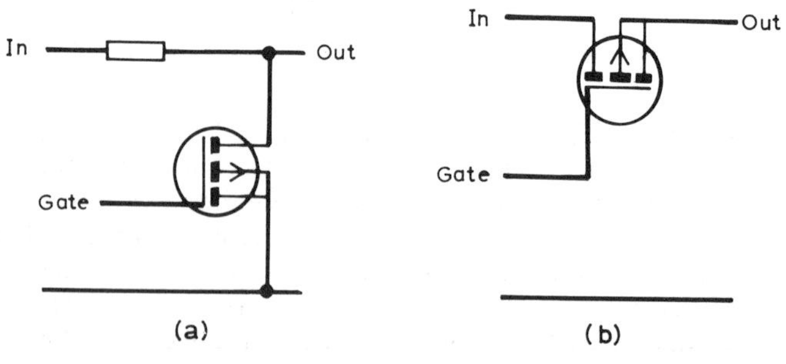
In
Out
Gate
In
Out
Gate
(a)
(b)

Figure 2.25. Using FETs as switches: (a) shunt, (b) series

Other switching devices

Unijunctions have two base contacts and an emitter contact, forming a device with a single junction which does not conduct until the voltage between the emitter and base contact 1 (*Figure 2.26*) reaches a specified level. At this level, the whole device becomes conductive. The unijunction

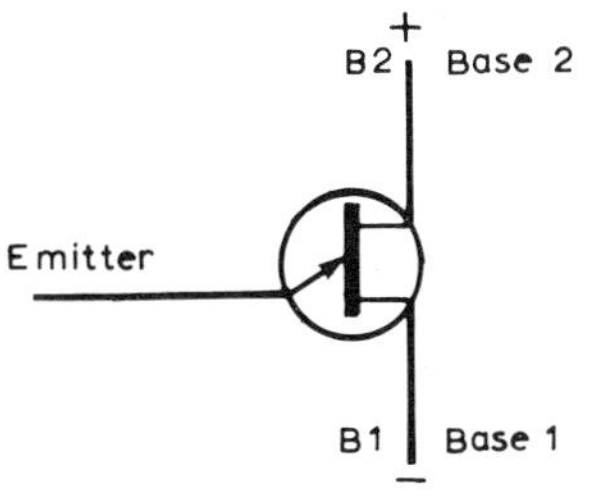
+
B2
Base 2
Emitter
B1
Base 1
−

Figure 2.26. Unijunction symbol

is used to generate short pulses, using circuits such as that of *Figure 2.27*. The frequencies of operation of this circuit are not noticeably affected by supply voltage changes, since the unijunction fires (becomes conductive) at a definite fraction of the supply voltage. The intrinsic stand-off ratio, *n*, is defined as

$$\frac{\text{firing voltage } (e - b_1)}{\text{supply voltage } (b_2 - b_1)}$$

and has values ranging typically from 0.5 to 0.86. Pulse repetition rates up to 1 MHz are obtainable.

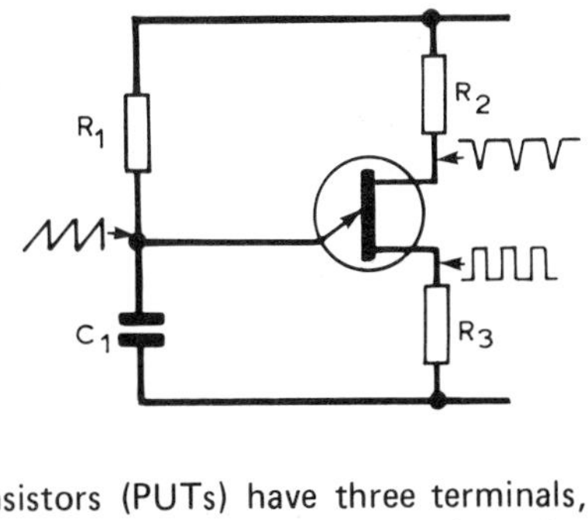
R1
R2
C1
R3

Figure 2.27. A unijunction oscillator. R_2, R_3 are about 100 ohms each, and the frequency of oscillation is determined by the time constant RC

Programmable unijunction transistors (PUTs) have three terminals, one of which is used to set the value of intrinsic stand-off ratio, *n*, by connection to a potential divider (*Figure 2.28*). Firing will occur at the programmed voltage; the frequency range is generally up to 10 kHz.

Thyristors are controlled silicon diodes which are switched into conduction by a brief pulse or a steady voltage at the gate terminal. Voltage of 0.8 to 1.5 V and currents of a few μA up to 30 mA are needed at the gate, according to the current rating of the thyristor.

Figure 2.28. A programmable unijunction transistor (PUT). The firing voltage between anode and cathode is selected by the voltage applied to the third electrode

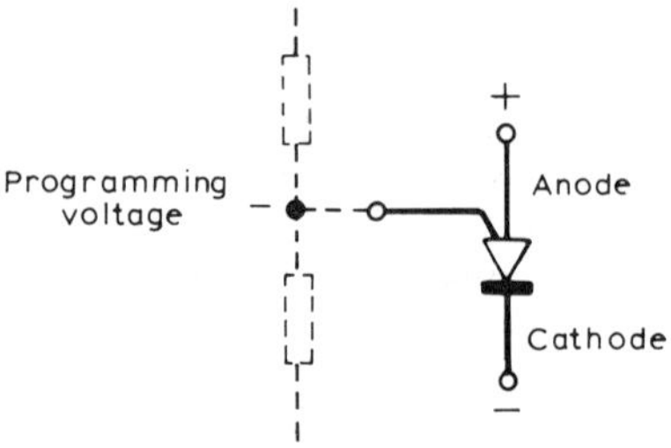
Programming voltage
+
Anode
Cathode
−

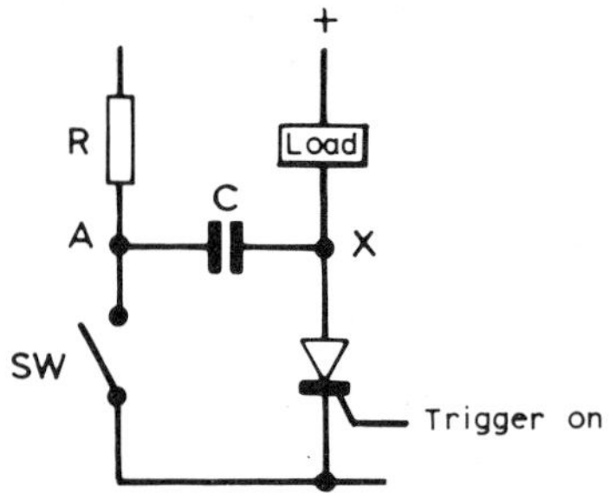
+
R
Load
C
A
X
SW
Trigger on

Figure 2.29. A capacitor turn-off circuit for a thyristor. When the switch is momentarily closed, the sudden voltage drop at A will cause an equal drop at X, turning off the thyristor until it is triggered again

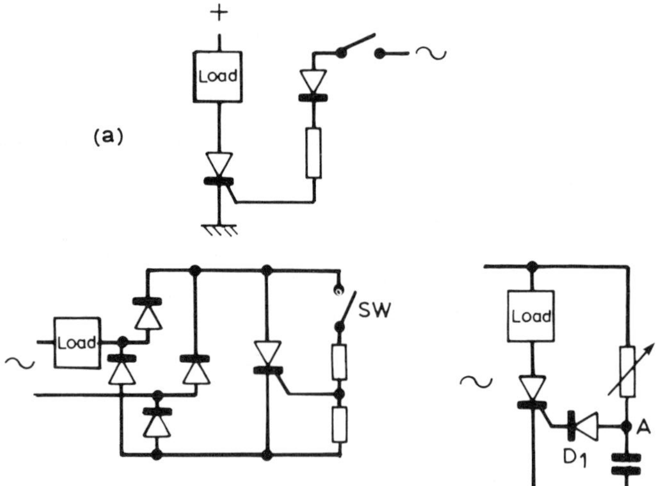
+
Load
(a)
Load
SW
(b)
Load
A
D_1
(c)

Figure 2.30. A.C. thyristor circuits. (a) Basic half-wave a.c. relay circuit. (b) A full-wave relay circuit. (c) Basic phase-control circuit. In the half-cycle during which the thyristor can conduct, the gate is activated only when the voltage at A has risen enough to cause the trigger diode D_1 to conduct. The time in the cycle at which conduction starts is controlled by the setting of the variable resistor

The thyristor ceases to conduct only when the voltage between anode and cathode falls to a low value (about 0.2 V) or when the current between anode and cathode becomes very low (less than 1 mA). D.C. switching circuits need some form of capacitor discharge circuit (*Figure 2.29*) to switch off the load. A.C. switching circuits, using a.c. or full wave rectified waveforms, are switched off by the waveform itself on each cycle. A few typical a.c. thyristor circuits are shown in *Figure 2.30*. Note that the gate signal may have to be applied through a pulse transformer, particularly when the thyristor switches mains currents, to avoid connecting the firing circuits to the gate.

Triacs are two-way thyristors whose terminals are labelled MT1, MT2 and gate. For reliable firing, the pulse at the gate should be of the same polarity as MT2 (some circuits are shown in *Figure 2.31*).

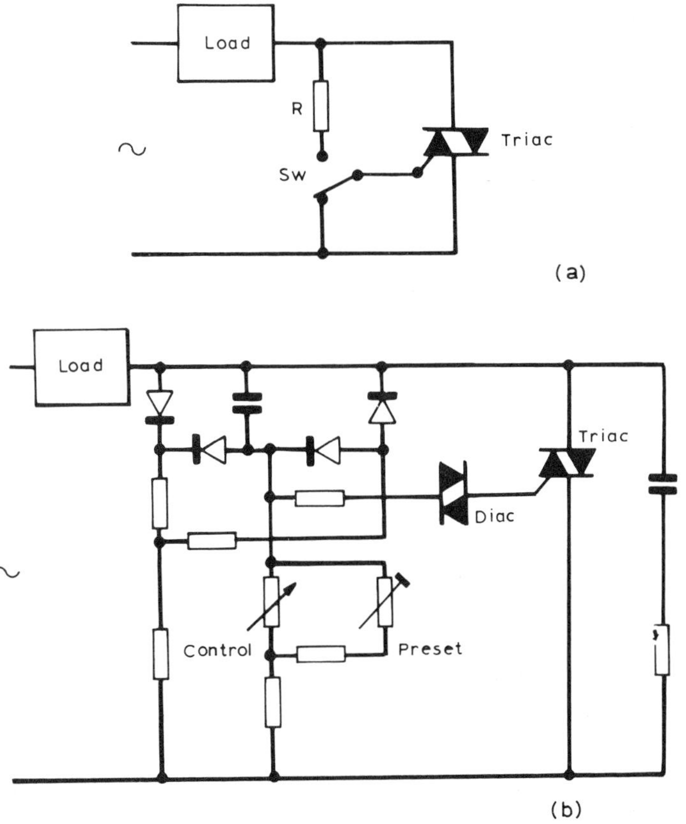
Load
R
~
Triac
Sw
(a)
Load
Triac
Diac
~
Control
Preset
(b)

Figure 2.31. Triac circuits. (a) Basic full-wave relay circuit. (b) Power regulator circuit, using a diac trigger diode, and radio interference suppression circuit across the triac

Firing pulses for thyristors and triacs can be obtained from unijunctions or from other types of trigger device such as diacs, silicon bidirectional switches, four-layer diodes or silicon unidirectional switches. The diac, or bidirectional trigger diode, is non-conductive in either direction until its breakdown voltage is exceeded, after which the device conducts readily until the voltage across its terminals (either polarity) is low. Firing voltages of 20 to 36 V are typical, and the 'breakback' voltage (at which the device ceases to conduct) is typically 6 V. Brief peak currents of 2 A are possible. The silicon bidirectional switch also uses a gate electrode, but operates with one polarity only. Four-layer diodes have lower firing and breakback voltages than diodes, but essentially similar characteristics.

The silicon controlled switch (SCS) is a useful device with four electrodes which can be used, according to connections, either as a programmable unijunction or as a low-power thyristor. The connections are referred to as anode, cathode, gate-anode and gate-cathode. If the gate-cathode is used together with the anode and cathode, thyristor operation (at low currents) is obtained; if the gate-anode is used, the

Table 2.1 PRO-ELECTRON CODING

The first letter indicates the semiconductor material used:

A	Germanium
B	Silicon
C	Gallium arsenide and similar compounds
D	Indium antimonide and similar compounds
R	Cadmium sulphide and similar compounds

The second letter indicates the application of the device:

A	Detector diode, high speed diode, mixer diode
B	Variable capacitance (varicap) diode
C	A.F. (not power) transistor
D	A.F. power transistor
E	Tunnel diode
F	R.F. (not power) transistor
G	Miscellaneous
L	R.F. power transistor
N	Photocoupler
P	Radiation detector (photodiode, phototransistor, etc.)
Q	Radiation generator (LED etc.)
R	Control and switching device (such as thyristor)
S	Switching transistor, low power
T	Control and switching device (such as a triac)
U	Switching transistor, high power
X	Multiplier diode (varactor or step diode)
Y	Rectifier, booster or efficiency diode
Z	Voltage reference (zener), regulator or transient suppressor diode.

The remainder of the code is a serial number. For consumer applications, such as radio, TV, hi-fi, this has three figures. For industrial and telecommunications use, a letter W, X, Y or Z and two figures are used.

device behaves as a PUT. The unused electrode is generally left open circuit.

Table 2.1 shows the European Pro-Electron coding used for semiconductors. The US (JEDEC) 1N and 2N numbers are registration numbers only (this also applies to the Japanese 2SA, 2SB etc. system), and the function of a semiconductor cannot be guessed from the number.

Chapter 3

Discrete Component Circuits

This Chapter illustrates a selection of well established circuits and data, and comments are reduced to a minimum so as to include the greatest possible number of useful circuits. The common-emitter and a few other amplifier circuits have already been dealt with in Chapter 2.

Where several different types of circuits are shown (as for oscillators) practical considerations may dictate the choice of design. For example,

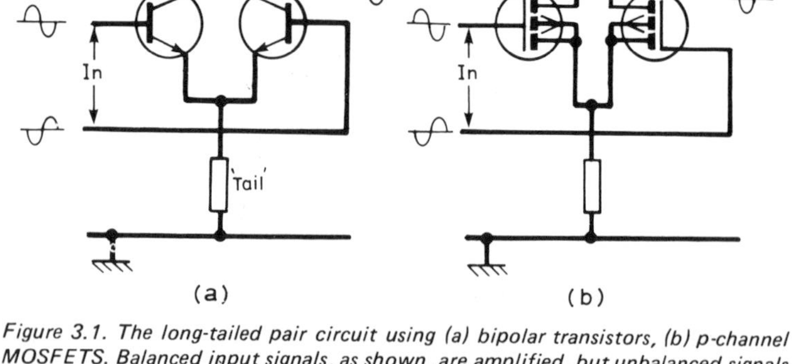

+
Out
In
'Tail'
(a)
+
Out
In
(b)

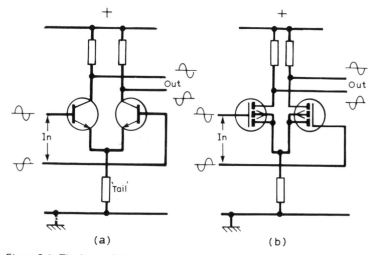

Figure 3.1. The long-tailed pair circuit using (a) bipolar transistors, (b) p-channel MOSFETS. Balanced input signals, as shown, are amplified, but unbalanced signals (in the same phase at each input) are attenuated

a Hartley oscillator uses a tapped coil, but the arrangements for frequency variation may be more convenient than those for a Colpitts oscillator, which uses a capacitive tapping. Some crystal oscillator circuits are not always self-starting, particularly with 'difficult' crystals. For this reason, as many variations on basic circuits have been shown as is feasible within the space.

The long-tailed pair, shown both in bipolar and in FET form in *Figure 3.1* is the most versatile of all transistor circuits, which is why it is so extensively used for linear i.c.s. A common-mode signal is a signal applied in the same phase to both bases or gates. Any amplification of such a common mode signal can only be caused by lack of balance between the transistors, so that this value of gain is low. The difference signal is amplified with a considerably greater gain. The long-tailed pair is most effective when used as a balanced amplifier, with balanced input and output, but single-ended inputs or outputs can be provided, as shown in *Figure 3.2a* and *b.* The overall voltage gain of a long-tailed pair circuit is about half the gain that would be obtained from one of the transistors, using the same load and bias conditions.

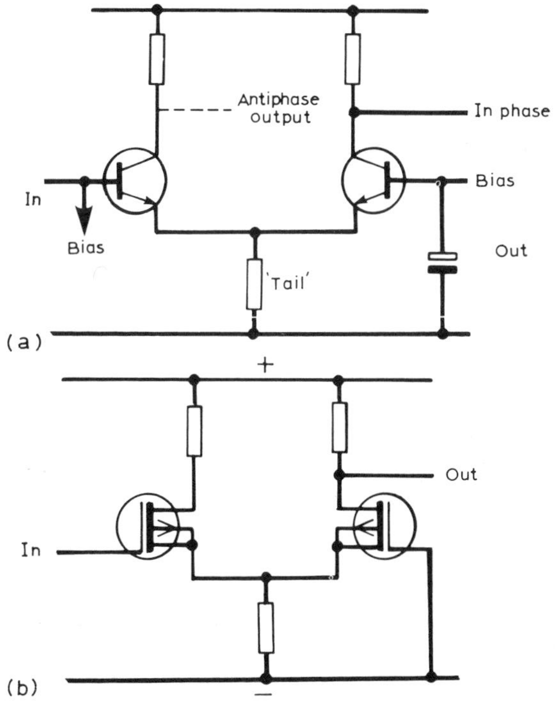
Antiphase output
In phase
In
Bias
Bias
Out
'Tail'
(a)
+
Out
In
(b)
−

Figure 3.2. Single-ended inputs and outputs on a long-tailed pair circuit. The second input is earthed to signals — no bias arrangements have been shown. (a) Bipolar transistors, (b) p-channel MOSFETs

Figure 3.3 ilustrates a magnetic pickup preamplifier circuit. The problem here is to apply the frequency correction (equalisation) needed for disc replay. Discs are recorded to the RIAA standard (see BS 1928: 1965), in which bass frequencies are attenuated (to prevent excessive cutter movement) and treble frequencies are boosted (to have an amplitude well above that of surface noise). This deliberate distortion must be corrected at playback by *CR* networks; to achieve this correctly, three time constants, 75 μs, 318 μs and 3180 μs, are used. The 75 μs time constant is of a treble cut filter, taking effect (that is, with its turnover point) at 2.12 kHz; the 318 μs kHz is a bass boost starting at 500 Hz, and the 3180 μs time constant is a final stage of bass cut at 50 Hz and under.

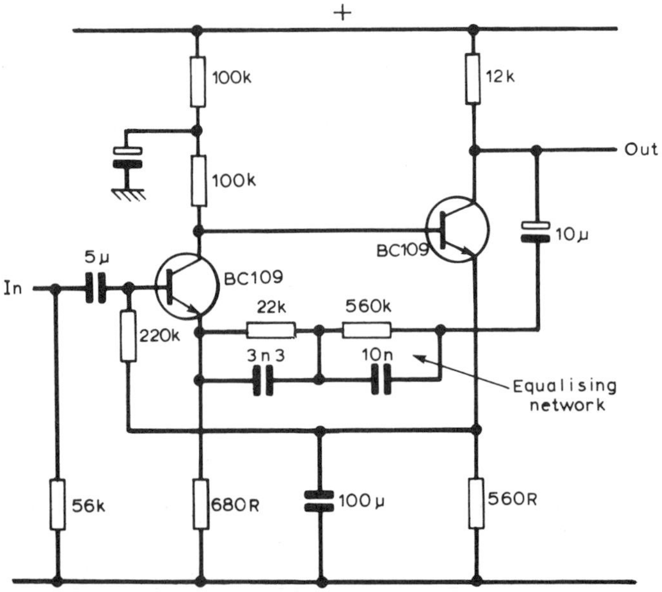
+
100k
12k
Out
100k
10μ
BC109
5μ
In
BC109
22k
560k
220k
3n3
10n
Equalising network
56k
680R
100μ
560R

Figure 3.3. An input stage for magnetic pickup cartridges. This stage combines high gain with equalisation, and a fairly high voltage supply (40 V or more) is needed to avoid overloading on transients

The output from most magnetic cartridges, other than the moving-coil type, is around 5 mV at the standard conditions of 5 cm/s stylus velocity and 1 kHz signal. This corresponds to a signal amplitude of close to the minimum most amplifiers need to give full output for 2.5 mV input at full volume setting. The preamplifier should present an input resistance of around 50 kΩ and should be capable of accepting considerable overloads at the input, 50 mV or more, without noticeable distortion. The trend at the time of writing is away from the feedback-loop type of equalisation, and towards an equalising network of purely passive type applied after a 'flat' preamplifier stage — this is held to be advantageous because of the effects of transients on feedback amplifiers, particularly when momentarily overloaded.

Figure 3.4 shows some tape/cassette input circuits. Once again, equalisation is needed, but the time constants are not quite so universally agreed; they appear to change almost annually as new tape materials and new types of tape head construction appear. In addition to these 'standard' corrections, individual tape decks may need further corrections, a multiplex filter may be included to remove f.m. stereo subcarrier

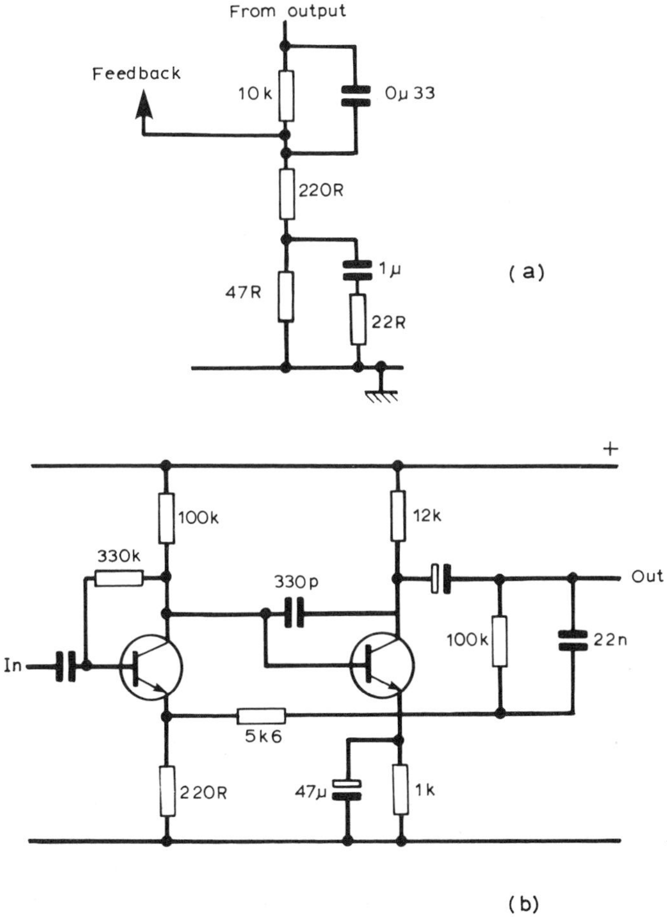
From output
Feedback
10 k
0μ 33
220R
1 μ
47R
22R
(a)
+
100k
12k
330k
330 p
Out
100 k
22n
In
5k6
220R
47μ
1 k
(b)

Figure 3.4. Tape equalisation and input stages. (a) One form of feedback network which can be used for reel-to-reel tape equalisation. (b) A cassette recorder input stage, using rather different time constants in the equalisation networks

signals, and noise-cancelling circuits, such as the Dolby circuits, may be used. At the last count, the equalisation frequencies being used on replay were 3180 μs for all tapes and either 70 μs or 120 μs for chrome and ferric tapes respectively; ferrichrome and pure iron particle tapes are replayed at 70 μs. The equalisation needed for recording amplifiers is too specialised to include here partly because recording equalisation time constants depend much more on individual needs.

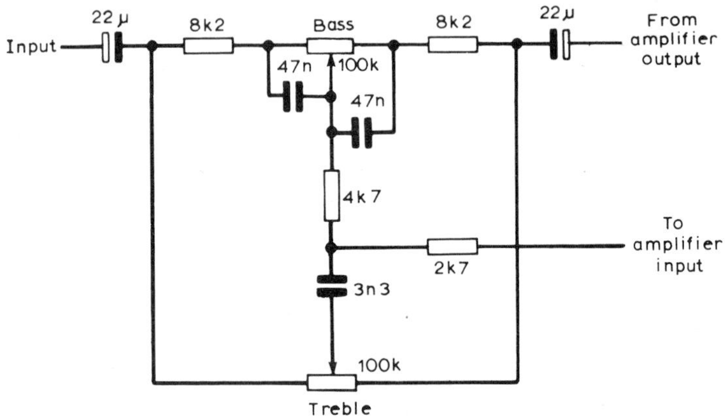
Input
22 μ
8k2
Bass
8k2
22 μ
From amplifier output
47n
100k
47n
4k7
To amplifier input
2k7
3n3
100k
Treble

Figute 3.5. A Baxandall type of tone control circuit. This circuit is normally located between two voltage amplifier transistors which provide the necessary gain

Figure 3.5 shows a version of the Baxandall tone control circuit, which is virtually the standard method of tone control used nowadays. There is very little interaction between the treble and the bass controls, low distortion, and a good range of control; about 20 dB of boost or cut. The Baxandall circuit is usually located between bipolar transistors, but several designs claim significant improvements by using a FET at the output of the control stage.

Figure 3.6 deals with active filters. These designs use only resistors and capacitors, together with semiconductors, and are considerably simpler to design than *LC* filters. Low-pass, high-pass, bandpass and notch filters are illustrated in *Figure 3.6*. The filters generally have a slope of 12 dB per octave (meaning that the response changes by 12 dB for each doubling or halving of frequency).

Figure 3.7 shows a typical input preamplifier stage for a moving-coil microphone. The particular features here are low noise operation, matching a fairly low input resistance, high gain, and hum rejection. The low output and low resistance of the moving-coil microphone requires the use of a microphone transformers. If a balanced layout is possible, hum pickup can be greatly reduced.

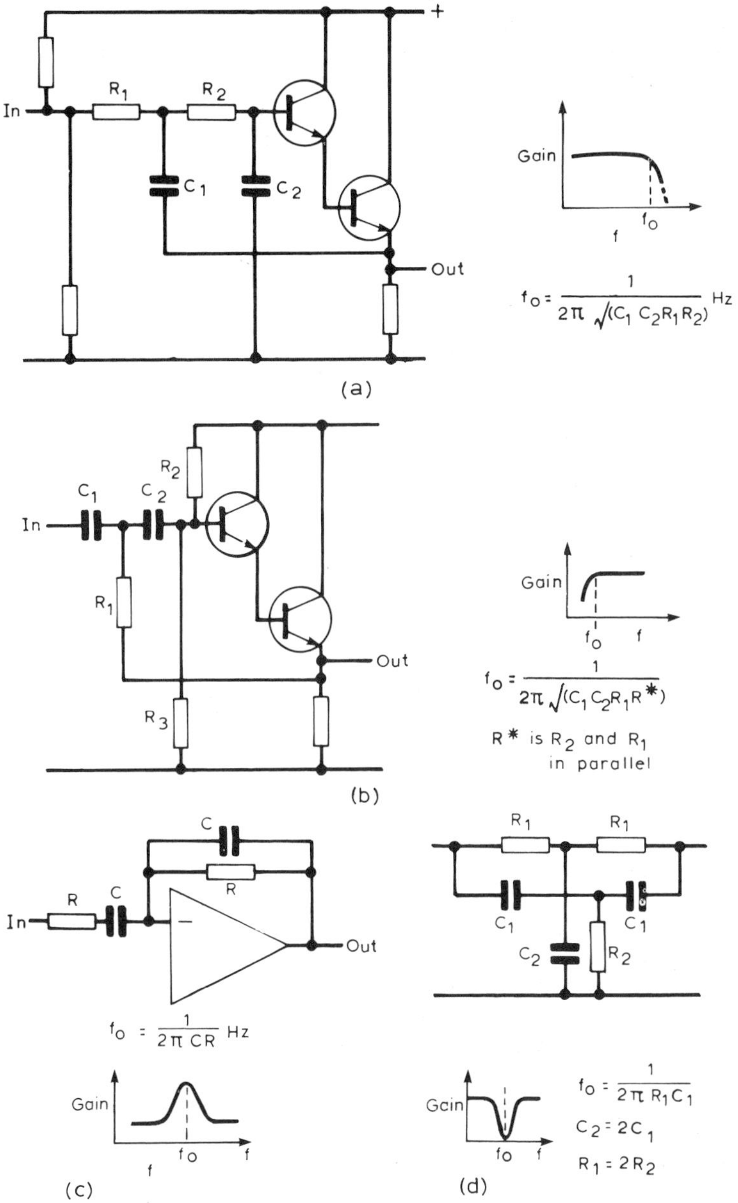
+
In
R1
R2
C1
C2
Out
Gain
f
fo
fo = 1 / 2π √(C1 C2 R1 R2) Hz
(a)
C1
C2
R2
In
R1
R3
Out
Gain
fo
f
fo = 1 / 2π √(C1 C2 R1 R*)
R* is R2 and R1
in parallel
(b)
C
R
R
C
In
Out
fo = 1 / 2π CR Hz
Gain
fo
f
(c)
R1
R1
C1
C1
C2
R2
Gain
fo
f
fo = 1 / 2π R1 C1
C2 = 2C1
R1 = 2R2
(d)

Figure 3.6. Active filter circuits. (a) Low pass, (b) High pass. These two types are Sallen & Key filters, named after the inventors. Circuit (c) is a bandpass filter using the Wien bridge network and an inverting amplifier. The gain of the amplifier determines the effective Q of the circuit, so that higher gain causes narrower bandwidth. Circuit (d) is a passive twin-T notch filter. The selectivity can be improved by using positive feedback

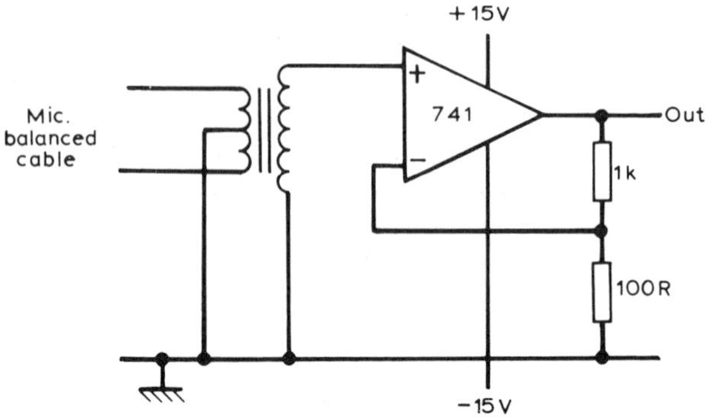
+15V
Mic.
balanced
cable
741
Out
1k
100R
−15V

Figure 3.7. A moving-coil microphone input circuit. A suitable input transformer is essential if high-quality results are expected. The transformer should preferably be supplied by the makers of the microphone

The next three sets of circuits deal with audio output stages. Class A stages are those in which the transistor(s) are always biased on and never saturated (bottomed). A Class A stage may use a single transistor (a single-ended stage) or two transistors which share the current in some way (a push-pull stage), but the efficiency is low. % Efficiency is defined as

$$\frac{\text{power dissipated in the load} \times 100}{\text{total power dissipated in the output stage}}$$

and is always less than 50% for Class A operation.

A Class A stage should pass the same current when no signal is applied as when maximum signal is applied. Because of this, the dissipation is large, so that large-area heatsinks are needed for the output transistors.

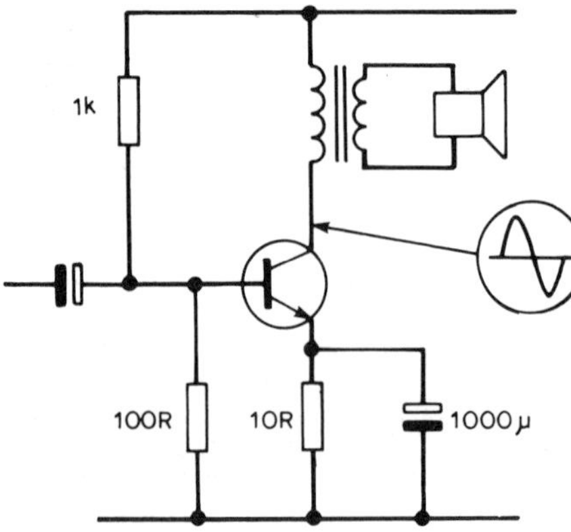
1k
100R
10R
1000μ

Figure 3.8. A Class A single-ended output stage. Good heatsinking is essential

Class B audio operation uses two (or more) transistors biased so that one conducts on one half of the waveform and the other on the remaining half. Some bias must be applied to avoid 'crossover distortion' due to the range of base-emitter voltage for which neither transistor would conduct in the absence of bias. Class B audio stages can have efficiency figures as high as 75%, though at the expense of rather higher distortion than a Class A stage using the same layout. The higher efficiency enables greater output power to be obtained with smaller heatsinks, and the use of negative feedback can, with careful design, reduce distortion to negligible levels.

Figure 3.8 shows a Class A single-ended power output stage, suitable for general-purpose use such as car radio operation. *Figure 3.9* shows the totem-pole or single-ended push-pull circuit, which can be used for either Class A or Class B operation according to the bias level. This version uses complementary symmetry – the output transistors are

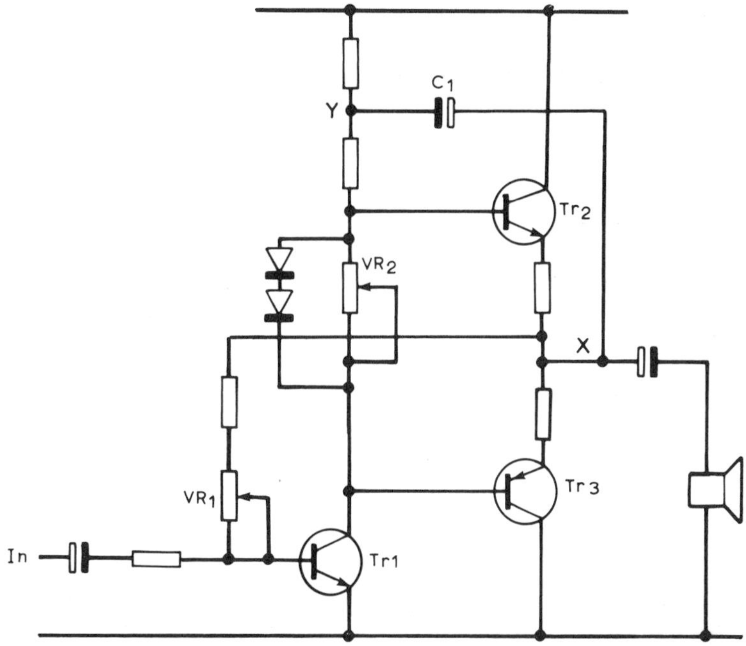

C1
Y
Tr2
VR2
X
VR1
Tr3
In
Tr1

Figure 3.9. A single-ended push-pull (totem-pole) Class B output stage, using complementary power transistors. VR_1 sets the voltage at point X to half of the supply voltage, VR_2 sets the quiescent (no signal) current through the output transistors. C_1 is a 'bootstrap' capacitor which feeds back in-phase signals to point Y, increasing input impedance. Oscillation is avoided because the gain of Tr_2 is less than 1

pnp and npn types. When complementary output transistors cannot be obtained, a pseudo-complementary circuit, such as that of *Figure 3.10*, can be used, though this is not truly symmetrical.

Figures 3.11, 3.12 and *3.13* illustrate some of the circuits used for wideband voltage amplification. *Figure 3.11* deals with methods of frequency compensation using inductors or capacitors to compensate for the shunting effect of stray capacitances. *Figure 3.12* shows a circuit which uses feedback to reduce the gain and so extend the flat portion

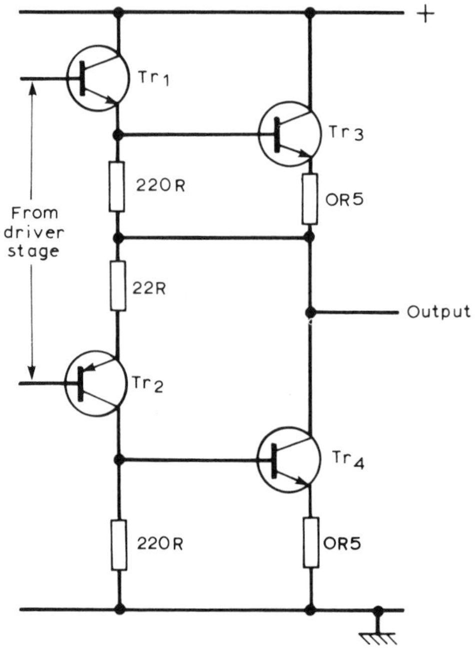
+
Tr1
Tr3
220R
OR5
From
driver
stage
22R
Output
Tr2
Tr4
220R
OR5

Figure 3.10. A quasi-complementary output stage. The low power complementary transistors Tr_1, Tr_2 drive the high-power output pair, Tr_3, TR_4, which are not complementary types. The circuit is not symmetrical, and can cause considerable distortion when overdriven

of the frequency range – this is a useful basic circuit for video frequencies. *Figure 3.13* shows a cascode amplifier, a type of construction which was widely used in the days of valves, but which has been strangely neglected in transistor circuits. The advantage of the cascode is stability, because there is practically no feedback from output to input and high gain over a large bandwidth. FET cascodes and combinations of FET and bipolar transistors can also be used.

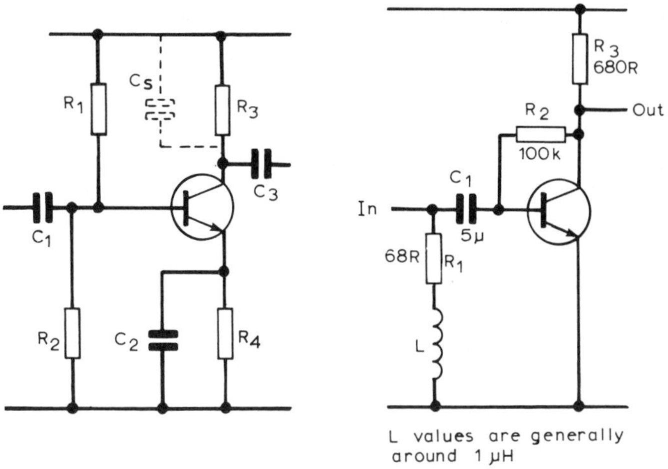
R1
Cs
R3
C3
C1
R2
C2
R4
R3
680R
R2
Out
100k
C1
In
5µ
68R
R1
L
L values are generally around 1 µH

Figure 3.11. Frequency compensation for wideband amplifiers. (a) Capacitive compensation. The value of C_2 is chosen so that R_4 is progressively decoupled at high frequencies. As a rough guide, C_sR_3 should equal C_2R_4. (b) Inductive shunt compensation. The value of L is chosen so as to resonate with the input capacitance of the transistor at a frequency above that of the uncompensated 3 dB point. These compensation methods are useful, but cannot compensate for low gain caused by an unsuitable transistor type. Transistors capable of amplification at high frequencies must be used in these circuits

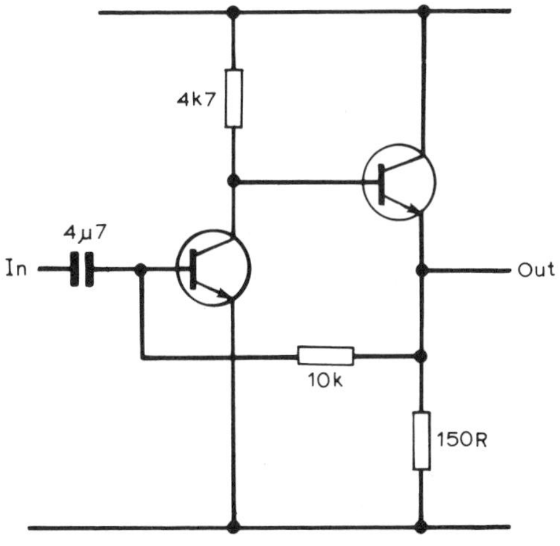
4k7
4µ7
In
Out
10k
150R

Figure 3.12. A feedback pair circuit which is capable of wideband amplification when suitable transistors are used. Bandwidths of 5 MHz or more are obtainable

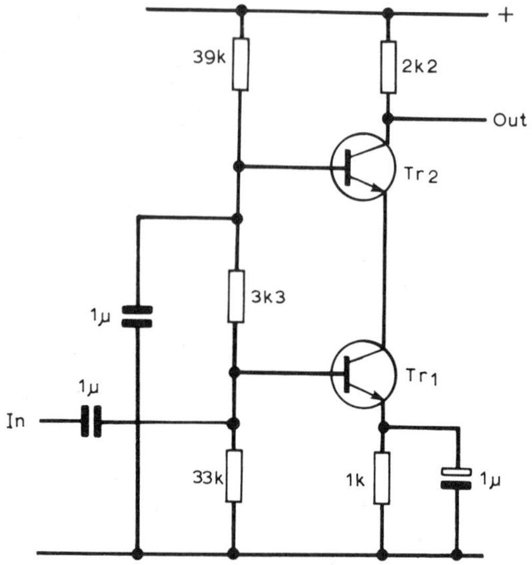
+
39k
2k2
Out
Tr2
3k3
1μ
1μ
Tr1
In
33k
1k
1μ

Figure 3.13. The cascode circuit, using, in this example, bipolar transistors. This is a exceptionally stable circuit, because of the isolation between input and output, and has high gain and a large bandwidth

The circuits of *Figures 3.14* to *3.16* are of sine wave oscillators which operate at radio frequencies. The Hartley type of oscillator (*Figure 3.14*) uses a tapped coil; the Colpitts type (*Figure 3.15*) uses a capacitor tap. Though these are not the only r.f. oscillator circuits, they are the circuits most commonly used for variable frequency oscillators. *Figure 3.16* illustrates some crystal controlled oscillator circuits. The frequency of the output need not be the fundamental crystal frequency, since most crystals will oscillate at higher harmonics (overtones) and harmonics can be selected at the output. Frequency multiplier stages (see *Figure 3.26*) can then be used to obtain still higher frequencies.

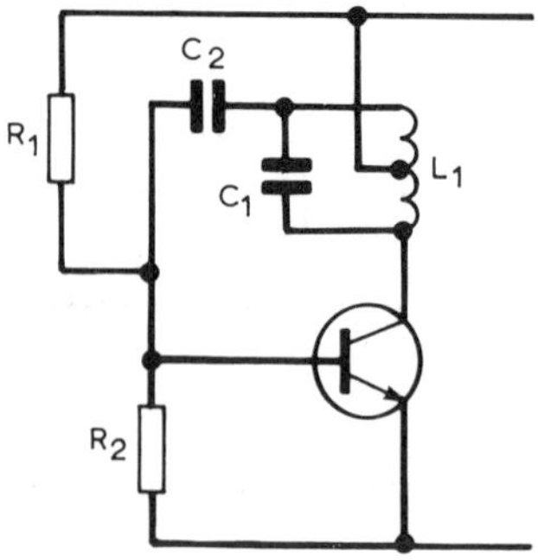
C2
R1
L1
C1
R2

Figure 3.14. The Hartley oscillator. The resonant circuit is L_1C_1, and the value of C_2 should be chosen so that the amount of positive feedback is not excessive, since this causes a distorted waveform. R_1 should be chosen so that the transistor is just drawing current when C_1 is short-circuited

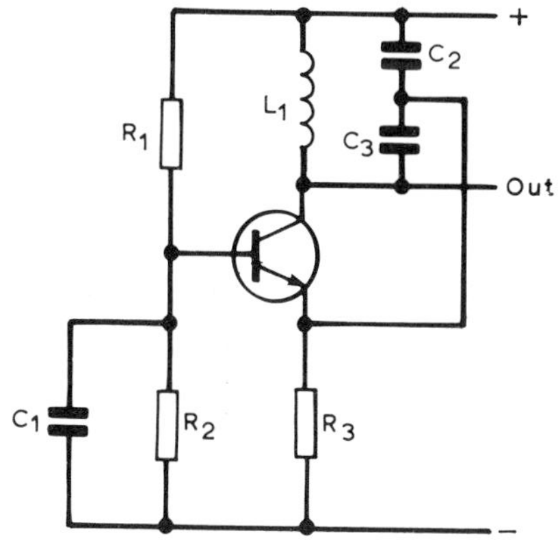
+
C2
L1
R1
C3
Out
C1
R2
R3
–

Figure 3.15. The Colpitts oscillator. The tapping is provided by C_2, C_3, and it is the series combination of these two capacitors which tunes L_1. As a rough guide, the value of C_3 should be about ten times the value of C_2 to avoid over-driving. R_1 is chosen to give about 1 mA of collector current. C_1 must not be omitted, because oscillation is impossible (because of negative feedback to the base) unless the base is decoupled

For low frequencies, oscillators such as the Wien bridge (*Figure 3.17*) or twin-T types are extensively used. Usable frequency ranges are from 1 Hz, or lower, to around 1 MHz.

Untuned or aperiodic oscillators are important as generators of square and pulse waveforms. *Figure 3.18* shows the familiar multivibrator (astable) together with modifications which improve the shape of the waveform. The less familiar serial multivibrator is shown in *Figure 3.19;* this circuit is a useful source of narrow pulses. When a pulse of a determined, or variable, width is required from any input (trigger)

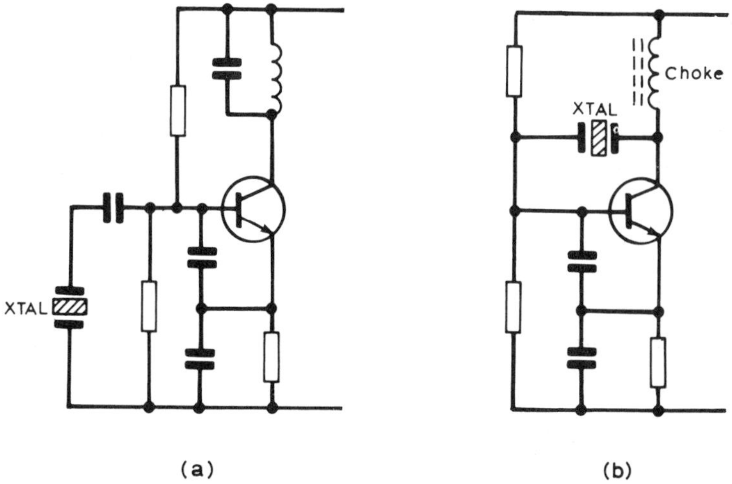
XTAL
Choke
XTAL
(a)
(b)

Figure 3.16. Two versions of a Colpitts crystal oscillator, showing alternative positions for the cystal

pulse, a monostable circuit must be used. *Figure 3.20* shows a monostable circuit, with a block diagram to illustrate how a combination of astable and monostable can form a useful pulse generator.

Figure 3.21 shows the basic bistable circuit, now rather a rarity in the discrete form thanks to the low price of i.c. versions. The Schmitt trigger is illustrated in *Figure 3.22*; its utility is as a comparator and

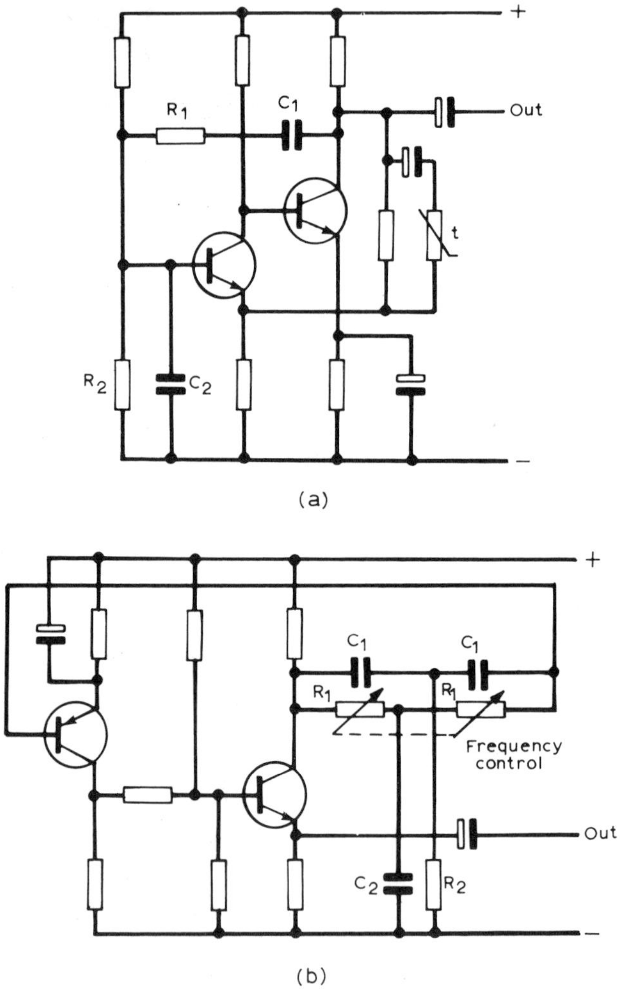

+
R1
C1
Out
t
R2
C2
–
(a)
+
C1
C1
R1
R1
Frequency control
Out
C2
R2
–
(b)

Figure 3.17. Low-frequency oscillators. (a) Wien bridge, (b) twin-T. The Wien bridge circuit uses a thermistor to keep the amplitude of the output signal constant. R_1, R_2 may be a ganged variable if a variable frequency output is needed

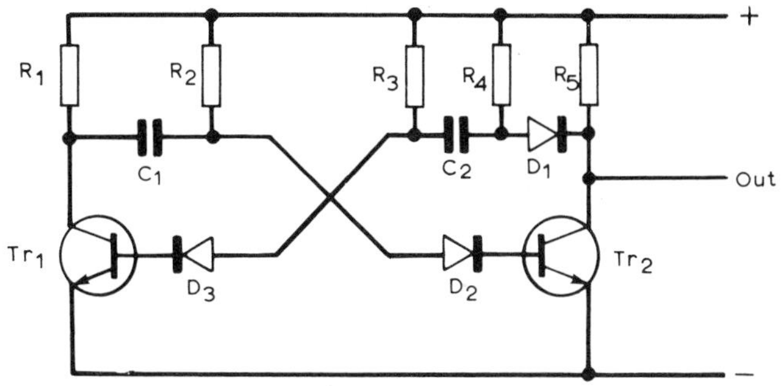

+
R1
R2
R3
R4
R5
C1
C2
D1
Out
Tr1
D3
D2
Tr2
–

$$T_1 = 0{\cdot}7\, C_1 R_2$$
$$T_2 = 0{\cdot}7\, C_2 R_3$$
$$f = \frac{1}{T_1 + T_2}$$

Figure 3.18. The astable multivibrator. The frequency of operation is given by the formula shown. The diodes D_2, D_3 prevent breakdown of the base-emitter junctions of the transistors when the transistors are turned off, and D_1 isolates the collector of Tr_2 from C_2 when Tr_2 switches off. In this way, a fast-rising waveform can be obtained

trigger stage which gives a sharply changing output from a slowly changing input. The hysteresis (voltage difference between the switching points) is a particularly valuable feature of this circuit. A circuit with hysteresis will switch positively in each direction with no tendency to 'flutter' or oscillation, so that Schmitt trigger circuits are used extensively where electronic sensors have replaced purely mechanical devices such as thermostats.

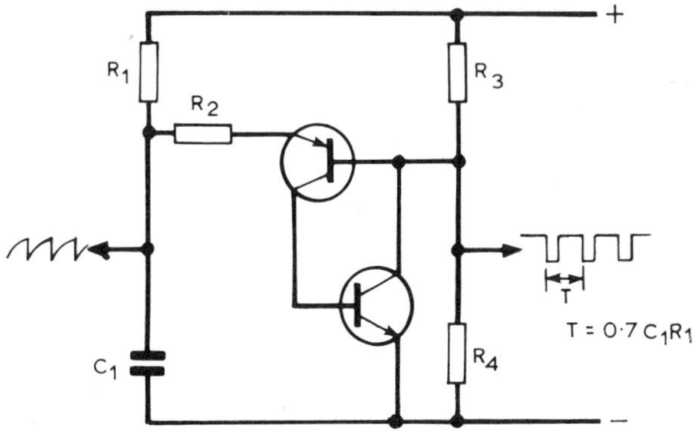

+
R1
R2
R3
C1
R4
T
T = 0·7 C1R1
–

Figure 3.19. The serial astable. Only one time constant is needed, and the outputs are a sawtooth and a pulse as shown. Usually $R_3 = R_4$; values of around 10 kΩ are usual

Radio-frequency circuits are represented here by only a few general examples, because the circuits and design methods that have to be used are fairly specialised, particularly for transmission, and the reader who wishes more information on purely r.f. circuits is referred to the excellent amateur radio publications. *Figure 3.23* shows an a.m./f.m. i.f. amplifier for 470 kHz and 10.7 MHz such as would be used in a.m./f.m. receivers. This design uses a common emitter amplifier, since the

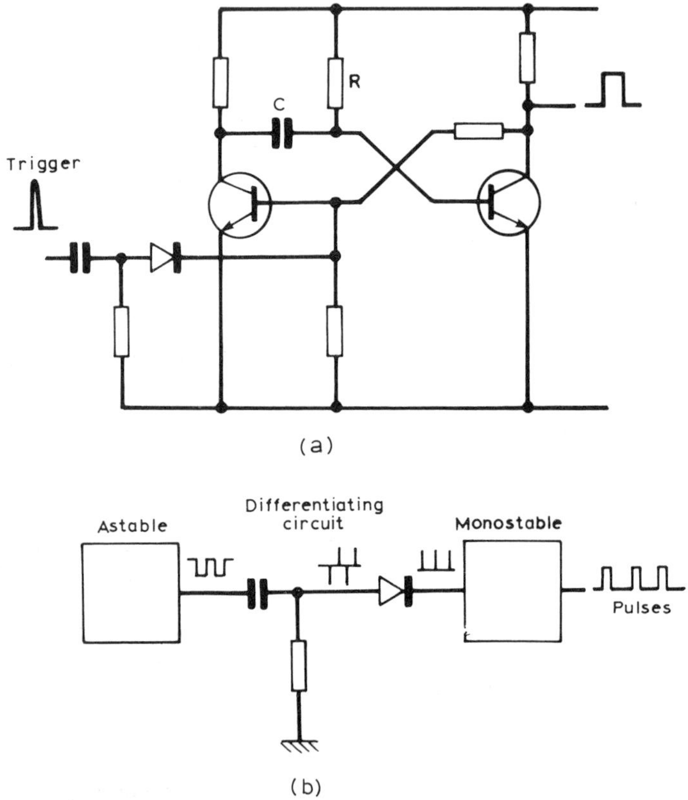
R
C
Trigger
(a)
Astable
Differentiating circuit
Monostable
Pulses
(b)

Figure 3.20. The monostable (a). The pulse width of the output pulse is determined by the time constant CR. The block diagram (b) shows how an astable (to determine frequency) and a monostable (to determine pulse width) can be combined to form a precise pulse generator

operating frequency is well below the turnover frequency, f_r (at which gain is unity) for the transistor. For v.h.f. use, common base stages (*Figure 3.24*) and dual-gate FETs (*Figure 3.25*) are extensively used. *Figure 3.26* shows frequency multiplier and intermediate stages for transmitters, and *Figure 3.27* a selection of low-pair output (power

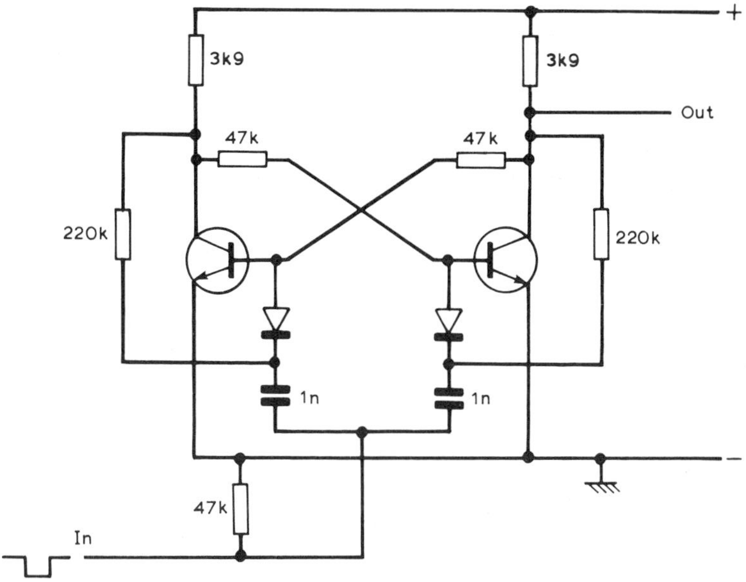
3k9
3k9
Out
47k
47k
220k
220k
1n
1n
47k
In
+
–

Figure 3.21. The bistable, or flip-flop. The output changes state (high to low or low to high) at each complete input pulse

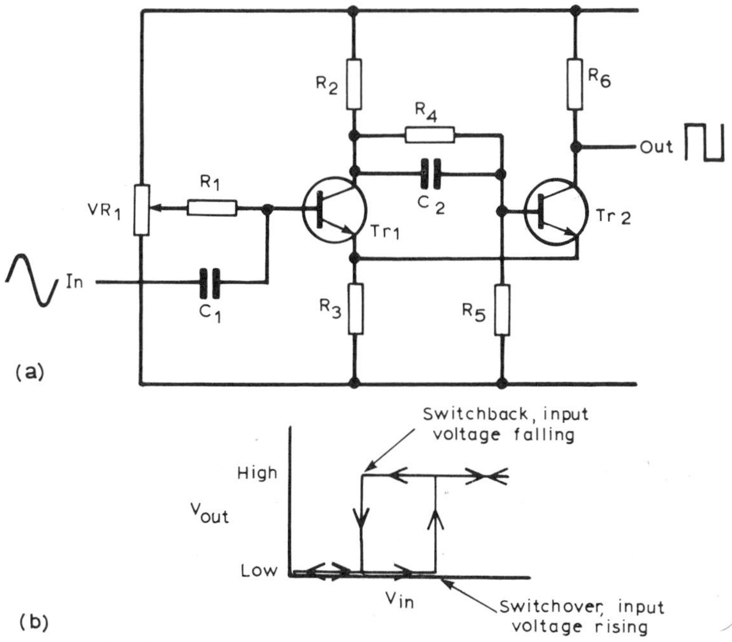
R2
R6
R4
Out
R1
C2
VR1
Tr1
Tr 2
In
C1
R3
R5
(a)
Switchback, input voltage falling
High
Vout
Low
Vin
Switchover, input voltage rising
(b)

Figure 3.22. The Schmitt trigger circuit. (a) This is a switching circuit whose output is always flat-topped and steep-sided whatever the input waveform. The characteristic (b) shows hysteresis – a difference between the switching voltages depending on the direction of change of the input voltage

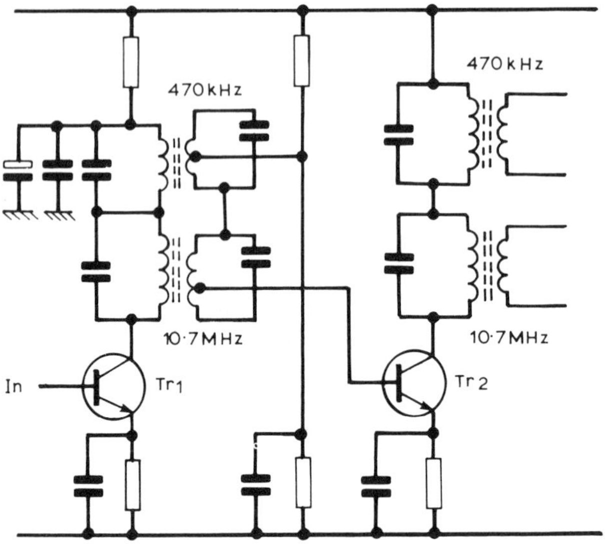
470kHz
470kHz
10·7MHz
10·7MHz
In
Tr1
Tr2

Figure 3.23. An i.f. amplifier typical of a.m./f.m. receivers. The frequency difference between the i.f.s is so great that no special filtering is needed, but the 10.7 MHz transformers must be located close to their transistors

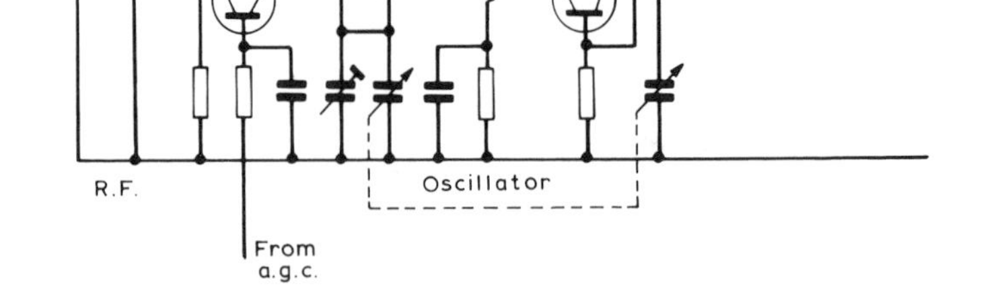
In
Tr1
Tr2
I.F.
10·7MHz
R.F.
Oscillator
From
a.g.c.

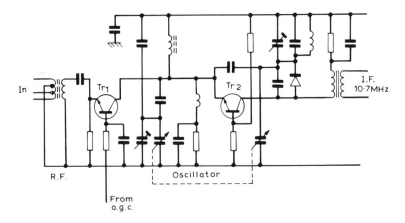

Figure 3.24. A typical f.m. 'front-end', using a common-base r.f. amplifier and a common-base oscillator. All transistors will operate at higher frequencies in the common-base connection than in the common-emitter connection

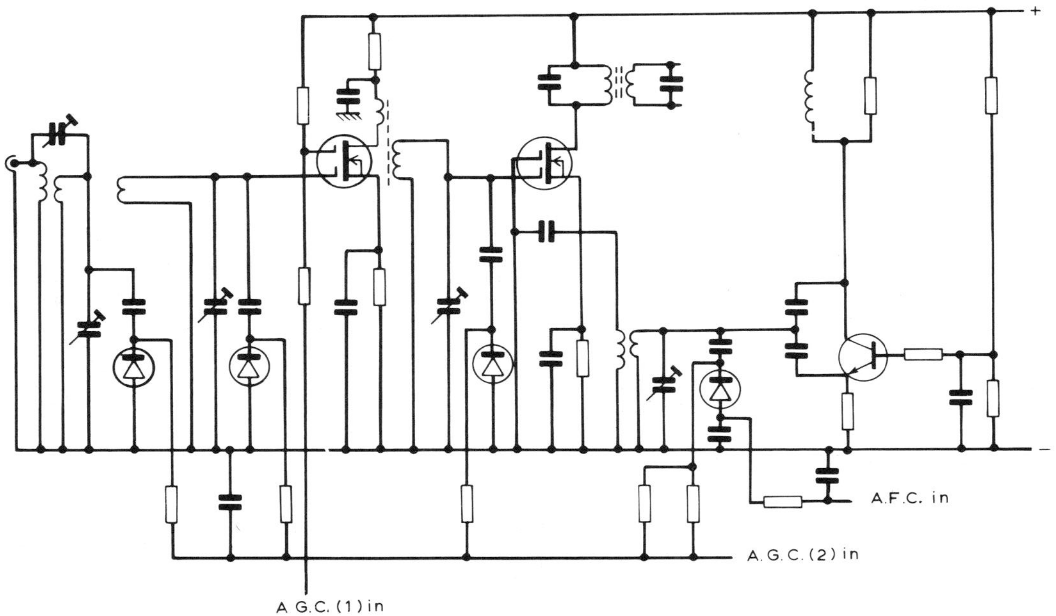
+
–
A.F.C. in
A.G.C. (2) in
A G.C. (1) in

Figure 3.25. An f.m. front end of a high-quality tuner (Armstrong), using dual-gate MOSFETs. Using both gates enables the r.f. a.g.c. to be completely isolated from the signal input, and the mixer signal input to be isolated from the oscillator input

amplifier or p.a.) stages. Transmitters which use variable frequency oscillators (v.f.o.) will require broadband output stages as distinct from sharply tuned stages, and this precludes the use of Class C amplifiers (in which the transistor conducts only on signal peaks). Without a sharply tuned, high Q, load, Class C operation introduces too much distortion (causing unwanted harmonics) and so Class B is preferable.

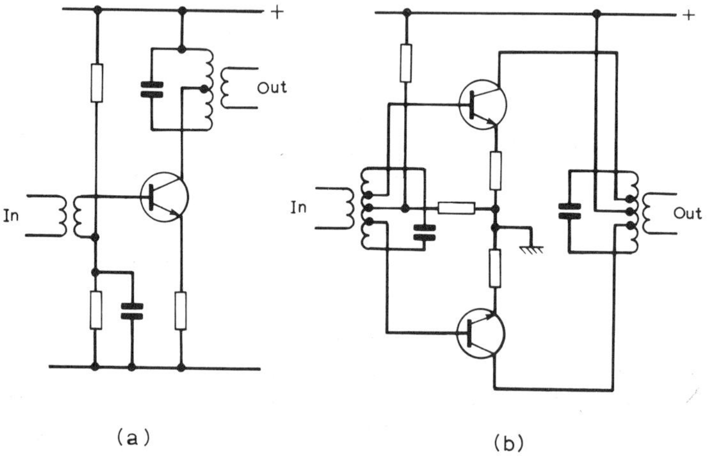
+
Out
In
(a)
+
In
Out
(b)

Figure 3.26. Frequency multipliers for transmitters. (a) A single transistor multiplier for even or odd multiples, (b) a push-pull multiplier for odd multiples. In each type of circuit, the output is tuned to a frequency which is a multiple of the input frequency. Other techniques not shown here include push-push, in which two transistors have antiphase signals at their collectors, but both feed the same output at the collectors. This circuit is used for even multiples. Varactor diodes are also extensively used as multipliers at low power levels

To carry information by radio or by digital signals requires some form of modulation and demodulation. For radio use amplitude modulation and frequency modulation are the most common techniques. Straightforward amplitude modulation produces two sidebands, with only one third of the total power in the sidebands, so that double sideband a.m. is used virtually only for medium and long wave broadcasting. Short wave communications use various forms of single sideband or suppressed carrier a.m. systems; v.h.f. radio broadcasting uses wideband f.m. and other v.h.f. communications use narrow-band f.m.

Figure 3.28 shows two simple modulator circuits, excluding specialised types. Carrier suppression can be achieved by balanced modulators in which the bridge circuit enables the carrier frequency to be balanced out while leaving sideband frequencies unaffected.

Sideband removal can be achieved by crystal filters, a fairly straightforward technique which is applicable only if the transmitting frequency is fixed, or by a phase-shift modulator which makes use of the phase shift which occurs during modulation. Frequency modulation, unlike amplitude modulation, is carried out on the oscillator itself, so requiring

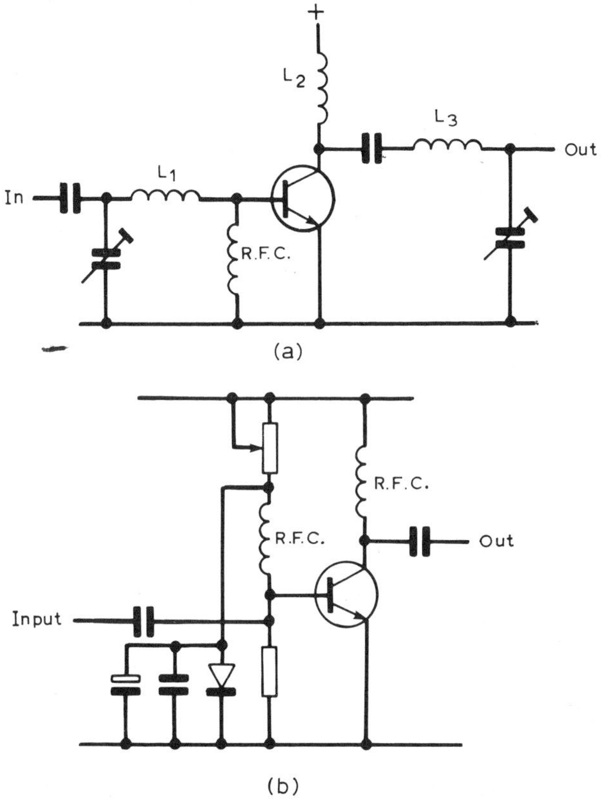
+
L2
L3
Out
L1
In
R.F.C.
(a)
R.F.C.
Out
R.F.C.
Input
(b)

Figure 3.27. Power amplifiers for transistor transmitters. (a) A Class C single transistor p.a. stage, (b) a Class B design, necessary for single-sideband transmitters. Tuning inductors have been omitted for clarity. In both circuits some decoupling capacitors have not been shown — complete decoupling is essential. At the higher frequencies, circuit layout is critical, and the circuit diagram becomes less important than the physical layout

reasonably linear operation of the stages following the oscillator. *Figure 3.29* illustrates some types of discrete component demodulators.

Pulse modulation systems are used extensively in applications ranging from data processing to radar. Pulse amplitude modulation and frequency modulation is essentially similar in nature to a.m. and f.m.

of sine waves, and will not be considered here. Forms of modulation peculiar to pulse operation are pulse width modulation (p.w.m.), pulse position modulation (p.p.m.) and pulse-code modulation (p.c.m.). A technique which is not a pulse modulation system but which is

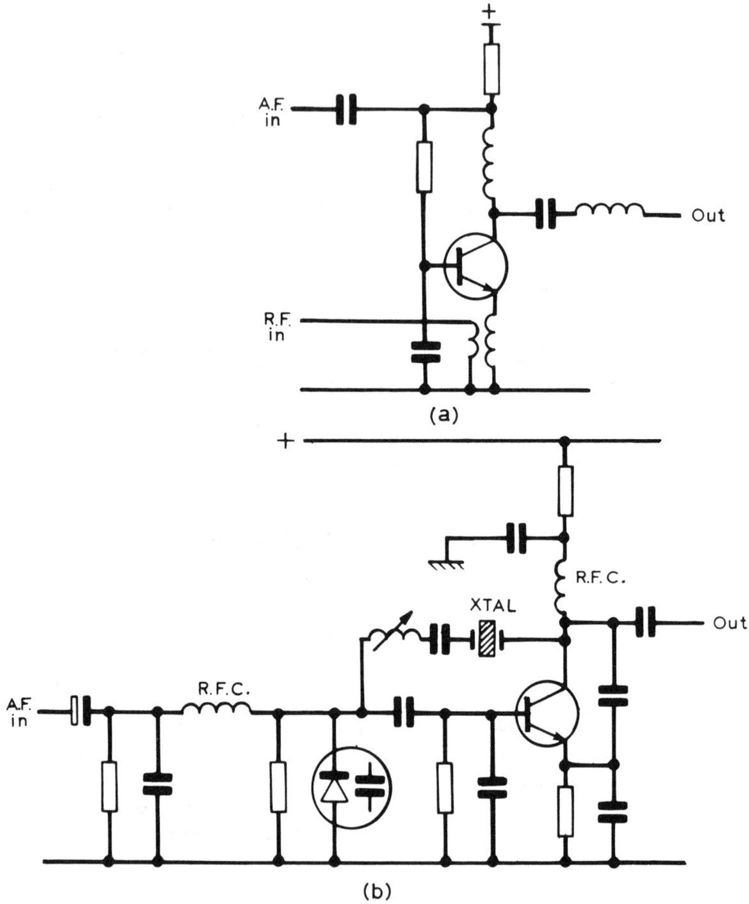
A.F. in
R.F. in
Out
(a)
A.F. in
R.F.C.
XTAL
R.F.C.
Out
(b)

Figure 3.28. Two simple modulator circuits. (a) A collector modulated stage for an a.m. transmitter, (b) a varactor diode f.m. modulator

extensively used for coding slow pulse information is frequency shift keying (f.s.k.) in which the high (logic 1) and low (logic 0) voltages of a pulse are represented by different audio frequencies.

Figure 3.30 is concerned with optical circuits, including LED devices and light detectors.

The circuits of *Figure 3.31* deal with power supply units. *Figure 3.31* shows the no-load voltage output, and the relationship between d.c. load voltage, minimum voltage, and a.c. voltage at the transformer. Only capacitive input circuits have been shown, since choke-input filters are by now rather rare. The relationship between the size of the reservoir capacitor and the peak-to-peak ripple voltage is given approximately by

$$V = \frac{I_{dc} \times t}{C}$$

with I_{dc} equal to load current (amperes), *t* in seconds the time between voltage peaks and *C* the reservoir capacitance in farads.

A more convenient set of units is I_{dc} in mA, *t* in ms, and *C* in μF, using the formula unchanged. *V* is then the peak-to-peak ripple voltage in volts.

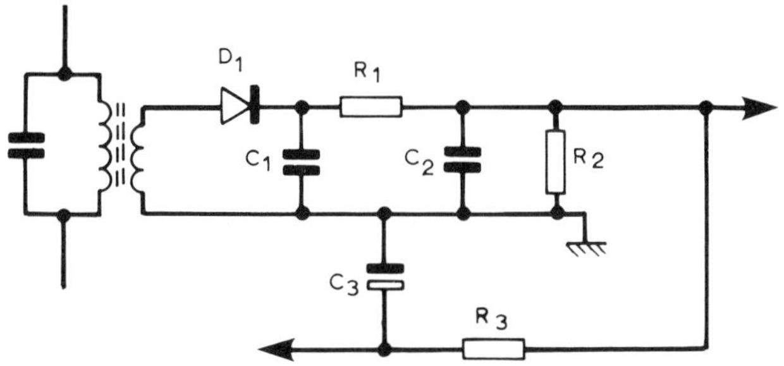
D1
R1
C1
C2
R2
C3
R3

(a)

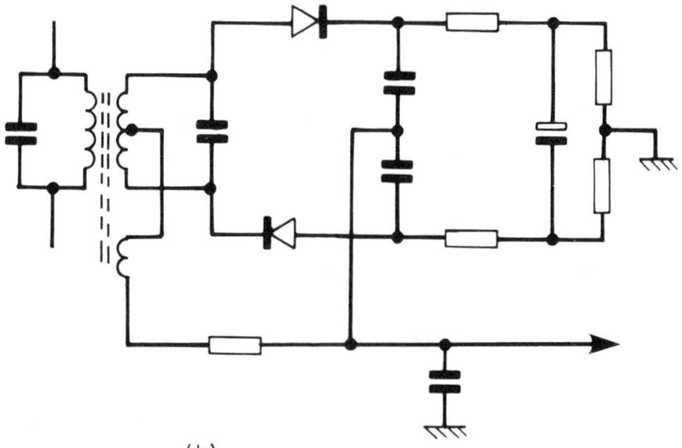

(b)

Figure 3.29. Demodulators. The a.m. demodulator (a) uses a single diode. The time constant of C_1 with $R_1 + R_2$ must be long compared with the time of a carrier wave, but short compared with the time of the highest-frequency audio wave. The f.m. demodulator (b) is a ratio detector

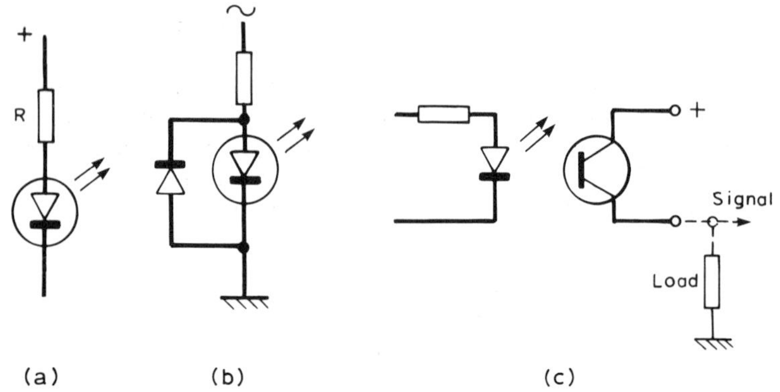

R
Signal
Load
(a)
(b)
(c)

Figure 3.30. Optoelectronic circuits. (a) Driving a single LED – a current-limiting resistor must be used. (b) When an LED is operated from a.c., a diode must always be included to protect the LED from reverse voltage. (c) Optocoupler, used to couple signals at very different d.c. levels. This is useful for triac firing, or for modulating the grid of a c.r.t., since d.c. signals can be transferred, which is not possible using a transformer

All power supplies which use the simple transformer-rectifier-capacitor circuit will provide an unstabilised output, meaning that the output voltage will be affected by fluctuations in the mains voltage level and also by changes in the current drawn by the load. The internal resistance of the power supply unit causes the second effect and can be the reason for instability in amplifier circuits, or of misfiring of pulse circuits. A stabiliser circuit provides an output which, ideally, remains constant despite any reasonable fluctuation in the mains voltage and

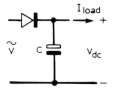

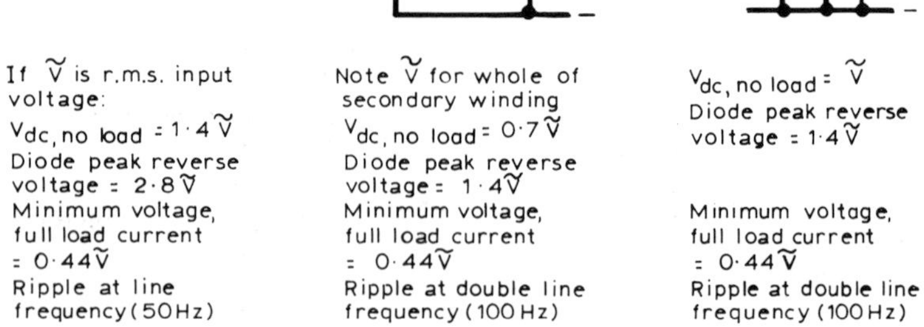

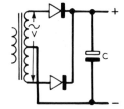

I_{load}
$\tilde{V}$
C
V_{dc}
C
If $\tilde{V}$ is r.m.s. input voltage:
$V_{dc, no\ load} = 1 \cdot 4 \tilde{V}$
Diode peak reverse voltage = $2 \cdot 8 \tilde{V}$
Minimum voltage, full load current = $0 \cdot 44 \tilde{V}$
Ripple at line frequency (50 Hz)
Note $\tilde{V}$ for whole of secondary winding
$V_{dc, no\ load} = 0 \cdot 7 \tilde{V}$
Diode peak reverse voltage = $1 \cdot 4 \tilde{V}$
Minimum voltage, full load current = $0 \cdot 44 \tilde{V}$
Ripple at double line frequency (100 Hz)
$V_{dc, no\ load} = \tilde{V}$
Diode peak reverse voltage = $1 \cdot 4 \tilde{V}$
Minimum voltage, full load current = $0 \cdot 44 \tilde{V}$
Ripple at double line frequency (100 Hz)

Figure 3.31. Rectifier circuits in detail

has zero internal resistance so that the output voltage is unaffected by the load current.

Stabilisation is achieved by feeding into the stabiliser circuit a voltage which is higher than the planned output voltage even at the worst combination of circumstances – low mains voltage and maximum load current. The stabiliser then controls the voltage difference between input and output so that the output voltage is steady.

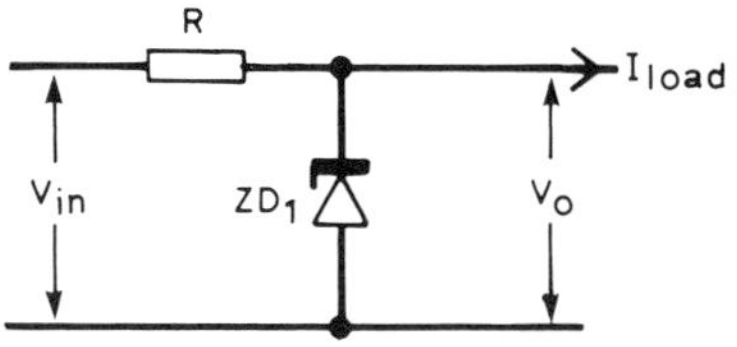
R
Iload
Vin
ZD1
Vo

Design data: Allow 2mA minimum current through zener diode

Maximum current $= (I_{load} + 2)$mA

Diode dissipation, max. $= V_o(I_{load} + 2)$mW

Resistor dissipation, max $=$
$(V_{in} - V_o)(I_{load} + 2)$ mW

Figure 3.32. A simple zener-diode stabiliser

Figure 3.32 shows a simple zener diode stabiliser suitable for small scale circuits taking only a few milliamps. This is a shunt stabilising circuit, so called because the stabiliser (the zener diode) is in parallel (shunt) with the load. The value of the resistor *R* is such that there will be a 'holding' current of 2 mA flowing into the zener diode even at the lowest input voltage and maximum signal current. The circuit of *Figure 3.33* (sometimes known as the 'amplified zener') is a shunt stabiliser which does not depend on dissipating power in the zener when the load current drops.

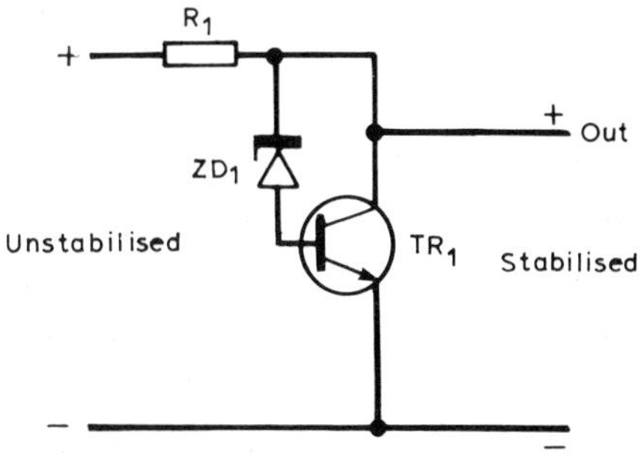
R1
+
+
Out
ZD1
Unstabilised
TR1
Stabilised
−
−

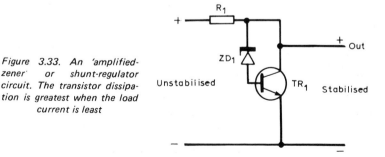

Figure 3.33. An 'amplified-zener' or shunt-regulator circuit. The transistor dissipation is greatest when the load current is least

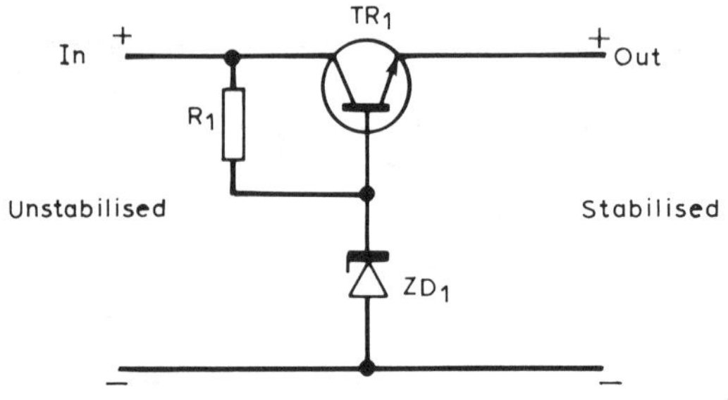
TR1
In
+
+
Out
R1
Unstabilised
Stabilised
ZD1
—
—

Figure 3.34. A simple series stabiliser. The transistor dissipation is greatest when the load current is maximum

Figure 3.34 shows a simple series stabiliser, using a zener diode to set the voltage at the base of an emitter follower. A more elaborate negative feedback circuit is shown in *Figure 3.35,* with provision for altering the stabilised voltage. The circuit of *Figure 3.36* also provides automatic shut-off (with the shut-off current alterable by setting the

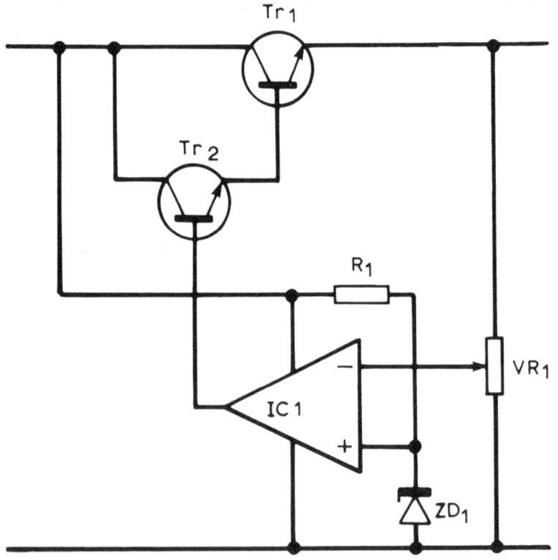
Tr1
Tr2
R1
VR1
IC 1
ZD1
—
+

Figure 3.35. A variable-voltage series stabilised supply circuit. Tr_1 is usually a 2N3055, and Tr_2 a lower power general purpose transistor. The most suitable i.c. is the LM3900, because it continues to amplify even when the output voltage is close to either supply voltage, unlike the 741

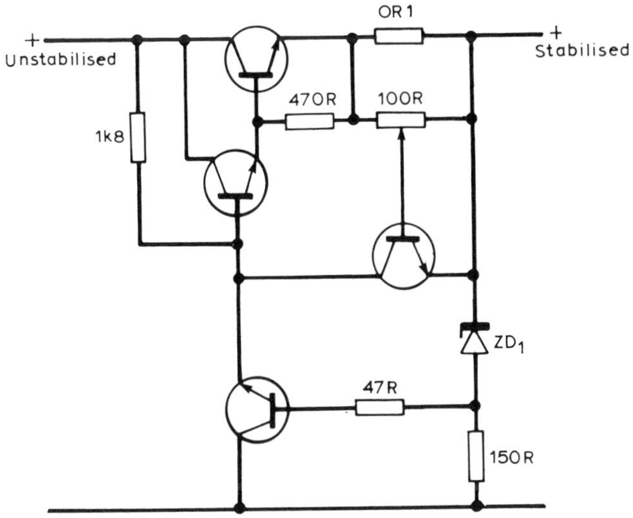
OR1
+
Unstabilised
+
Stabilised
470R
100R
1k8
ZD1
47R
150R

Figure. 3.36. A fixed-voltage stabiliser with over-current protection, limiting the current to an amount which is set by the potentiometer

100 Ω potentiometer) when excessive load current flows. More elaborate circuits are not considered here because of the extensive use of i.c. regulators (see Chapter 4).

Chapter 4

Linear I.C.s

Linear i.c.s are single-chip arrangements of amplifier circuits that are intended to be biased and operated in a linear way. This definition is usually extended to include i.c.s which have a comparatively slow switching action, such as the 555 timer.

The most important class of linear amplifier i.c. is the operational amplifier which features high gain, high input resistance, low output resistance and a narrow bandwidth extending to d.c. Such amplifiers are almost invariably used in negative feedback circuits, and make use of a balanced form of internal circuit (*Figure 3.1*) so that power supply hum and noise picked up by stray capacitance are both discriminated against.

The 741 is typical of operational amplifiers generally, so that the design methods, circuits and bias arrangements which are used for this i.c. can be used, with small modifications, for other types. Referring to the pinout diagram and symbol of *Figure 4.1*, the 741 uses two inputs marked + and –. These signs refer to the phase of the output signal relative to each input, so that feedback directly from the output to the + input is positive, and feedback directly from the output to the – input is negative.

The circuit arrangement of the 741 is such that, using balanced power supplies, the d.c. level at the output ought to be at zero volts when both inputs are connected to zero volts. This does not generally happen because of slight differences in internal components, so that an *input offset voltage* is needed to restore the output to zero voltage. Alternatively, the offset can be balanced out by a potentiometer connected as shown in *Figure 4.2*. Once set in this way so that the

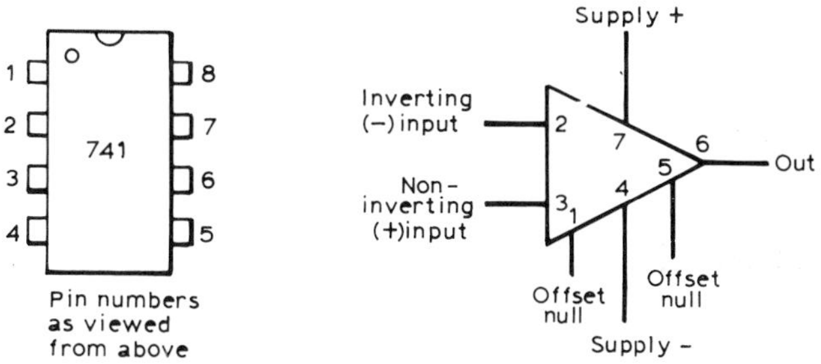
1
2
3
4
8
7
6
5
741
Pin numbers as viewed from above
Supply +
Inverting (−) input
Non-inverting (+) input
2
3
1
7
4
5
6
Out
Offset null
Offset null
Supply −

Maximum ratings

Positive supply +18V
Negative supply −18V
Differential input voltage ±30V
Common-mode input voltage ±15 V
Load current 10mA
Operating temperature 70°C
Storage temperature 140°C
Dissipation 400mW

Figure 4.1. 741 operational amplifier outline, with pin numbering (a) and the connections (b). The offset-null pins are used only for d.c. amplifier applications

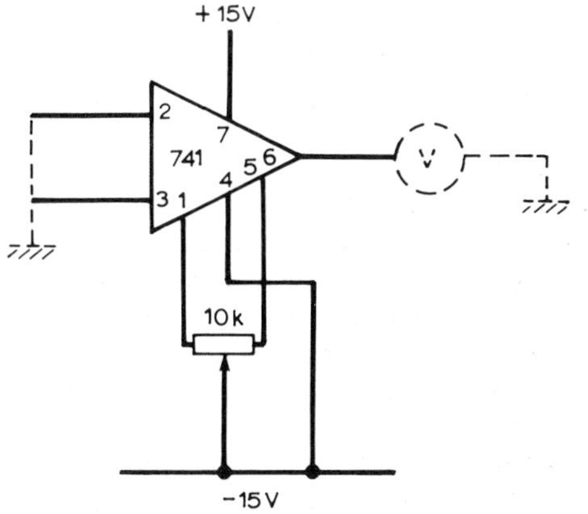
+15V
2
7
741
6
5
4
3
1
V
10k
−15V

Figure 4.2. Using an offset-null control. With the inputs both earthed (balanced power supplies) and a voltmeter connected to the output (dotted lines), the 10 KΩ potentiometer is adjusted so that the output voltage is zero

output is at zero volts (with the inputs earthed), the output voltage will then slowly change (drift). The drift may be caused by temperature changes, by supply voltage changes, or simply by old age. Drift is a problem which mainly affects high-gain d.c. coupled amplifiers and long time-constant integrators; a.c. amplifier circuits, and circuits which can use d.c. feedback bias are not affected by drift.

Bias Methods

For linear amplification, both inputs must be biased to a voltage which lies approximately halfway between the supply voltages. The output voltage can then be set to the same value by:

(a) making use of an offset-balancing potentiometer, or,

(b) connecting the output to the (–) input through a resistor, so making use of d.c. feedback.

Method (a) is seldom used, and the use of d.c. feedback is closely tied up with the use of a.c. feedback; the two will be considered together.

The power supply may be of the balanced type, such as the ±15 V supply, or unbalanced, provided that the bias voltage of input and output is set about midway between the limits (+15 and –15, or +V and 0) of supply voltages. Bias voltages should not be set within three volts of supply voltage limits, so that when a ±15 V supply is used, the input or output voltages should not exceed +12 V or –12 V. This limitation applies to bias (steady) voltage or to instantaneous voltages. If a single-ended 24 V power supply is used, the input and output voltages should not fall below 3 V nor rise above 21 V. Beyond these limits, the amplifying action may suddenly collapse because there is not sufficient bias internally.

Basic circuits

Figure 4.3 shows the circuits for an inverting amplifier, using either balanced or unbalanced power supplies. The d.c. bias conditions are set by connecting the (+) input to mid-voltage (which is earth voltage when balanced power supplies are used) and using 100% d.c. feedback from the output to the (–) input. The gain is given by:

$$G = R_1/R_2$$

Note that a capacitor C_1 is needed when a single-ended power supply is used to prevent the d.c. bias voltage from being divided down in the same ratio as the a.c. bias. When balanced power supplies are used, direct coupling is possible provided that the signal source is at zero d.c. volts.

The input resistance for these circuits is simply the value of resistor R_2, since the effect of the feedback is to make the input resistance at the (—) input almost zero; this point is referred to as a 'virtual earth' for signals. The output resistance is typically about 150 ohms.

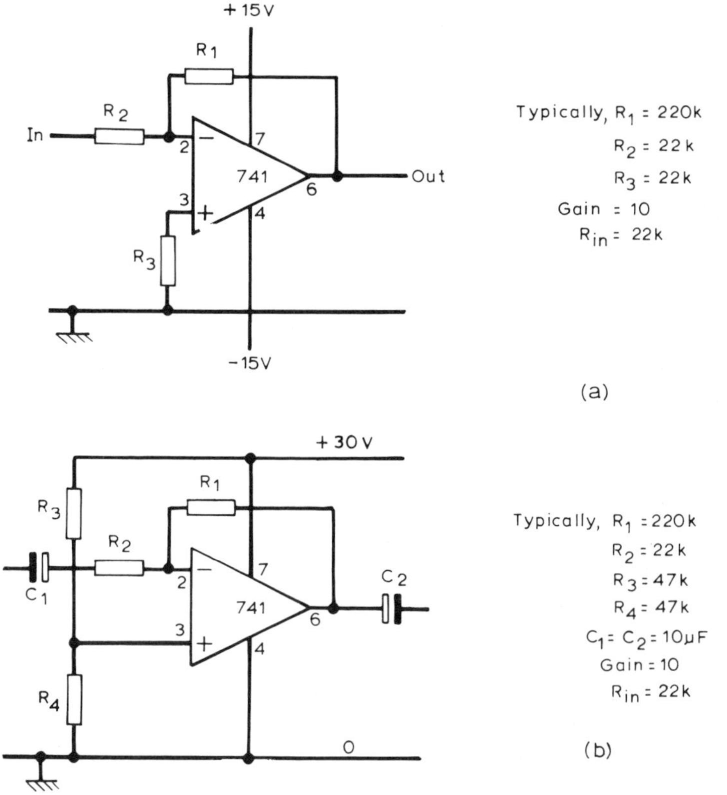
+15V
R1
R2
In
741
Out
2
3
7
6
4
R3
-15V
Typically, R1 = 220k
R2 = 22k
R3 = 22k
Gain = 10
Rin = 22k
(a)
+30V
R1
R3
R2
C1
741
C2
2
3
7
6
4
R4
0
Typically, R1 = 220k
R2 = 22k
R3 = 47k
R4 = 47k
C1 = C2 = 10µF
Gain = 10
Rin = 22k
(b)

Figure 4.3. Inverting amplifier configuration. (a) uses balanced power supplies. Ideally, R_3 should equal R_2, though differing values are often used. The gain is set by the ratio R_1/R_2, and the input resistance is equal to R_2. Using an unbalanced power supply (b), the + input is biased to half the supply voltage (15 V in this example) by using equal values for R_3 and R_4. The gain is again given by R_1/R_2. Coupling capacitors are needed because of the d.c. bias conditions

Circuits for non-inverting amplifiers are shown in *Figure 4.4*. Non-inverting amplifiers also make use of negative feedback to stabilise the working conditions in the same way as the inverting amplifier circuits, but the signal input is now to the (+) input terminal. The gain is

$$G = \frac{R_1 + R_2}{R_2}$$

and the circuit is sometimes referred to as the 'voltage-follower with gain'. The input resistance is high, usually around 1 MΩ, for the dual-supply version, though the bias resistors (*Figure 4.4b*) reduce this to a few hundred kΩ.

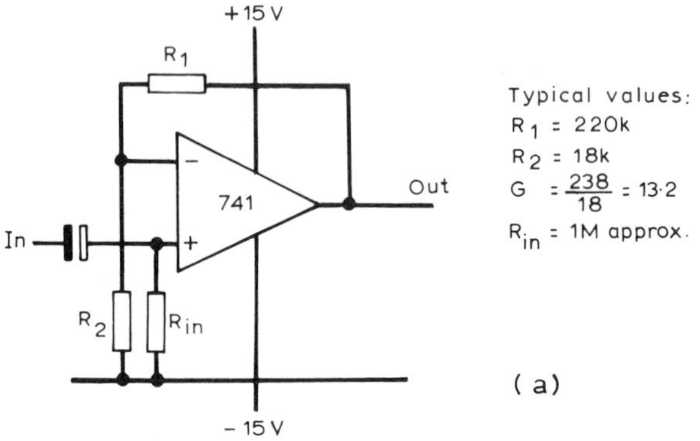
+15 V
R1
741
Out
In
R2
Rin
– 15 V
Typical values:
R1 = 220k
R2 = 18k
G = 238/18 = 13·2
Rin = 1M approx.
(a)

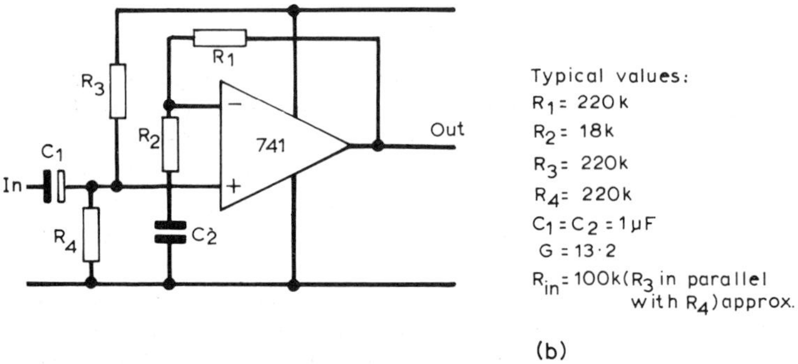
R1
R3
R2
741
Out
C1
In
R4
C2
Typical values:
R1 = 220 k
R2 = 18k
R3 = 220k
R4 = 220k
C1 = C2 = 1 μF
G = 13·2
Rin = 100k(R3 in parallel with R4) approx.
(b)

Figure 4.4. Non-inverting amplifiers. Using a balanced power supply (a), only two resistors are needed, and the voltage gain is given by $\frac{R_1 + R_2}{R_2}$*. The input resistance is very high. When an unbalanced supply (b) is used, a capacitor* C_2 *must be connected between* R_2 *and earth to ensure correct feedback of signal without disturbing bias. The input resistance is now lower because of* R_3 *and* R_4 *which as far as signal voltage is concerned are in parallel*

Figure 4.5 shows the 741 used as a differential amplifier, though with a single-ended output. The gain is set by the ratio R_1/R_2 as before – note the use of identical resistors in the input circuits to preserve balance.

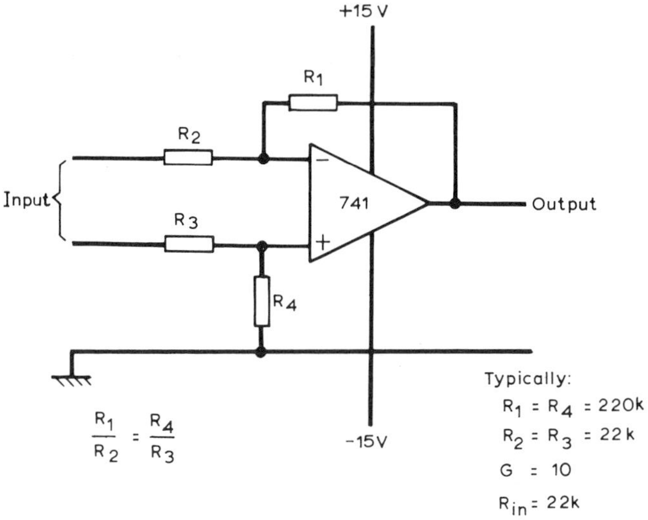
+15 V
R1
R2
R3
R4
741
Input
Output
−15V
R1/R2 = R4/R3
Typically:
R1 = R4 = 220k
R2 = R3 = 22k
G = 10
Rin = 22k

Figure 4.5. Differential amplifier application. Both inputs are used for signals which must be in antiphase (balanced about earth). Any common-mode signals (in phase at both inputs) are greatly attenuated

General notes on op-amp circuits

The formulae for voltage gain hold for values of gain up to several hundred times, because the gain of the op-amp used *open-loop* (without feedback) is very high, of the order of 100 000 (100 dB). The maximum load current is about 10 mA, and the maximum power dissipation 400 mW. The 741 circuit is protected against damage from short-circuits at the output, and the protection circuits will operate for as long as the short-circuit is maintained.

The frequency range of an op-amp depends on two factors, the gain-bandwidth product for small signals, and the slew rate for large signals. The gain-bandwidth product is the quantity, $G \times B$, with G equal to voltage gain (not in dB) and B the bandwidth upper limit in Hz. For the 741, the G–B factor is typically 1 MHz, so that, in theory, a bandwidth of 1 MHz can be obtained when the voltage gain is unity, a bandwidth of 100 kHz can be attained at a gain of ten, a bandwidth of 10 kHz at a gain of 100 times, and so on. This trade-off is usable only for small signals, and cannot necessarily be applied to all types of operational amplifiers. Large amplitude signals are further limited by the slew rate of the circuits within the amplifier. The slew rate of an amplifier is the value of maximum rate of change of output voltage

$$\text{slew rate} = \frac{\text{maximum voltage change}}{\text{time needed}}$$

Units are usually volts per microsecond.

Because this rate cannot be exceeded, and feedback has no effect on slew rate, the bandwidth of the op-amp for large signals, sometimes called the power bandwidth, is less than that for small signals. The slew rate limitation cannot be corrected by the use of negative feedback; in fact negative feedback acts to increase distortion when the slew rate limiting action starts, because the effect of the feedback is to increase the rate of change of voltage at the input of the amplifier whenever the rate is limited at the output. This accelerates the overloading of the amplifier, and can change what might be a temporary distortion into a longer-lasting overload condition.

The relationship between the sine wave bandwidth and the slew rate, for many types of operational amplifier is:

$$\text{Max. slew rate} = 2\,\pi\, f_{\text{max}}\, E_{\text{peak}}$$

or

$$f_{\text{max}} = \frac{\text{max. slew rate}}{2\,\pi\, E_{\text{peak}}}$$

where slew rate is in units of volts per second (*not* V/μs), f_{max} is the maximum full-power frequency in Hz, and E_{peak} is the peak voltage of the output sine wave.

This can be modified to use slew rate figures in the more usual units of V/μs, with the answer in MHz. For example, a slew rate of 1.5 V/μs corresponds to a maximum sine wave frequency (at 10 V output) of

$$f_{\text{max}} = \frac{1.5}{2\,\pi \times 10}\ \text{MHz} = 0.023\ \text{MHz or } 23\ \text{kHz.}$$

Slew rate limiting arises because of internal stray capacitances which must be charged and discharged by the current flowing in the transistors inside the i.c.: improvement is obtainable only by redesigning the internal circuitry. The 741 has a slew rate of about 0.5 V/μs, corresponding to a power bandwidth for 12 V peak sine wave signals of about 6.6 kHz. The slew rate limitation makes op-amps unsuitable for applications which require fast-rising pulses, so that a 741 should not be used as a signal source or feed (interface) with digital circuitry, particularly TTL circuitry, unless a Schmitt trigger stage is also used.

Higher slew rates are obtainable with more modern designs of op-amps; for example, the Fairchild LS201 achieves a slew rate of 10 V/μs.

Other operational amplifier circuits

Figures 4.6 to *4.9* illustrate circuits other than the straightforward voltage amplifier types. *Figure 4.6* shows two versions of follower circuit with no voltage gain, but with useful characteristics. The unity gain inverter will provide an inverted output of exactly the same amplitudes as the input signal, subject to slew rate limitations. The non-inverting circuit, or voltage follower, performs the same action as the familiar emitter-follower, having a very high input resistance and a low output resistance. For this type of circuit, the action of the feedback causes both inputs to change voltage together, as a common-mode signal

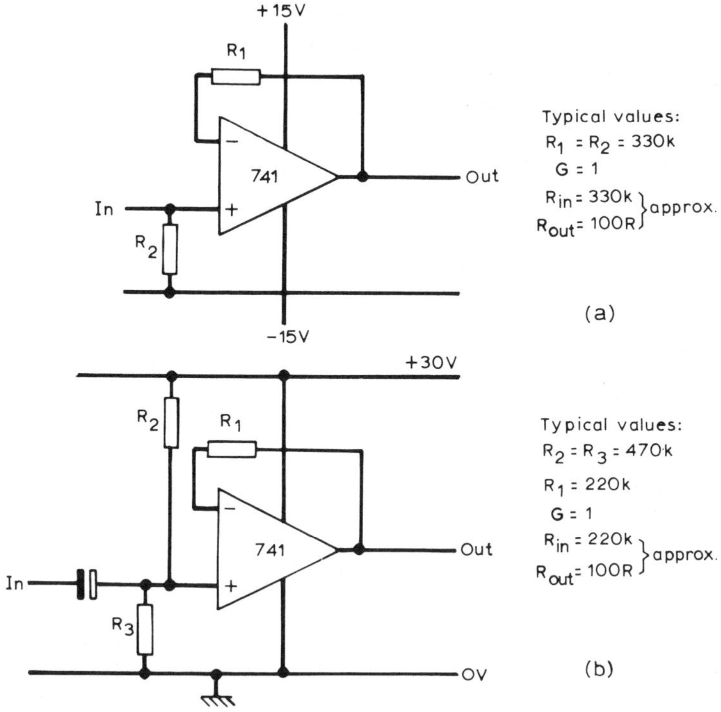

+15V
R1
741
Out
In
R2
−15V
Typical values:
R1 = R2 = 330k
G = 1
Rin = 330k
Rout = 100R
approx.
(a)
+30V
R2
R1
741
Out
In
R3
0V
Typical values:
R2 = R3 = 470k
R1 = 220k
G = 1
Rin = 220k
Rout = 100R
approx.
(b)

Figure 4.6. The voltage follower. The gain is unity, with high input resistance and low output resistance. The dual-voltage supply version (a) uses only two resistors whose values are not critical. Ideally, R_1 should equal R_2, and both should be high values so that the input resistance is high, equal to R_2. The single supply voltage version (b) uses three resistors, with $R_2 = R_3$ and R_1 made equal to $R_2/2$, which is also the value of input resistance

would, so that any restrictions on the amplitude of common-mode signals (see the manufacturer's sheets) will apply to this circuit. *Figure 4.7* shows two examples of a 741 as it is used in a variety of 'shaping' circuits in which the gain/frequency or gain/amplitude graph is intended to be non-linear.

The use of op-amps for switching circuits is limited by the slew rate, but the circuits shown in *Figures 4.8* and *4.9* are useful if fast-rising or falling waveforms are not needed.

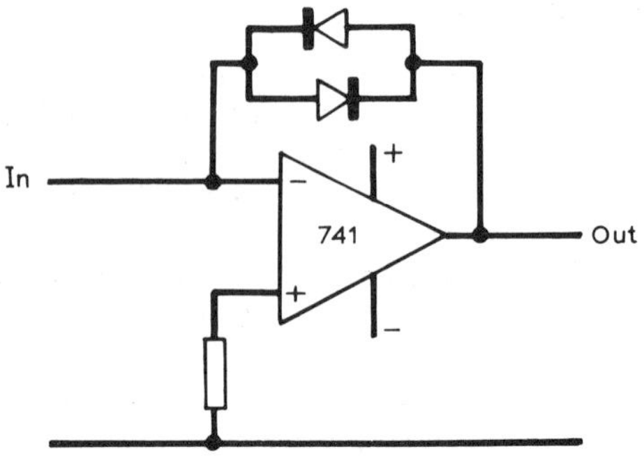
In
741
Out

(a)

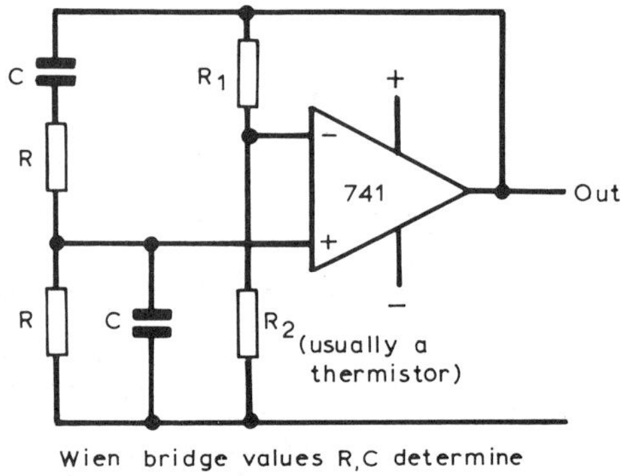
C
R
R
C
R1
R2 (usually a thermistor)
741
Out
Wien bridge values R,C determine frequency

(b)

Figure 4.7. Using the 741 in circuits which are not linear amplifiers. (a) A constant-output amplifier. Because the diodes will permit feedback of voltages whose amplitude is enough to allow the diodes to conduct, the output voltage is limited to about this amplitude, but without excessive clipping. The gain is very large for small input signals, and very small for large input signals. (b) The Wien bridge in the feedback network causes oscillation. The waveform is a sine wave only if the gain is carefully controlled by making $R_1/R_2 = 3$*, and this is usually done by making* R_2 *a thermistor whose resistance value decreases as the voltage across it increases. The frequency of oscillation is given by*

$$f = \frac{1}{2\pi RC}$$

Current differencing amplifiers

A variation on the op-amp circuit uses current rather than voltage input signals, and is typified by the National Semiconductor LM3900. In this i.c., which contains four identical op-amps, the + and − inputs are current inputs, whose voltage is generally about +0.6 V when correctly biased. A single-ended power supply is used, and the output voltage can reach to within a fraction of a volt of the supply limits. The output voltage is proportional to the difference between the *currents* at the

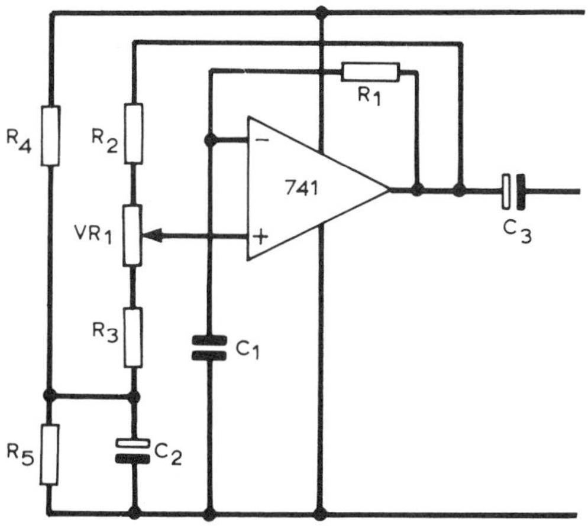
R1
R4
R2
741
VR1
C3
R3
C1
R5
C2

Figure 4.8. A 741 used as an astable multivibrator; a single power supply voltage version is illustrated. The use of R_2, VR_1, R_3 in the positive feedback path sets the + input to a definite fraction of the output voltage above or below the normal half-supply-voltage. When the output changes over, C_1 charges through R_1 until the − input reaches the same voltage as the + input upon which the circuit switches over. The voltage change at the + input is then rapid, but the voltage at the − input cannot change until C_1 has charged again. The frequency is therefore set by the time constant C_1R_1 and also by the setting of VR_1

two inputs, so that bias conditions are set by large value resistors. *Figure 4.10* shows a typical amplifier circuit, in which the current into the (+) input is set by R_1, whose value is 2.2 MΩ. Because the ideal bias voltage for the output is half of supply voltage, a 1 MΩ resistor is used connected between the output and the (−) input. In this way, the currents to the two inputs are identical, and the amplifier is correctly biased. The advantages of this type of op-amp are now being recognised, and an equivalent to the LM3900 is now available from Motorola.

Other linear amplifier i.c.s

A very large variety of i.c.s intended for a.f., i.f. and r.f. amplifiers can be obtained. For any design work, the full manufacturer's data sheet pack must be consulted, but a few general notes can be given here. A.F. i.c. circuits use direct coupling internally, because of the difficulty of fabricating capacitors of large value on to silicon chips, but the high gains which are typical of operational amplifiers are not necessary for most a.f. applications. Faster slew rates and greater open-loop bandwidths can therefore be attained than is practicable using op-amps.

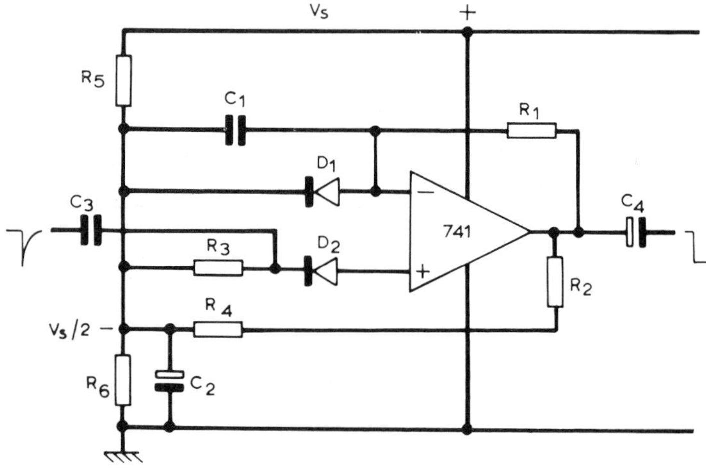
Vs
+
R5
C1
R1
D1
C3
C4
741
R3
D2
R2
R 4
Vs /2
R6
C2

Figure 4.9. A 741 monostable. With no input, the output voltage is high, which causes the + input voltage to be higher than the voltage, $V_s/2$ set by R_5 and R_6 (equal values). Because of D_1, the – input voltage cannot rise to the same value as the + input. A negative pulse at the + input causes the output voltage to drop rapidly, taking the + input voltage low. The – input voltage then drops at a rate determined by the time constant C_1R_1. When the – input voltage equals the + input voltage, the circuit switches back, and the diode D_1 conducts to 'catch' the – input voltage and so prevent continuous oscillation

Many a.f. i.c.s use separate chips for preamplifier and for power amplifier uses, with separate feedback loops for each. Frequency correcting networks composed of resistors and capacitors are usually needed to avoid oscillation, and heatsinks will be needed for the larger power amplifier i.c.s. The need for external volume, stereo balance, bass and treble controls, along with feedback networks, makes the circuitry rather more involved than some other i.c. applications.

Figure 4.11 shows two a.f. circuits examples. Note that the stability of these audio i.c.s is often critical, and decoupling capacitors, as specified by the manufacturers, must be connected as close to the i.c.

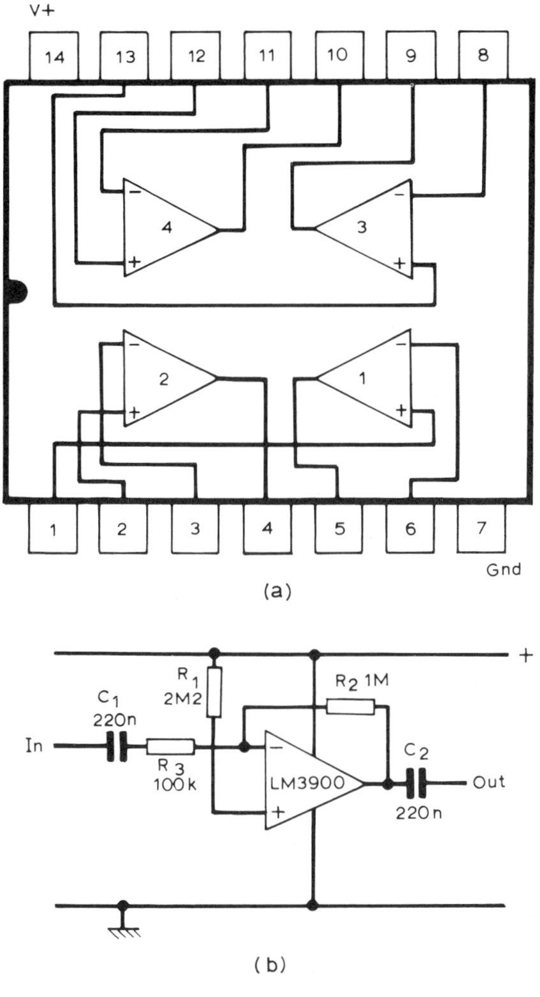
V+
14
13
12
11
10
9
8
4
3
2
1
1
2
3
4
5
6
7
Gnd
(a)
+
R_1
2M2
R_2 1M
C_1
220n
In
C_2
R_3
100k
LM3900
Out
220n
(b)

Figure 4.10. The current differencing amplifier, or Norton Op-amp. (a) Pinout for the LM3900, which contains four amplifiers in a single fourteen-pin package. (b) Typical amplifier circuit. Note the high resistor values

pins as possible. For stability reasons, also, stripboard construction is extremely difficult with some i.c. types, and suitable printed-circuit boards should be used.

I.F. and r.f. amplifier circuits contain untuned wideband amplifier circuits to which tuning networks, which may be *LC* circuits or trans-filters, may be added. It is possible to incorporate r.f., mixer, i.f. and

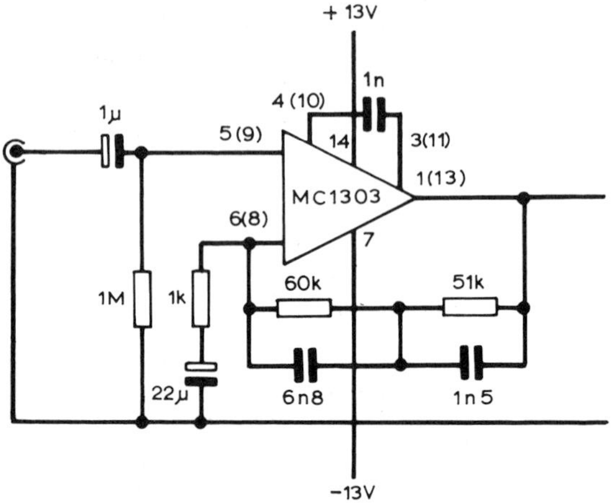
+ 13V
1n
4 (10)
1μ
5 (9)
14
3(11)
1(13)
MC1303
6(8)
7
60k
51k
1M
1k
22μ
6n8
1n5
−13V

(a)

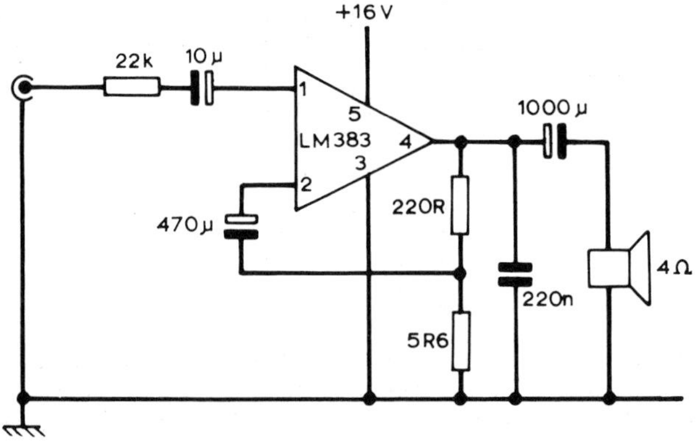
+16 V
22k
10 μ
1
5
1000 μ
LM383
4
3
2
220R
470μ
4Ω
220n
5R6

(b)

Figure 4.11. Audio amplifier i.c.s. (a) The MC1303 preamplifier is a dual unit for stereo use – the pin numbers in brackets are for the second section. Inputs up to 5 mV can be accepted, and the circuit here is shown equalised for a magnetic pickup. The output is 250 mV with a 5 mV input at a distortion level of about 0.1%. (b) The LM383 power amplifier uses a five-pin TO 220 package. The power output is 7 W into 4 ohms, with a distortion level of 0.2% at 4 W output. The maximum power dissipation is 15 W when a 4 °C/W heatsink is used

demodulator stages into a single i.c., but generally only when low-frequency r.f. and i.f. are used. A very common scheme for f.m. radio receivers is to use a discrete component tuner along with i.c. i.f. and demodulator stages, using the usual 10.7 MHz i.f. *Figure 4.12* shows an example of such an i.f. stage. Once again, when a large amount of gain

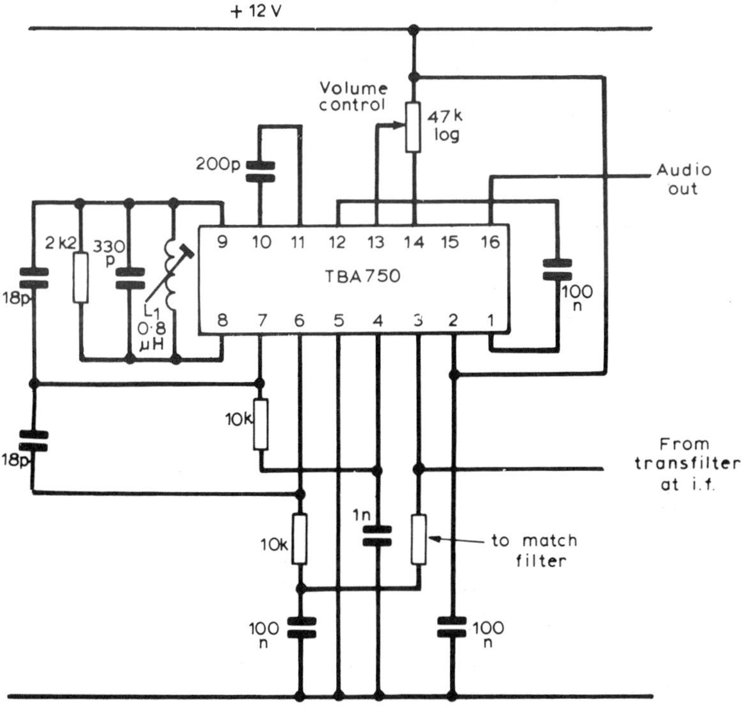
+ 12 V
Volume control
47 k log
200p
Audio out
2 k2
330 p
18p
L1 0·8 μH
9 10 11 12 13 14 15 16
TBA750
8 7 6 5 4 3 2 1
100 n
10k
18p
From transfilter at i.f.
1n
10k
to match filter
100 n
100 n

Figure 4.12. An i.f. detector i.c. for use in 10.7 MHz stereo f.m. i.f. stages. The minimum input for limiting is 100 μV, and the volume control range (operating on d.c.) is 80 dB. The audio output is 1.4 V r.m.s. with a signal of 15 kHz deviation

is attained in one i.c. stability is a major problem, and the manufacturer's advice on decoupling must be carefully followed. At the higher frequencies, the physical layout of components is particularly important, so that p.c.b.s intended for the TBA750 i.c. (and similar) should be used rather than stripboards.

Phase locked loops

The phase-locked loop is a type of linear i.c. which is now being used to a considerable extent. The block diagram of the circuit is outlined in

Figure 4.13 and consists of a voltage-controlled oscillator, a phase-sensitive detector, and comparator units. The oscillator is controlled by external components, so that the frequency of oscillation can be set by a suitable choice of these added components. An input signal to the Phase Locked Loop (PLL) is compared in the phase-sensitive detector to the frequency generated in the internal oscillator, and a voltage output obtained from the phase-sensitive detector. Provided that the input frequency is not too different from the internally generated

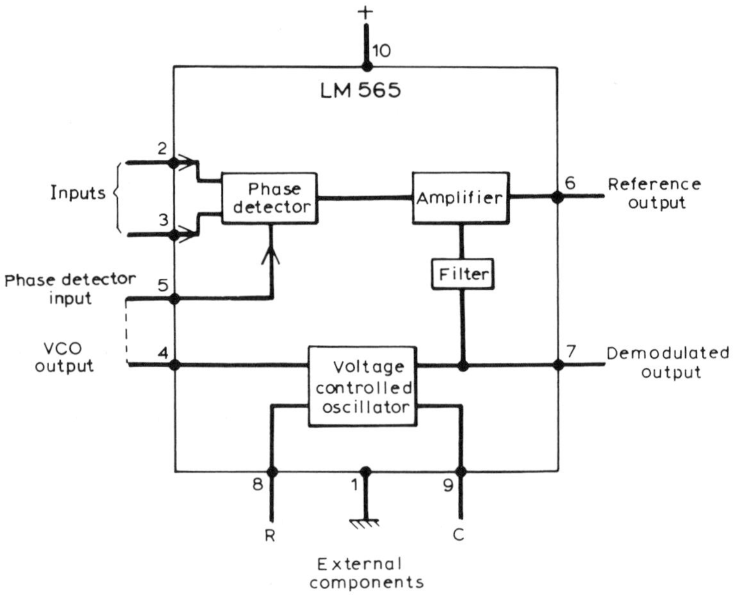
+
10
LM 565
2
Inputs
3
Phase detector
Amplifier
6
Reference output
Filter
Phase detector input
5
VCO output
4
Voltage controlled oscillator
7
Demodulated output
8
1
9
R
C
External components

Figure 4.13. The phase-locked loop (p.l.l.) block diagram. The pin numbering is for the LM565. The signal input can be to pin 2 or 3 in this i.c., and in normal use pins 4 and 5 are linked

frequency (within the 'pull-in' range), the voltage from the phase-sensitive detector can then be used to correct the oscillator frequency until the two signals are at the same frequency and in the same phase. Either the oscillator or the correcting voltage may be used as an output. The circuit can be used, for example, to remove any traces of amplitude modulation from an input signal, since the output (from the internal oscillator) is not affected by the amplitude of the input signal, but is locked to its frequency and phase. The circuit may also be used as an f.m. demodulator, since the control voltage will follow the modulation of an f.m. input in its efforts to keep the oscillator locked in phase.

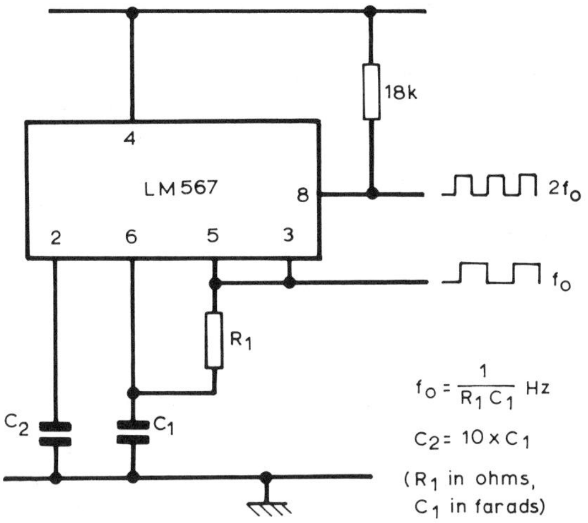

18k
4
LM567
8
2fo
2
6
5
3
fo
R1
$f_o = \frac{1}{R_1 C_1}$ Hz
C2
C1
$C_2 = 10 \times C_1$
(R_1 in ohms,
C_1 in farads)

(a)

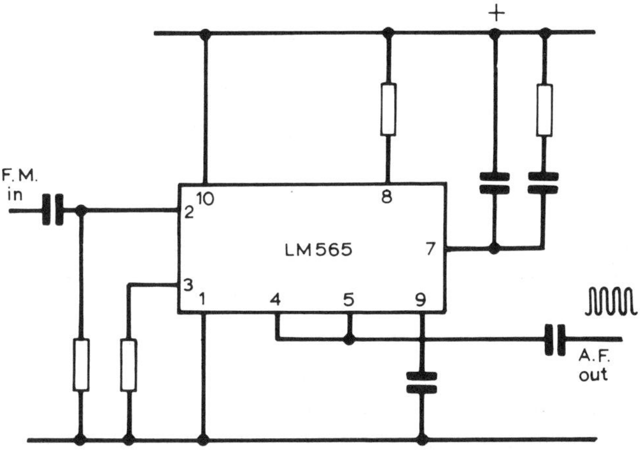

+
F.M.
in
10
2
8
LM565
7
3
1
4
5
9
A.F.
out

(b)

Figure 4.14. P.L.L. circuits. (a) Oscillator with fundamental and second harmonic outputs. (b) F.M. demodulator — the component values must be calculated with reference to the i.f. frequency which is used — this cannot be as high as the normal 10.7 MHz because the operating frequency limit of this i.c. is 500 kHz

Voltage stabiliser i.c s

The ease with which zener diode junctions and balanced amplifiers may be constructed in integrated form, together with the increasing demand for stabilised supplies and the steady increase in the power which can be dissipated from i.c.s due to improved heat-sinking methods, has led to the extensive use of i.c. voltage stabiliser circuits, to the extent that discrete component stabilisers are almost extinct. The decisive factor is that i.c.s can include such features as overload and short-circuit protection at virtually no extra cost, using an elaborate circuit such as that of the SGS-ATES TBA625A shown in *Figure 4.15*. The overload

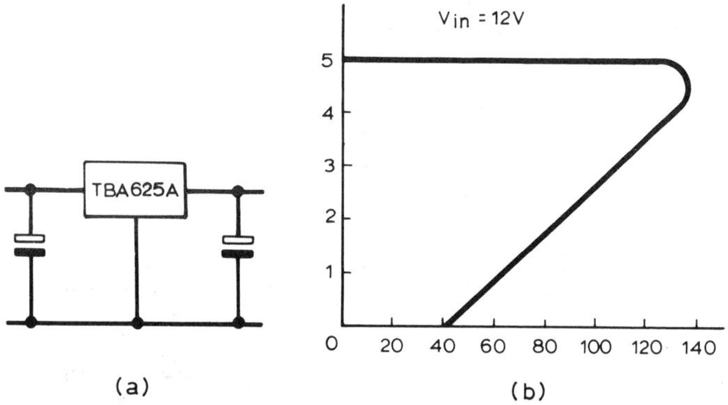
TBA625A
(a)
Vin = 12V
5
4
3
2
1
0
20
40
60
80
100
120
140
(b)

Figure 4.15. Voltage regulation. (a) Use of an i.c. regulator, the TBA625A, is illustrated. (b) Foldback overcurrent protection — at excessive currents the voltage output and current output both drop as indicated in the graph

protection is usually of the 'foldback' type, illustrated by the characteristic of *Figure 4.15b*, in which excessive current causes the output voltage to drop to zero with a much smaller current flowing. *Figure 4.16* shows circuit applications — note that a fixed voltage regulator can be used to provide an adjustable output, and higher current operation can be achieved by adding power transistors to the circuit. Voltage stabiliser i.c.s are available for all the commonly used voltage levels.

Motor-speed controllers are a more specialised form of stabiliser circuit, and are used to regulate the speed of d.c. motors in record players, tape and cassette recorders, and model motors. A typical application is shown in *Figure 4.17*.

One very important class of linear i.c.s is concerned with television circuitry. The development of linear i.c.s has been such that virtually every part of a TV circuit with the exceptions of the tuner head and the horizontal output stage can now be obtained in i.c. form. Because of the specialised nature of such circuits, the reader is referred to the

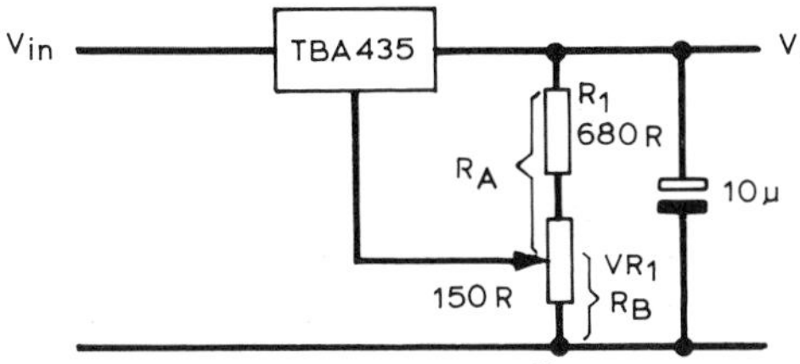
Vin
TBA435
Vo
R1
680R
RA
10µ
VR1
150R
RB

$V_O = 18\left(1 + \frac{R_A}{R_B}\right)$ (approx)

(a)

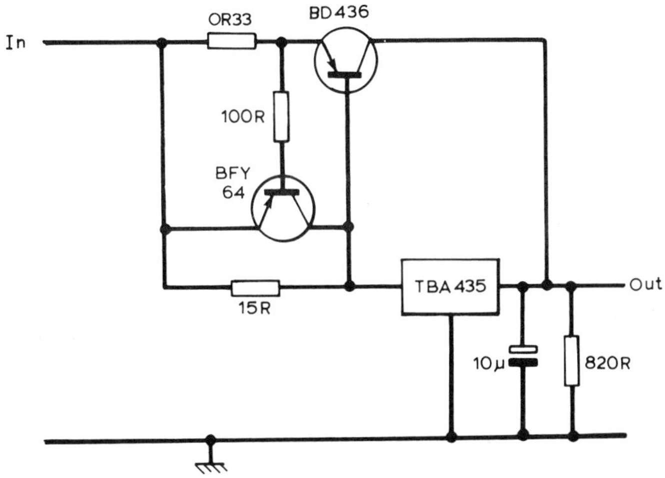
OR33
BD436
In
100R
BFY
64
15R
TBA435
Out
10µ
820R

(b)

Figure 4.16. Extending the range of fixed-voltage regulators. (a) Extending voltage range, (b) extending current range

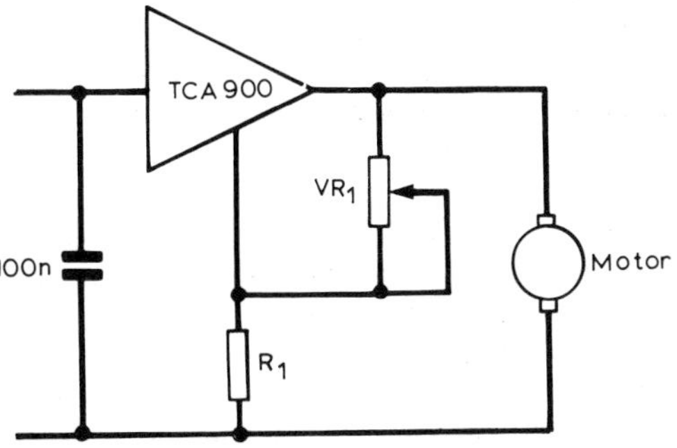
TCA900
VR1
100n
Motor
R1

Figure 4.17. A motor-speed control i.c. The values of VR_1 and R_1 must be calculated with reference to the resistance of the motor windings

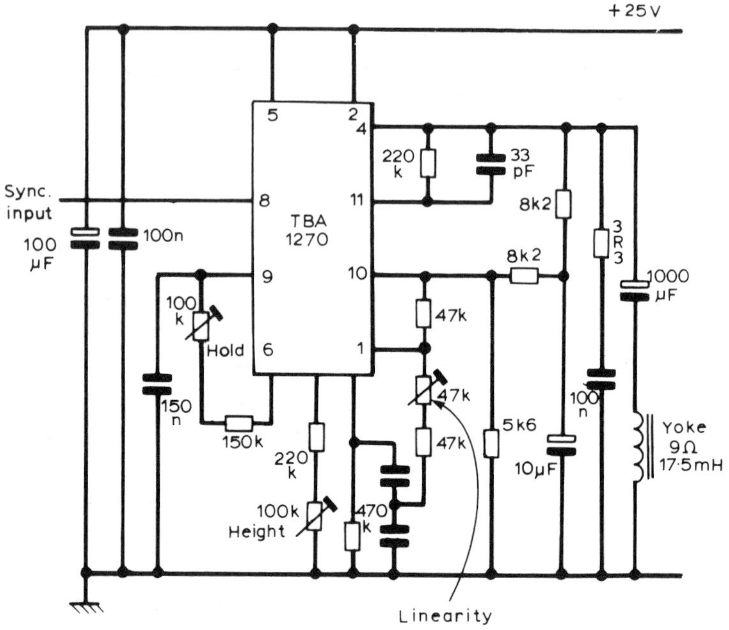
+25V
5
2
4
220
k
33
pF
Sync.
input
8
11
TBA
1270
8k2
100
μF
100n
3
R
3
8k2
9
10
1000
μF
100
k
47k
Hold
6
1
47k
150
n
100
n
5k6
Yoke
9Ω
17·5mH
150k
47k
220
k
10μF
100k
Height
470
k
Linearity

Figure 4.18. The TDA 1270 used for vertical deflection in a 12 in portable TV receiver

manufacturer's handbooks for further information, but as an example of the uses of such i.c.s, *Figure 4.18* shows a TDA1270 (SGS-ATES) which performs the functions of vertical oscillator, timebase generator and output stage.

The 555 timer

This circuit is generally classed among linear circuits because it uses op-amp circuits as comparators. The purpose of the timer is to generate time delays or waveforms which are very well stabilised against voltage changes. A block diagram of the internal circuits is shown in *Figure 4.19.*

A negative-going pulse at the trigger input, pin 2, makes the output of comparator (2) go high. The internal resistor chain holds the (+) input of comparator (2) at one third of the supply voltage, and the (−) input of comparator (1) at two-thirds of supply voltage, unless pin 5 is connected to some different voltage level. The changeover of comparator (2) causes the flip-flop to cut off Tr_1, and also switch the output stage to its high-voltage output state. With Tr_1 cut off, the external capacitor *C* can charge through *R* (also external) until the voltage at pin

6 is high enough, equal to 2/3 of the supply voltage, to operate comparator (1). This resets the flip-flop, allows Tr_1 to conduct again, so discharging *C*, and restores the output to its low voltage state. Resetting is possible during the timing period by applying a negative pulse to the reset pin, number 4.

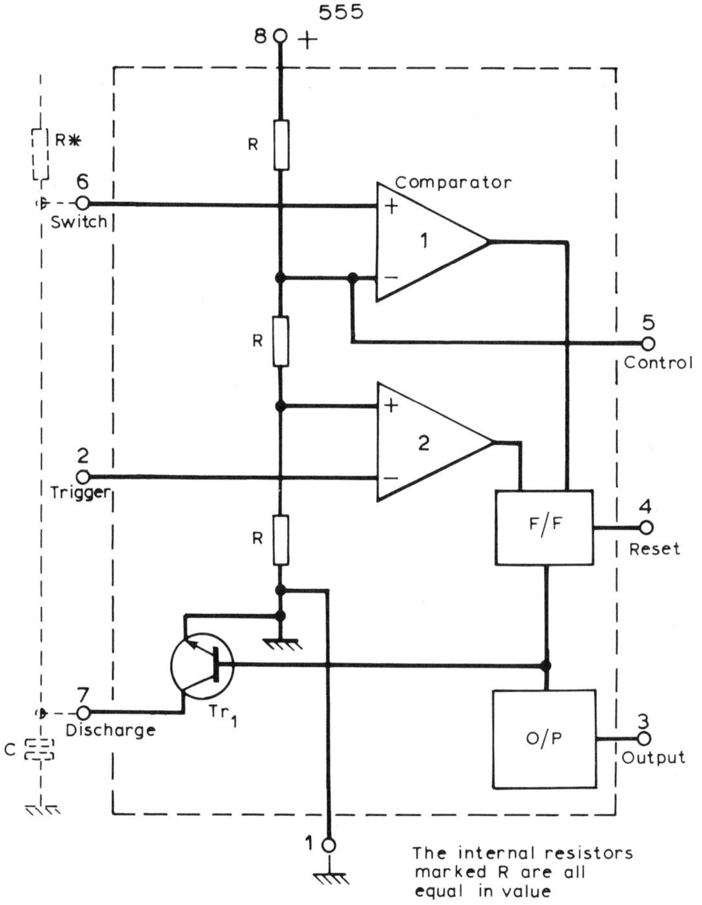
555
8
+
R*
6
Switch
R
Comparator
1
R
5
Control
2
2
Trigger
R
F/F
4
Reset
7
Discharge
C
Tr1
O/P
3
Output
1
The internal resistors
marked R are all
equal in value

Figure 4.19. The 555 timer block diagram. R and C are external components which are added in most applications of the timer*

The triggering is very sensitive, and some care has to be taken to avoid unwanted triggering pulses, particularly when inductive loads are driven. Retriggering caused by the back-e.m.f. pulse from an inductive load is termed 'latch-up', and can be prevented by the diode circuitry shown in *Figure 4.20*.

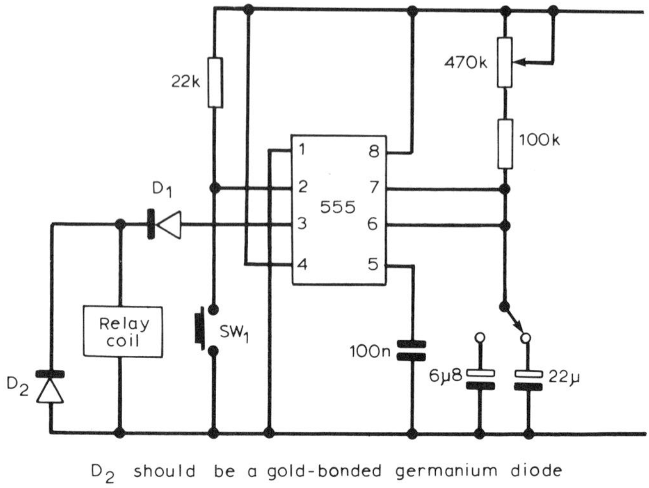
22k
470k
100k
1
2
3
4
8
7
6
5
555
D1
Relay coil
SW1
100n
6µ8
22µ
D2
D2 should be a gold-bonded germanium diode

Figure 4.20. A relay-timer circuit using the 555. On pressing the switch, the relay is activated for a time determined by the setting of the 470 kΩ variable and the capacitor value selected by the switch. Note the use of diodes to prevent latch-up and damage to the i.c. when the relay is switched off

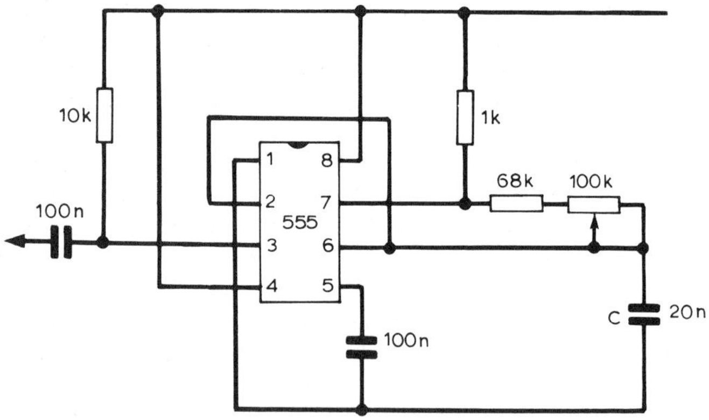
10k
1k
1
2
3
4
8
7
6
5
555
68k
100k
100n
100n
C
20n

Figure 4.21. An astable pulse generator, with variable frequency output controlled by the 100 kΩ potentiometer. The capacitor C can be a switched value if desired

Two typical circuits are shown in *Figures 4.21* and *4.22*. The timer is available from several manufacturers, all using the same 555 number though prefixed with different letter combinations which indicate the manufacturer.

Chapter 5

Digital I.C.s

Basic logic notes

Unlike linear circuits, digital circuits process signals which consist of only two significant voltage levels, labelled logic 0 and logic 1. Most logic systems use positive logic, in which logic 0 is represented by zero volts, or a low voltage, below 0.5 V; and logic 1 is represented by a higher voltage. Changes of level from 0 to 1 or 1 to 0 are made as quickly as possible, since slow changes can cause faulty operation of some types of logic circuits.

The use of two logic levels naturally leads to the use of a scale-of-two or binary scale for counting. In a binary scale, the only digits used are 1 and 0, with the placing of the 1 indicating what power of 2 is represented. *Table 5.1* shows methods of converting to and from binary numbers and decimal numbers.

In addition, because large binary numbers are awkward to handle, and difficult to copy without error, hexadecimal (or hex) numbers are used for many applications, particularly in microprocessor machine code (see later). Hex coding is used when binary numbers occur in groups of four, eight (called a byte) or multiples, and the conversions are shown in *Table 5.2*. The use of hex coding makes the tabulation of binary numbers considerably simpler, but the circuits to which the hex codes refer will still operate in binary.

Note in this context that a three-state or tri-state device does not imply a third logic voltage. This description refers to a circuit which can be switched to a high impedance at an input or output (or both) so that it does not affect or respond to the voltages of signals connected to it.

Table 5.1. CODE CONVERSION

DECIMAL-TO-BINARY CONVERSION

Write down the decimal number. Divide by two, and write the result underneath, with the remainder, 0 or 1, at the side. Now divide by two again, placing the new (whole number) result underneath, and the remainder, 0 or 1, at the side. Repeat until the last figure (which will be 2 or 1) has been divided, leaving zero.

Now read the remainders in order from the foot of the column upwards to give the binary number.

Example: Convert decimal 1065 into binary:

1065	
532	1
266	0
133	0
66	1
33	0
16	1
8	0
4	0
2	0
1	0
0	1

Binary number is 10000101001

BINARY-TO-DECIMAL CONVERSION

Using the table of binary powers, which shows the values of successive powers of two, starting from the right-hand side of the binary number (the least significant figure). Write down the decimal equivalent for each 1 in the binary number, and then add.

Example: 1 0 0 0 0 1 0 1 0 0 1

	Decimal
	1
	8
	32
	1024
Total	1065

Power of two	*Decimal*	*Power of two*	*Decimal*
0	1	10	1 024
1	2	11	2 048
2	4	12	4 096
3	8	13	8 192
4	16	14	16 384
5	32	15	32 768
6	64	16	65 536
7	128	17	131 072
8	256	18	262 144
9	512	19	524 288

Table 5.2. BINARY, HEXADECIMAL AND DECIMAL

4-bit Binary	*Decimal*	*Hexadecimal*
0000	0	0
0001	1	1
0010	2	2
0011	3	3
0100	4	4
0101	5	5
0100	6	6
0111	7	7
1000	8	8
1001	9	9
1010	10	A
1011	11	B
1100	12	C
1101	13	D
1110	14	E
1111	15	F

To convert binary to hex: arrange the binary numbers in four-bit groups starting from the right-hand side (the least significant digit). Convert each group to hex, using the table above.

Examples: 01011101 converts to 5D, 11100011 converts to E3

To convert hex to binary: write down the equivalent for each number, using the table above.

Examples: E1 = 11100001, 6F = 01101111

Combinational logic circuits are those in which the output at any time is determined entirely by the combination of input signals which are present at that time. A combinational logic circuit can be made to produce a logic 1 at its output only for some set pattern of 1s and 0s at various inputs, just as a combination lock will open only when the correct pattern of numbers has been set. One of the major uses for

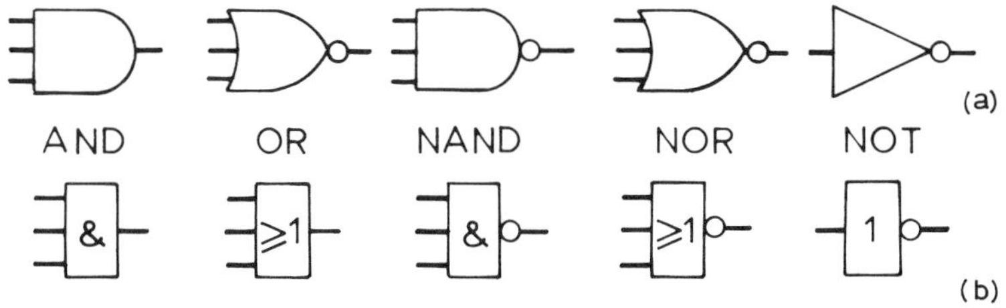
(a)
AND
OR
NAND
NOR
NOT
&
≥1
&
≥1
1
(b)

Figure 5.1. Gate symbols. (a) International (U.S. MIL) symbols, used in magazine articles, most books and manufacturer's manuals, (b) British Standard (B.S.) symbols used for TEC and C & G examinations and textbooks

combinational logic circuits is in recognising patterns, whether these be binary numbers or other information in binary form.

Any required combinational logic circuit can be built up from a few basic circuits, the choice of which is dictated more by practical than by theoretical considerations. The three most basic circuits are those of the

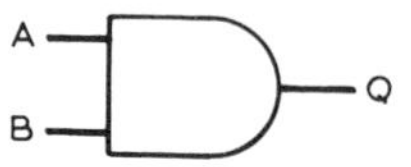
A
B
Q

Truth table

A	B	Q
0	0	0
0	1	0
1	0	0
1	1	1

Boolean :

A • B = Q

Read as A and B gives Q

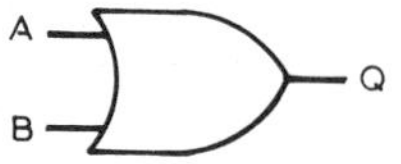
A
B
Q

Truth table

A	B	Q
0	0	0
0	1	1
1	0	1
1	1	1

Boolean:

A + B = Q

Read as A or B gives Q

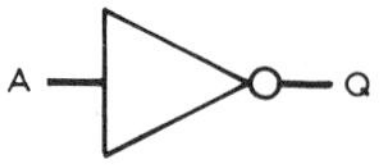
A
Q

A	Q
1	1
1	0

Boolean :

$\overline{A}$ = Q

Read as inverse (or complement) of A is Q,
or as Q = NOT A

Figure 5.2. Truth tables and Boolean expressions for three basic gates

AND, OR and NOT gates; the NOT gate is also referred to as an inverter or complementer. These are represented in circuits by the symbols shown in *Figure 5.1*; the international symbols are much more commonly used. The internal circuitry of the i.c.s is not usually shown, since the action of the circuits is standardised. All that need be known about the internal circuits is the correct level of power supplies, driving signals and output signals, together with any handling precautions.

The action of a logic gate or circuit can be described by a truth table or by a Boolean expression. A truth table shows each possible input to the logic circuit with the output which such a set of inputs produces; examples are shown in *Figure 5.2*. A Boolean expression is a much shorter way of showing the action, using the symbols + to mean OR and . to mean AND. The action of a four-input AND gate, for example, can be written as A.B.C.D = 1; a truth table for this gate would take up half a page, because the number of lines of truth table is equal to 2^n, where n is the number of inputs.

A
B
Q
A
B
Q

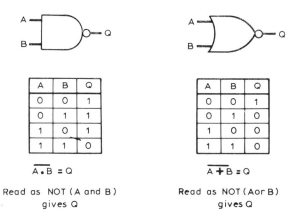

A	B	Q
0	0	1
0	1	1
1	0	1
1	1	0

$\overline{A . B} = Q$

Read as NOT (A and B) gives Q

A	B	Q
0	0	1
0	1	0
1	0	0
1	1	0

$\overline{A + B} = Q$

Read as NOT (A or B) gives Q

Figure 5.3. NAND and NOR gate truth tables and Boolean expressions

The NAND and NOR gates shown in *Figure 5.3* combine the actions of the AND and OR gates with that of an inverter. These gates are simpler to produce, and either can be used as an inverter so that for practical purposes these are more common than the ordinary AND and OR gates. Another type of circuit which is used to such an extent that it is available in i.c. form is the exclusive-OR (X-OR) gate. The truth table for a two-input X-OR gate is shown in *Figure 5.4*, in comparison

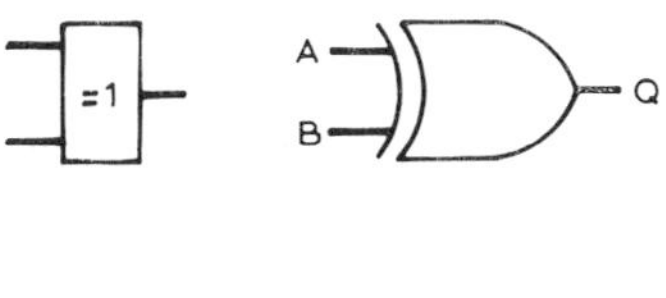
=1
A
B
Q

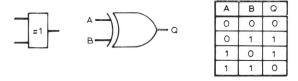

A	B	Q
0	0	0
0	1	1
1	0	1
1	1	0

Boolean: $(\overline{A . B}) . (A + B) = Q$

Figure 5.4. Exclusive-OR (X-OR) gate. Symbols, truth table and Boolean expression

with that for a simple OR-gate. The Boolean expression is $Q = (A + B).\overline{(A.B)}$, where the bar above the brackets indicates NOT (= inverse). The expression A.B will be 1 when A = 1 and B = 1, and the inverse bar indicates that the output must be zero for this state. This quantity is ANDED with (A or B), so as to make the output zero for A = 1 and B = 1.

The advantages of setting out logic circuit requirements as Boolean expressions rather than in truth tables are:

(1) The Boolean expressions are considerably quicker to write.

(2) The Boolean expressions can be simplified (often) by applying a set of rules.

(3) The Boolean expressions can usually indicate what combination of single gates will be needed for the circuit.

The rules of Boolean algebra (invented, incidentally, long before digital logic circuits existed) are shown in *Table 5.3*. The usefulness of these

Table 5.3. BOOLEAN ALGEBRA

1. Definitions of gate action	
$0 + 0 = 0$	$0.0 = 0$
$0 + 1 = 1$	$0.1 = 0$
$1 + 0 = 1$	$1.0 = 0$
$1 + 1 = 1$	$1.1 = 1$
2. For gates to which one input is a variable A (which can be 0 or 1)	
$A + 0 = A$	$A.0 = 0$
$A + 1 = 1$	$A.1 = A$
$A + A = A$	$A.A = A$
$A + \overline{A} = 1$	$A.\overline{A} = 0$
$\overline{\overline{A}} = A$	
3. For more than one variable (each of which can be 0 or 1)	
$A + B = B + A$	$A.B = B.A$
$(A + B) + C = A + (B + C)$	$(A.B).C = A.(B.C)$
$(A.B) + (A.C) = A.(B + C)$	$(A + B).(A + C) = A + (B.C)$
4. De Morgan's theorem	
$\overline{A.B} = \overline{A} + \overline{B}$	$\overline{A + B} = \overline{A}.\overline{B}$

rules is that they may be applied to simplify an expression, so saving considerably in the number of logic gates that are needed to carry out a logic operation. For example, the proposed gate system in *Figure 5.5* carries out the process which in Boolean algebra becomes

$$(A \,.\, B \,.\, C) + A \,.\, (\overline{B} + \overline{C})$$

Leaving the first term unchanged, we can apply Rule 4 to the second term

$$(A \cdot B \cdot C) + A \cdot (\overline{B \cdot C})$$

Now taking out the common factor A changes the expression to

$$A \cdot (B \cdot C + \overline{B \cdot C}) \qquad \text{(Rule 3)}$$

$$= A \cdot 1 \qquad \text{(Rule 2)}$$

$$= A \qquad \text{(Rule 2)}$$

From this, it appears that only the A signal is needed. An experienced designer might, in fact, be able to deduce that the B and C signals were

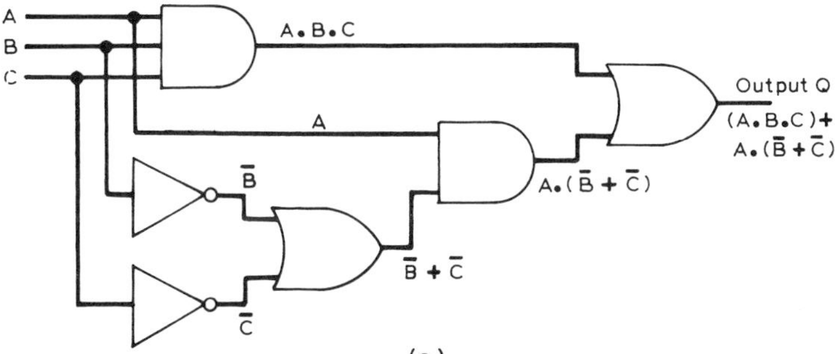

A
B
C
A•B•C
A
$\overline{B}$
$\overline{B}$ + $\overline{C}$
$\overline{C}$
A•($\overline{B}$ + $\overline{C}$)
Output Q
(A•B•C)+
A•($\overline{B}$ + $\overline{C}$)

(a)

A	B	C	A•B•C	$\overline{B}$	$\overline{C}$	$\overline{B}+\overline{C}$	A•($\overline{B}+\overline{C}$)	Q
0	0	0	0	1	1	1	0	0
0	0	1	0	1	0	1	0	0
0	1	0	0	0	1	1	0	0
0	1	1	0	0	0	0	0	0
1	0	0	0	1	1	1	1	1
1	0	1	0	1	0	1	1	1
1	1	0	0	0	1	1	1	1
1	1	1	1	0	0	0	0	1

(b)

Figure 5.5. Analysing the action of a gate system. (a) The text shows the method using Boolean algebra, the truth table shows that the Q output column corresponds exactly with the A input column. The truth table method is simple, but very tedious, because the number of lines of truth table are equal to 2^n, where n is the number of inputs (A, B, C, etc.)

redundant by an inspection of the gate circuit, but the Boolean algebra is often faster to write, if nothing else!

By way of another example, consider the gates of *Figure 5.6*, which implement the Boolean expression

$$Q = \overline{A} \cdot C \cdot + \overline{A} \cdot B \cdot D + A \cdot C \cdot D + A \cdot B \cdot D$$

Taking out the common factor D, this becomes

$$Q = D \cdot (\bar{A} \cdot C + \bar{A} \cdot B + A \cdot C + A \cdot B)$$

and taking out the next common factor (C + B), this transforms to

$$Q = D \cdot (\bar{A} \cdot (C + B) + A \cdot (C + B))$$

which is

$$Q = D \cdot (C + B) \cdot (\bar{A} + A)$$

$$= D \cdot (C + B), \text{ since } \bar{A} \text{ or } A \text{ must be logic 1.}$$

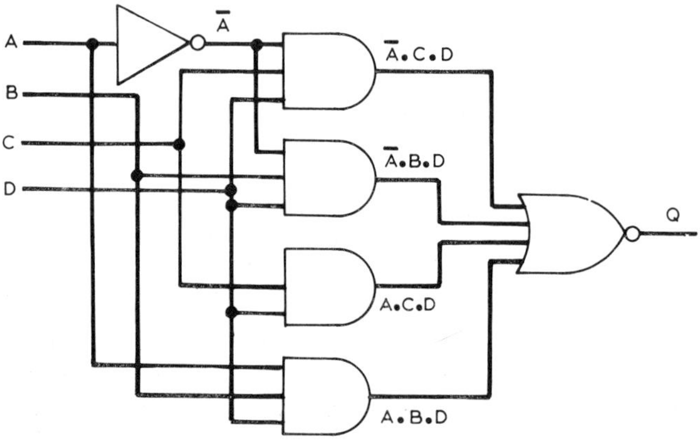
A
B
C
D
Ā
Ā•C•D
Ā•B•D
A•C•D
A•B•D
Q

(a)

A	B	C	D	Ā	Ā•C•D	Ā•B•D	A•C•D	A•B•D	Q
0	0	0	0	1	0	0	0	0	0
0	0	0	1	1	0	0	0	0	0
0	0	1	0	1	0	0	0	0	0
0	0	1	1	1	1	0	0	0	1
0	1	0	0	1	0	0	0	0	0
0	1	0	1	1	0	1	0	0	1
0	1	1	0	1	0	0	0	0	0
0	1	1	1	1	1	1	0	0	1
1	0	0	0	0	0	0	0	0	0
1	0	0	1	0	0	0	0	0	0
1	0	1	0	0	0	0	0	0	0
1	0	1	1	0	0	0	1	0	1
1	1	0	0	0	0	0	0	0	0
1	1	0	1	0	0	0	0	1	1
1	1	1	0	0	0	0	0	0	0
1	1	1	1	0	0	0	1	1	1

(b)

Figure 5.6. A gate system (a) whose action is analysed by Boolean algebra in the text. The truth table (b) shows that the output is logic 1 when D is 1 and C or B also 1

For really large and complex circuits, Boolean algebra can be tedious and difficult, but other approaches are equally tedious and difficult – there is no easy way of dealing with complex circuits.

Note that in circuits which use an elaborate sequence of gates, problems can arise because of the small but significant time delay which

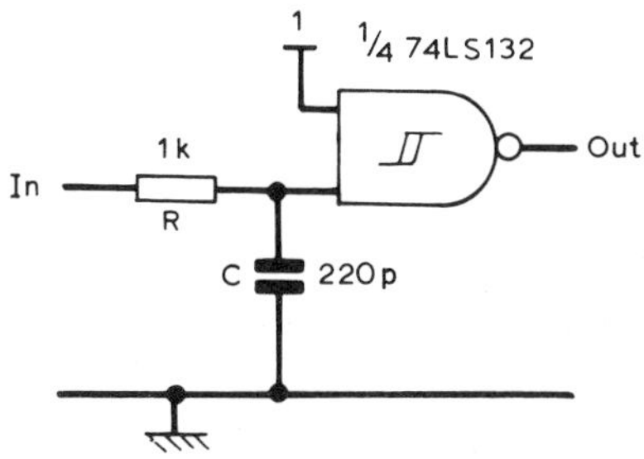

1
1/4 74LS132
1 k
In
R
Out
C
220 p

Figure 5.7. A circuit for suppressing unwanted short pulses. This also inverts, so that a second Schmitt inverter may be needed after the Schmitt gate

is introduced by each gate. These 'race hazard' problems arise when the final gate of a set is switched by signals which have passed through different numbers of gates in their different paths. A time delay between these paths can cause a brief unwanted pulse, of duration equal to the time delay between the input paths, at the output. This unwanted pulse may cause problems, particularly in a counting circuit which follows the gate circuits. Methods of eliminating race hazards are too complex to discuss in this book, but *Figure 5.7* shows a simple unwanted-pulse suppressor which can be used for comparatively slow logic systems. The output is integrated by *R*, *C*, using a time constant chosen so that a short pulse (50 ns or so) will have little effect, but the delay which the circuit causes to a longer pulse is not serious. The sharp rise and fall times of the logic signals then have to be restored by using a Schmitt NAND gate, IC1.

Sequential logic

Sequential logic circuits change output when the correct sequence of signals appears at the inputs. The simplest of the sequential logic circuits is the R-S (or S-R) flip-flop, or latch, circuit shown in symbol form in *Figure 5.8a* with one possible circuit using NAND gates. The truth table for this device is shown in *Figure 5.8b*, from which it can be seen that it is the sequence of signals at the two inputs rather than the signal levels themselves which decide the output. Note that the outputs Q and $\overline{Q}$ are intended to be complementary (one must be the inverse

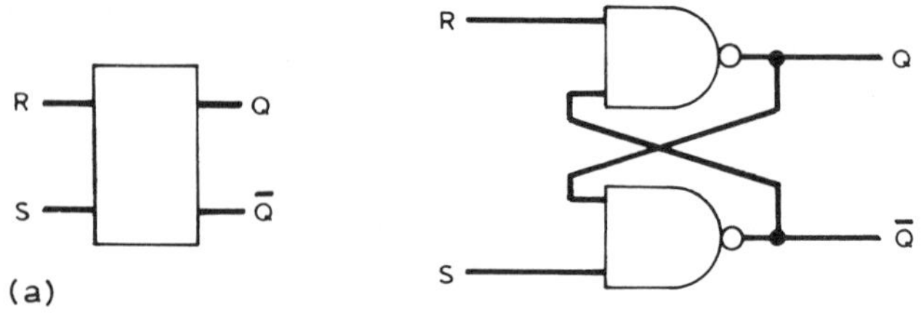
R
S
Q
$\overline{Q}$
(a)
R
S
Q
$\overline{Q}$

Truth table

R	S	Q	$\overline{Q}$
0	0	1*	1*
0	1	1	0
1	1	1	0
1	0	0	1
1	1	0	1

*Undesirable state

The R = 0, S = 0 state must be avoided because we normally want $\overline{Q}$ to be the inverse of Q

(b)

Figure 5.8. The R-S flip-flop. (a) Symbol and one possible circuit (NOR gates can also be used). (b) Truth table. The output for R=1, S=1 depends on the previous values of R and S

of the other), so that the state R = 0, S = 0 must never be allowed, since it results in Q = 1, $\overline{Q}$ = 1.

The R-S flip-flop is used to lock or latch information, one bit to each R-S latch, because the output of the latch is held fixed when both

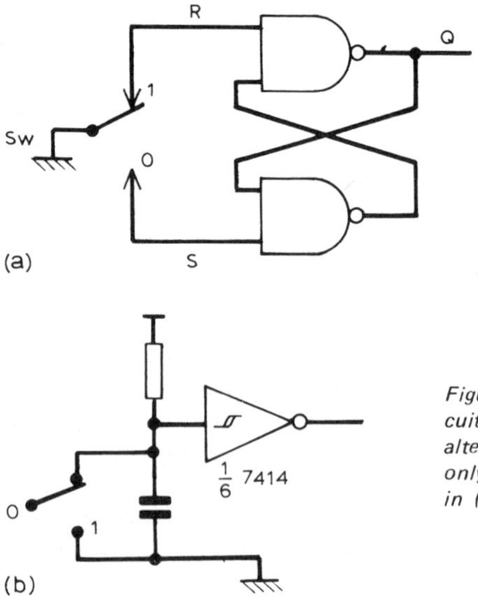
R
Q
1
Sw
0
(a)
S
$\frac{1}{6}$ 7414
0
1
(b)

Figure 5.9. A 'switch-debounce' circuit using an R-S flip-flop (a). An alternative arrangement which needs only an on/off switch type is shown in (b) — this uses a Schmitt inverter rather than the R-S flip-flop

inputs are at logic 1 in the state which existed just before the second input went high. One typical application is to suppress the effects of contact bounce in mechanical switches, as shown in *Figure 5.9*. The effect of bounce in this arrangement is to raise both inputs of the latch to logic 1, leaving the outputs unaffected.

The applications of the simple R-S latch are rather limited, and most sequential logic circuits make use of the principle of clocking. A clocked circuit has a terminal, the clock (CK) input, to which

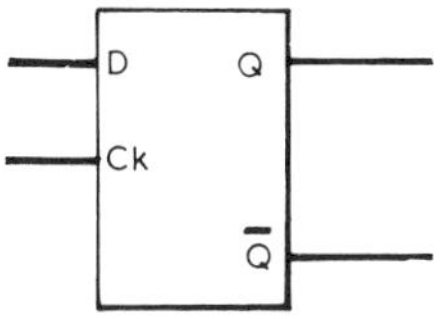
D
Q
Ck
$\bar{Q}$

D	Q_n	Q_n+1
0	0	0
0	1	0
1	0	1
1	1	1

Figure 5.10. Symbol for D-type flip-flop and truth table. In the truth table, Q_n denotes the output before the clock pulse, and Q_n+1 the output after the clock pulse

rectangular pulses, the clock pulses) can be applied. The circuit actions, which is determined by the other inputs, takes place only in the time of the clock pulse, and may be synchronised to the leading edge, the trailing edge, or the time when the clock pulse is at logic 1, according to the design of the circuit.

The D-type flip-flop, shown in *Figure 5.10*, with its truth table, is one form of D-type clocked flip-flop. The output at Q will become identical to the input at D (for DATA) only at the clock pulse, generally at the leading edge of the pulse, so earning the circuit the name 'edge-triggered'. By connecting the $\bar{Q}$ output back to the D-input, the flip-flop will act as a bistable counter, or divide-by-two circuit, because the voltage at Q will cause a change of state only when the leading edge of the clock-pulse occurs. The $\bar{Q}$ voltage itself will change only a little time after the leading edge of the clock pulse, so that there is no effect on the flip-flop. A few D-type flip-flops change state at the trailing edge of the clock pulse, and some allow the D-input to affect the Q-output for as long as the clock pulse is at logic 1. This latter type is sometimes called a 'transparent latch'.

The J-K flip-flop (*Figure 5.11a*) is a much more flexible design which uses a clock pulse along with two control inputs labelled J and K. The internal circuit makes use of two sets of flip-flops, designated master and slave respectively. At the leading edge of the clock pulse, the master flip-flop changes state under the control of the inputs at J and K. After this time, changes at J and K have no effect on the master flip-flop. There is no change at the output, however, until the trailing

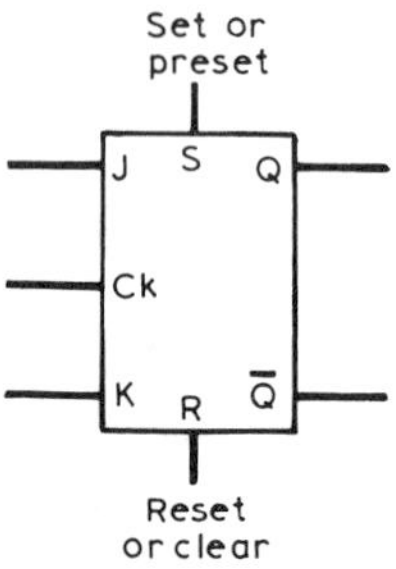
Set or
preset
J
S
Q
Ck
K
R
$\overline{Q}$
Reset
or clear

(a)

J	K	Qn	Qn+1
0	0	0	0
		1	1
0	1	0	0
		1	0
1	0	0	1
		1	1
1	1	0	1
		1	0

(b)

J	K	Qn	Qn + 1
0	0	X	no change
0	1	X	0
1	0	X	1
1	1	X	$\overline{X}$

(c)

Figure 5.11. The J-K flip-flop (a) and its truth table (b). The truth table can be shortened (c) by making use of X to mean 'don't care' – either 0 or 1

edge of the clock. The use of this system ensures a controllable time delay between the output and the input, which permits signals to be fed back from one flip-flop to another without the risk of instability, since an output signal fed back to a J or K input cannot have any effect until the clock leading edge, and will not itself change again until the clock trailing edge. This avoids the difficulty of having an input changing while the output which it controls also changes.

The remaining two signal inputs of the J-K flip-flop operate independently of the clock. The SET (or PRESET) terminal sets the Q output to logic 1, regardless of other signals, and the RESET or CLEAR terminal sets the output to logic 0, also regardless of other signals. The system must be arranged so that these two cannot operate together.

Figure 5.12 shows a type of binary counter variously named serial, ripple, or asynchronous. Using four stages of J-K flip-flops, the counter Q outputs will reach 1111 (decimal 15) before resetting. By using gates to detect 1010 (decimal ten), the counter can be forced to reset on the tenth input pulse, so that a decimal count is obtained. Such a set of gated flip-flops forms the basis of a BCD (Binary-Coded Decimal) counter such as the 7490.

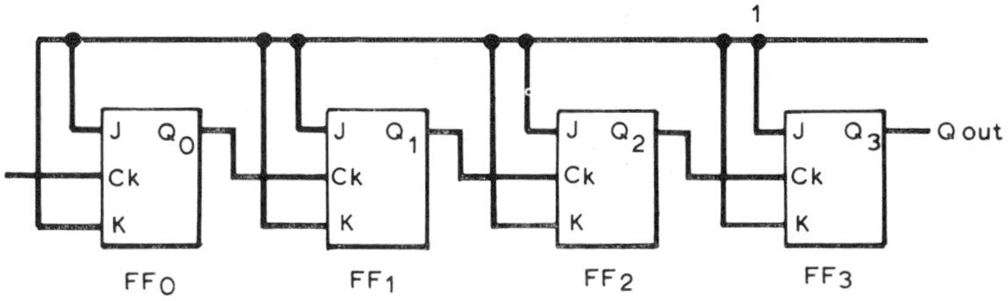
1
J
Ck
K
Q0
Q1
Q2
Q3
Q out
FF0
FF1
FF2
FF3

Figure 5.12. A ripple or asynchronous counter using J-K flip-flops. Note that the flip-flops (and the Q outputs) are labelled 0, 1, 2, 3 rather than 1, 2, 3, 4. This is because the first stage counts 2^0, the second stage 2^1, the third stage 2^2 and so on

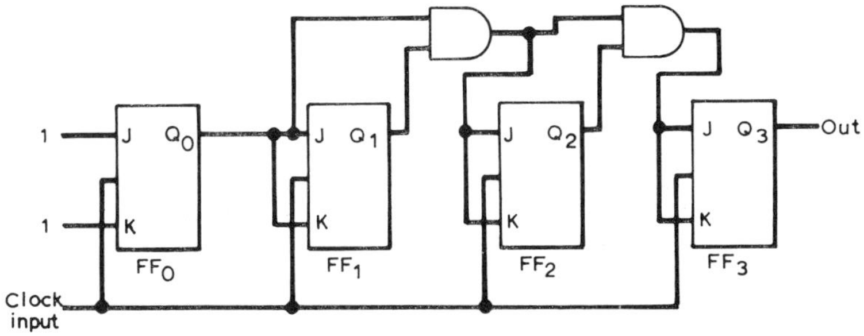
1
1
J
K
Q0
Q1
Q2
Q3
Out
FF0
FF1
FF2
FF3
Clock input

Figure 5.13. A synchronous counter. The input (clock) pulses are taken to each clock terminal, and the changeover of each flip-flop is controlled by the settings of the J, K terminals, which are gated

Figure 5.13 shows a synchronous counter, in which each flip-flop is clocked at the same rate, and counting of the clock pulses is achieved by using gated connections to the J and K inputs. The design of such synchronous counters for numbers which are not powers of two is complex and lengthy. The use of a technique called Karnaugh mapping simplifies the procedure to some extent for small count numbers, but is beyond the scope of this book. In any case, either ripple or synchronous counters can now be obtained as complete integrated circuits for most useful count values, making the design procedure unnecessary, and for complex applications, the use of a microprocessor is probably a simpler solution.

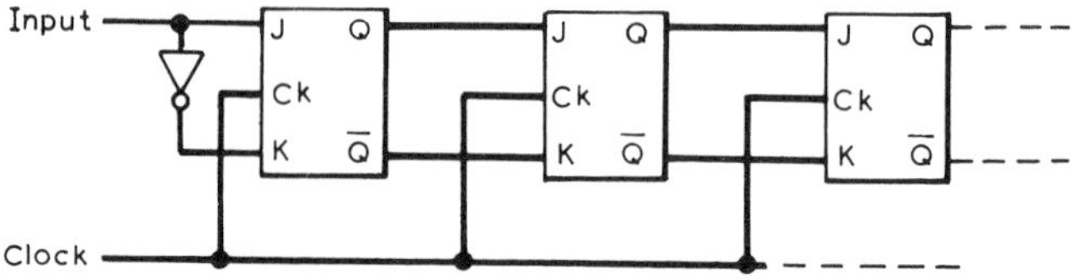
Input
J
Q
Ck
K
Q̄
Clock

Figure 5.14. Shift register. At each clock pulse, the binary digit at the Q output of each flip-flop is 'shifted' to the next flip-flop in line. The arrangement shown here gives right-shift; left-shift can be arranged by connecting $\bar{Q}$ to J and Q to K between flip-flops. For details and circuits, see Beginner's Guide to Digital Electronics

All counter circuits can be arranged so as to count in either direction up or down, and complete counters in i.c. form can be obtained which will change counting direction by altering a control voltage to 0 or 1. An example of such a counter is the 74192.

Shift registers, which can be formed from J-K flip-flops connected as shown in *Figure 5.14*, are also obtained in i.c. form. The action of a

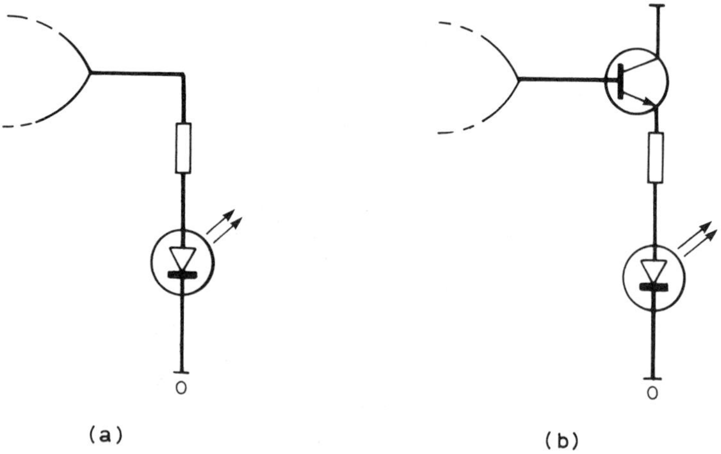
0
0
(a)
(b)

Figure 5.15. LED displays. TTL i.c.s can drive the LED directly (a) using a limiting resistor. CMOS i.c.s generally have low current outputs, and a transistor 'interface' circuit (b) is needed. Multiple transistors for this purpose can be obtained in i.c. form, such as the RCA CA8083 (5 NPN) and CA3082 (7 NPN transistors with a common collector connection)

shift register is to pass a logic signal (1 or 0) from one flip-flop to the next in line at each clock pulse. The input signals can be serial, so that one bit is shifted in at each clock pulse, or parallel, switched into each flip-flop at the same time. The output can similarly be serial, taken

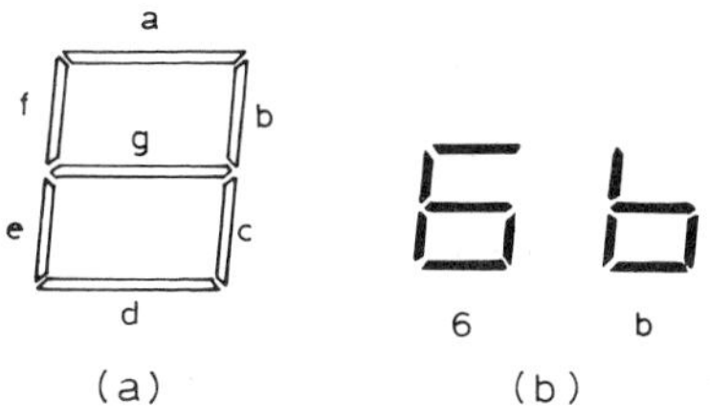

a
f
g
b
e
c
d
(a)
6
b
(b)

Count 0	Segments illuminated a b c d e f
1	b c
2	a b g e d
3	a b g c d
4	f g b c
5	a f g c d
6	a f e d c g
7	a b c
8	a b c d e f g
9	a b g f c
A	a b c f e
B	f e g c d
C	a f e d
D	g e d c b
E	a f g e d
F	a f g e

(c)

Figure 5.16. The seven-segment display (a) showing the segment lettering. An eighth segment, the decimal point, is often added, and may be on the right or left side. Hexadecimal letters as well as figures can be displayed provided that 6 and b are differentiated as shown (b) by the use of the a-segment. The truth table (c) shows the segments illuminated for each number in a hex count. Displays may be common anode or common cathode

from one terminal at each clock pulse, or parallel at each flip-flop output. The shift can be left, right or switched in either direction. Some circuits which make use of shift registers are discussed in detail in *Beginner's Guide to Digital Electronics.*

Displays and decoders

A binary number can be displayed using any visible on/off device; the most convenient is usually the LED (*Figure 5.15*). For decimal number and letter displays (alphanumeric displays), dot matrix displays are used. These consist of an array of LED dots, usually in a 7 X 5 arrangement, which can be illuminated to display a number or letter. A more common arrangement is the seven-segment display (*Figure 5.16*), consisting of bars which can be illuminated. Seven-segment displays may be common-anode, with a common positive connection and each bar illuminated by connecting its terminal to logic zero. The alternative is the common-cathode type, in which the common cathode terminal is earthed and each bar anode is illuminated by taking its voltage to logic 1. A current-limiting resistor must be wired in series with each bar terminal whichever type of system is used.

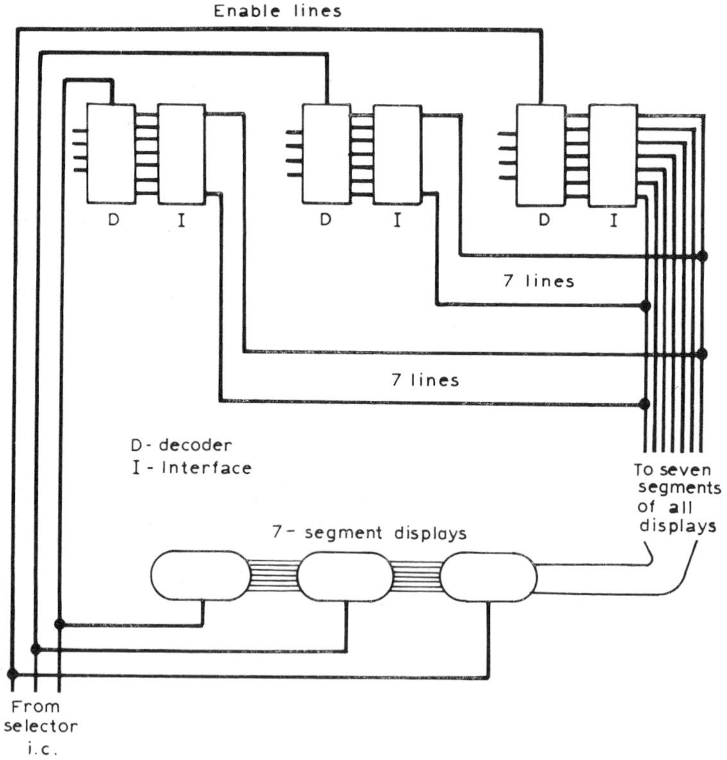
Enable lines
D
I
D
I
D
I
7 lines
7 lines
D - decoder
I - Interface
To seven segments of all displays
7- segment displays
From selector i.c.

Figure 5.17. A strobed or multiplexed display. The decoder i.c.s are operated in sequence by the enable lines (from a counter), which also select the correct display. Only seven segment connections are needed for all the displays, so that fewer pins are needed. The interface i.c.s must be tri-state so that when one is active, driving the segment lines, the others are not affected

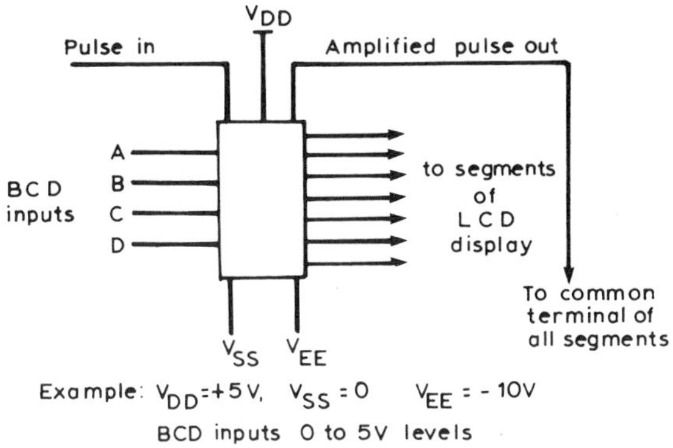

VDD
Pulse in
Amplified pulse out
A
B
C
D
BCD inputs
to segments of LCD display
To common terminal of all segments
VSS
VEE
Example: VDD=+5V, VSS=0 VEE= -10V
BCD inputs 0 to 5V levels

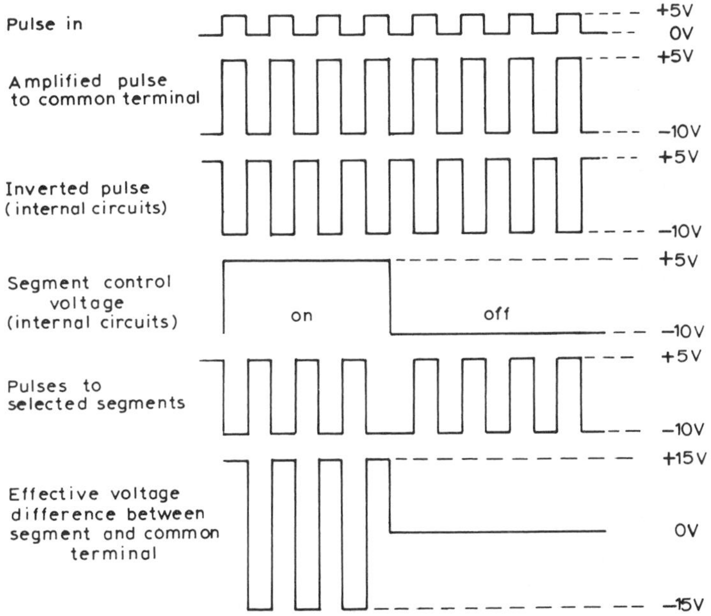

Pulse in
+5V
0V
Amplified pulse to common terminal
+5V
–10V
Inverted pulse (internal circuits)
+5V
–10V
Segment control voltage (internal circuits)
on
off
+5V
–10V
Pulses to selected segments
+5V
–10V
Effective voltage difference between segment and common terminal
+15V
0V
–15V

Figure 5.18. An LCD driving circuit and operating pulses. No d.c. must be applied to an LCD display, so that the pulses must be symmetrical about earth. The pulses are applied in phase to segments and common terminal (backplate) when the segments are inactive, but in antiphase to selected segments when a display is wanted

Whether the display is dot-matrix or seven-segment, a decoder i.c. is needed to convert from binary or decimal signals into the correct arrangement of bars or dots. A very common arrangement is to use binary coded decimal (BCD) in which a set of four bits represents each decimal digit, so that BCD-to-seven segment decoders are available; combined counter-decoders are also available. A few hexadecimal decoders exist for seven-segment drive, and Texas make combined decoder-display units. For dot-matrix displays, a read-only memory is usually used in place of a decoder, to take advantage of the much greater range of alphanumeric characters which can be obtained.

In some circuits, to save space or battery current, displays are strobed, meaning that the figures to be displayed are decoded and displayed one after the other, each in its correct place on the display (*Figure 5.17*).

Liquid crystal displays (LCD) are also used in battery-operated equipment. Their advantages include very low power consumption and better visibility under bright lighting, but at the expense of greater circuit complexity, higher cost, and the need for illumination in poor light. The signals from the decoder must be a.c., with no trace of d.c., at a frequency of a few hundred Hz. A typical LCD display circuit is shown in *Figure 5.18*.

I.C. types

The two main digital i.c. manufacturing methods are described as TTL (Transistor Transistor Logic) and MOS (Metal-Oxide-Silicon) – various types of MOS such as CMOS (complementary p and n channel used), PMOS (positive channel) and NMOS (negative channel) exist. TTL i.c.s use bipolar transistors in integrated form, with input and output circuits which are similar to those shown in *Figure 5.19*. Since the input is always to an emitter, the input resistance is low, and because the base of the input transistor is connected to the +5 V line, the input passes a current of about 1.6 mA when the input voltage is earth, logic 0. If an input is left unconnected, it will 'float' to logic 1, but can be affected by signals coupled by stray capacitance, so that such an input would normally be connected to +5 V through a 1 kΩ resistor. At the output, a totem-pole type of circuit is used. This can supply a current to a load which is connected between the output and earth (current sourcing), or can absorb a current from a load connected between the output and the +15 V line (current sinking). The normal TTL output stage can source or sink 16 mA, enabling it to drive up to ten TTL inputs. In the language of digital designers, the output stage has a fanout of ten.

TTL i.c.s which use these output stages must never be connected with the outputs of different units in parallel, since with one output stage at logic 1 and another at logic 0, large currents could pass,

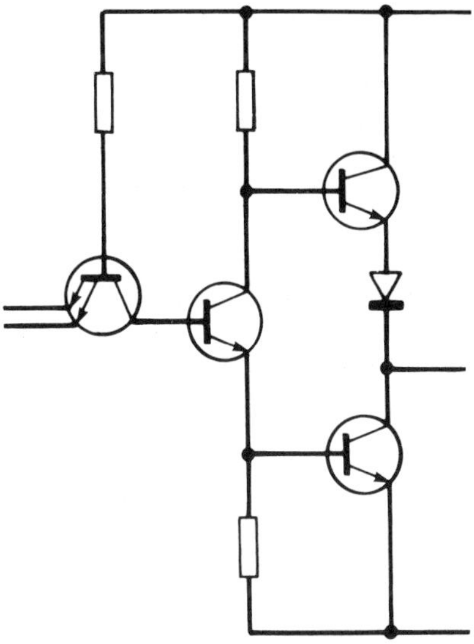

Figure 5.19. TTL circuitry. The example shown is a 2-input NAND gate. The inputs are to emitters of a transistor (in this example, a transistor with two emitters formed on to one base). The output is from a series transistor circuit so that rise and fall times are short

destroying the output stages. Modified output stages, which have open collector outputs, are available for connecting in parallel; an application which is called a wired OR, since the parallel connections create an OR gate at the output.

Inputs such as SET and RESET (or PRESET and CLEAR) normally have to be taken to logic 0 to operate, this is sometimes indicated by a small circle on the input at the symbol for the device. Clock pulses may affect the circuit either on the leading edge, trailing edge, both edges, or at logic 1; the characteristic sheet for the device must be consulted to make certain which clocking system is in use. The supply

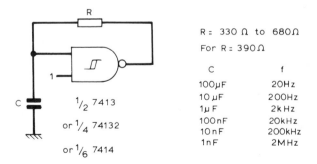

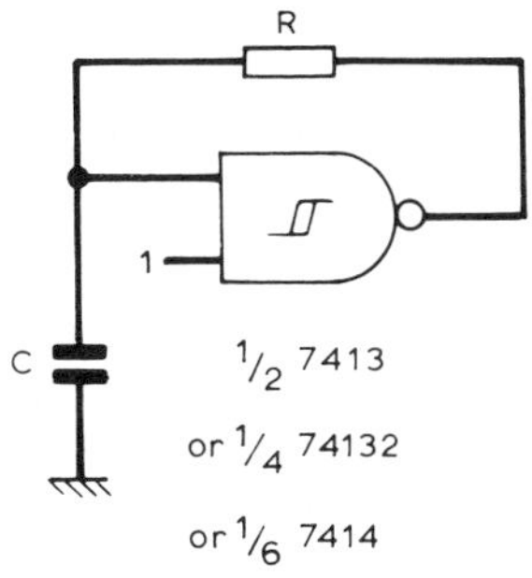
R
1
C
1/2 7413
or 1/4 74132
or 1/6 7414

R = 330 Ω to 680Ω

For R = 390Ω

C	f
100μF	20Hz
10 μF	200Hz
1μ F	2k Hz
100nF	20kHz
10 nF	200kHz
1nF	2MHz

Figure 5.20. Using a Schmitt trigger i.c. as an oscillator. Any inverting Schmitt circuit can be used. The table shows very approximately the values of frequency obtained using a 390 ohm resistor and various capacitor values

voltage must be +5.0 V ± 0.25 V, and large operating currents are needed if more than a few TTL i.c.s are in use. Each group of five TTL i.c.s should have its power supply lines decoupled to avoid transmitting pulses back through the supply lines (even when a stabilised supply is used). The operating pulses should have fast rise and fall times, of the order of 50 ns or less, since many TTL circuits can oscillate if they are linearly biased, as can happen momentarily during a slow rising or falling input. A useful tip is to use Schmitt trigger gates at the input to each circuit, so ensuring a fast rising and falling pulse from any input. These circuits can also be used as clock-pulse generators (*Figure 5.20*).

Table 5.4 (p. 142) shows the pin arrangements for a number of TTL i.c.s.

Most of the TTL range are now available as low power Schottky types. These i.c.s make use of a component called the Schottky diode, which can be built in i.c. form, and which limits the voltage between the collector and the base of a transistor when connected between these points. The use of Schottky diodes makes it possible to design gates which need much smaller currents than conventional TTL i.c.s and which, because the transistors never saturate, can switch as fast, or even faster. The typical input current which has to be sunk to keep a low-power Schottky gate input at logic 0 is around 0.4 mA, only a quarter of the amount needed for a conventional stage. These Schottky i.c.s are generally distinguished by the use of the letters LS in the type number, such as 74LS00, 74LS132, etc.

MOS circuits

Small-scale digital circuits, as well as large-scale microprocessor or calculator chips, can be made using MOS techniques, making use of n-channel, p-channel, or both. One very useful series, the 4000 series, uses both p and n channel FETs, and is known as CMOS (Complementary MOS). Typical input and output circuits are shown in *Figure*

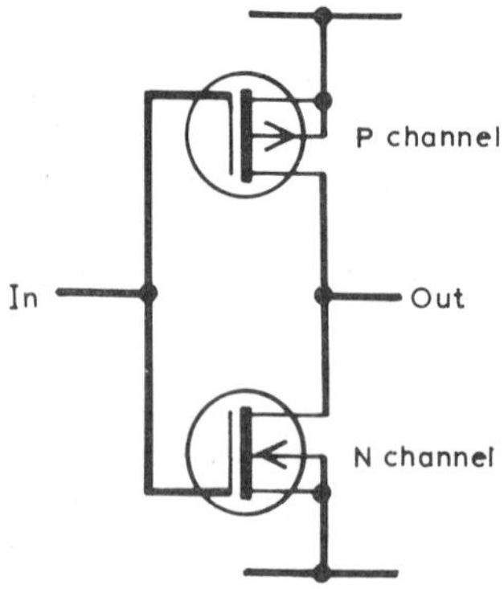
P channel
In
Out
N channel

Figure 5.21. Typical CMOS input and output circuits – the example is of an inverter. In practice, protection diodes are built in at the gate inputs to prevent excessive voltages from damaging the gates, so that nothing short of a visible spark at the input is likely to cause harm

5.21. Since the input is always to the gate of a FET, the input resistance is always very high, so that virtually no d.c. will flow in the input circuit, except through protective diodes. This feature makes the fanout capability for low frequency signals very large, though at higher operating speeds the fanout is limited by the need for the output currents to charge and discharge the stray capacitances at the gate inputs. Because of the high input resistance, the input gates are easily damaged by electrostatic charges. When the i.c. is in circuit, load resistors, along with built-in protection diodes, guard against overloads, but when the i.c. is not in a circuit protection against electrostatic damage can be assured only while all the pins are shorted together. For storage, the pins are often embedded in conductive foam, or the i.c. contained in a conductive plastic case. Particular care has to be taken when CMOS i.c.s are connected into circuit. In 'normal' domestic surroundings, the following precautions are adequate.

(1) The remainder of the circuit should be complete before the MOS i.c.s are added.

(2) Plugging into holders is less hazardous than soldering.

(3) The negative line of the circuit should be earthed when the i.c.s are plugged in.

(4) If soldering is used, the soldering iron must have an earthed bit.

(5) No input pins must ever be left unconnected in a circuit.

(6) The pins of the i.c. should not be handled.

For industrial conditions where low humidity and large insulating areas can present unusual electrostatic problems, the manufacturers' guides should be consulted.

The high input resistance of CMOS i.c.s makes some circuits, particularly oscillator circuits, much easier to implement, and the very low current consumption makes the use of battery operation possible for large circuits. The maximum operating voltage is +15 V, and satisfactory operation is possible at only 3 V; a safe range of operating voltages is 4 V to 12 V. The delay timer for a typical gate is greater than that of a comparable TTL i.c., but for many applications this is not important. Clock pulses with rise or fall times exceeding 5 μs should not be used.

Table 5.5 (p.177) shows the pin arrangement of a range of CMOS i.c.s.

Computers

The main circuit components of a digital computer of any size are registers and gates. The registers are, in the main, parallel in, parallel out types, but with provision for serial shifting. The digital computer

32		33	!	34	"	35	#	36	$
37	%	38	&	39	'	40	(	41	)
42	*	43	+	44	,	45	-	46	.
47	/	48	0	49	1	50	2	51	3
52	4	53	5	54	6	55	7	56	8
57	9	58	:	59	;	60	<	61	=
62	>	63	?	64	@	65	A	66	B
67	C	68	D	69	E	70	F	71	G
72	H	73	I	74	J	75	K	76	L
77	M	78	N	79	O	80	P	81	Q
82	R	83	S	84	T	85	U	86	V
87	W	88	X	89	Y	90	Z	91	[
92	\	93	]	94	^	95	_	96	`
97	a	98	b	99	c	100	d	101	e
102	f	103	g	104	h	105	i	106	j
107	k	108	l	109	m	110	n	111	o
112	p	113	q	114	r	115	s	116	t
117	u	118	v	119	w	120	x	121	y
122	z	123	{	124	¦	125	}	126	~
127	■								

Figure 5.22. The ASCII codes as they appear on a printer. Computers often use minor variations for some codes, particularly 123 to 127. There is no standard £ sign, for example, in the list

operates on binary numbers, and any information that is stored or used must be in binary form. Letters of the alphabet, for example, are coded as numbers, using the ASCII code system which is illustrated in *Figure 5.22.* Denary numbers are converted to binary form, using two different methods. One method is used for integers (whole numbers), and employs the methods that we have illustrated in *Table 5.1.* Numbers which are integers are stored in exact form, and the result of any arithmetic operation on an integer, with the exception of division, is also exact. The reason for excepting division is that division can result in a fraction, and integer numbers do not include fractions. Dividing the integer 5 by the integer 2, for example, gives the integer 2, because there is no way of expressing 0.5 in integer terms.

To avoid the limitations that are imposed by the use of integers, then, computers also allow for what are called 'real' or 'floating-point' numbers. This allows a large range of numbers, positive or negative, whole or fractional, to be used. The storage method is not so straightforward, however. Each real number is converted into a binary number, which can include a binary fraction. The number is then rearranged by shifting the binary point into a binary fraction and a power of two. These two are then stored and used for arithmetic. One common scheme is to store the binary fraction as a 32-bit number, and the power of two in eight bits. The snag in this scheme is that very few real numbers convert *exactly* into binary fraction form. This causes errors when real numbers are used, and the programming of the computer must be arranged so as to correct these errors by rounding numbers up or down where necessary.

The main computing actions then consist of operations on binary bits. These bits are stored in the memory of the computer, a set of miniature flip-flops or capacitors. Flip-flop memory is known as 'static' memory, and is arranged so that each group of flip-flops can be connected to by placing a binary-code address number on a set of lines. This method allows any part of the memory to be used either to store bits (writing) or to copy existing bits (reading). Memory like this is known as 'random-access memory (RAM)' to distinguish it from the older scheme which used large shift registers with the bits fed out serially. The form of RAM which uses miniature capacitors (formed as part of a MOS chip) is called 'dynamic RAM', and it uses the same principles of addressing. It is easier to construct, particularly if large memory capacity is needed, but retains data for only a matter of milliseconds. However, by arranging for the memory to be scanned at intervals and all charged capacitors recharged, the memory can be retained for as long as this scanning or 'refreshing' is applied.

The computing actions consist of addition and subtraction of binary numbers, gating under AND, OR or XOR rules, copying binary numbers from one memory address to another, and shifting or rotation actions within registers. All of these actions can be carried out by using gates and registers, and the main actions of logic gating and arithmetic are carried out by a section of the computer which is called the ALU, arithmetic and logic unit. These very simple actions are, remarkably enough, the basis of all computing operations.

Programmability

The feature that makes the computer so useful is programmability. When we construct a gate and register circuit, the circuit paths are normally fixed, and we can change them only by cutting tracks on the PCB and resoldering. Imagine, however, a circuit which contained gates and registers, but in which all the circuit paths could be changed by using gates in each connection (*Figure 5.23*). In the simplified example, the signal at S can be routed to register 1 or to register 2 according to the control voltage C. If C is at logic 1, then the signal at S reaches register 1.

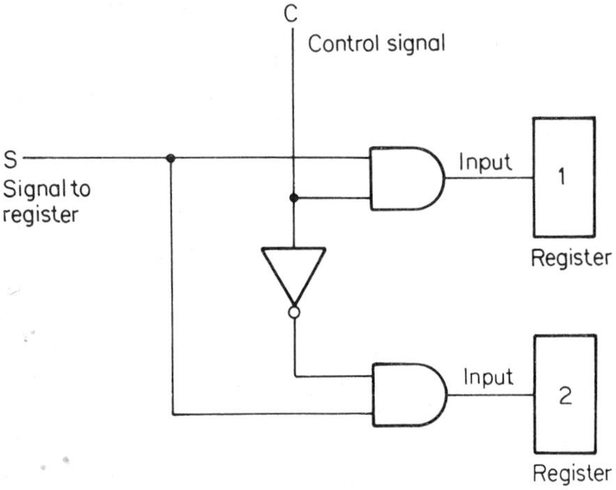
C
Control signal
S
Signal to register
Input
1
Register
Input
2
Register

Figure 5.23. How signal paths can be controlled by gates

If the control voltage at C is at logic 0, then the signal at S reaches register 2. If we provide enough of these gates, all of the signal paths in a circuit can be changed over simply by connecting a suitable logic voltage to a set of gates.

Now we can take this idea a step further. Suppose that the gates which control the connections are themselves operated by the bits stored in a register. By changing the bits that are stored in this register, we can change the connections in a circuit, and so alter the action of the circuit. This is how a circuit can be made programmable, and the program consists of the pattern of bits that is stored in the register that controls the internal circuit gates. In a computer, the register that is used in this way is called the 'instruction register'. All of the actions of the ALU, and of the other computer circuits, can be controlled by the binary number bits that are placed in this instruction register. In practice, each little section is controlled by its own instruction register, and all of these minor instruction registers are controlled by one main instruction register. The sets of numbers which are used in the minor instruction registers are called the 'microprogram' for the system. This whole set-up is known as the CPU, central processing unit, of the computer.

The action of the computer is then controlled by feeding numbers into the instruction register. One set of numbers sent to the instruction register, for example, may cause connections to be made so that a number is copied from memory, added to another number, and then stored back in memory. Another set may cause a number to be read from memory, shifted left, then stored back in memory. All of these operations are programmed by making use of the instruction register. The design of the registers is such that all actions are clocked, and the clock rate can be as fast as the type of semiconductors permits. If, for example, the clock rate is 1 MHz, then one million internal actions per second can be performed. This is *not* the same as one million additions per second, because one addition requires several internal actions of copying and gating to be carried out. A reasonable average is to assume that each computer operation will take from 3 to 12 clock pulses to complete. This is still a fairly fast rate, particularly when we consider that a 1 MHz clock rate is hopelessly slow by modern standards. Even microcomputers commonly use 4 MHz clock rates, and larger computers use rates of 100 MHz or more.

The 'smart' actions of computers are achieved by programming, feeding binary command numbers into the main instruction register as fast as they are needed. This cannot be done manually, because of the speed of the clock, and the method that is used is to store the correct sequence of command numbers in memory. The computer then reads the numbers as and when they are needed. This is possible because the memory is 'addressable', meaning that any part of the memory can be reached by putting a binary number in bit form on to a set of lines. These lines, called the address bus, are connected to all of the memory units, and another set of lines, the data bus, carry the signals which form the bits of numbers that the computer is working on, the data. The address bus obtains its signals from a register, the 'program counter', and the action of the CPU is that the program counter number is increased by one step each time a program command number has to be used. If these command numbers are stored in *exactly* the correct sequence in the memory, then, the whole action is automatic. If the CPU program counter register is loaded with the memory address of the first number of a program, this number will be copied into the instruction register, and the automatic action will start. If the instruction requires another number to be read, the program counter increments (its content has 1 added to it), and this next memory address is read. At the end of the first instruction, the program counter increments again to read the next instruction code. This will continue until one of the codes causes the CPU to stop. Even this term is misleading, because the presence of the clock pulses means that the CPU cannot stop. The nearest it can come to stop is to enter a 'loop'. This means executing the same instructions over and over again until something is done to start another program running. The instructions can be stored in the ordinary RAM memory, in which case they will be lost when power is switched off. An alternative is to store instructions in a different type of memory, ROM. ROM means 'read-only memory', and such a memory consists of a set of permanent connections, some to earth (logic 0) and some to a power supply pin (logic 1). Once again, these connections are reached by making use of address signals, so that for each address number, a different set of instruction bits is available. The advantage of using ROM is that the instructions are not lost when power is switched off, because the connections are permanent. The disadvantage, however, is that the ROM cannot be altered if another program is needed.

This brief description of CPU action nevertheless points out one very important feature. The CPU requires for its operation a set of binary numbers which act as its programming instructions. These numbers have to be stored in RAM (or ROM) memory, and they have to be stored in *exactly* the order that the CPU will need them. All the actions that are achieved by the computer are due to programs of this type stored in memory. Remember that the fundamental actions of the CPU may be no more than addition, subtraction, logic gating, and shifting. Anything else that the CPU can accomplish has to be done by a program in the memory. The set of number codes that dictates the action of the CPU is called 'machine-code', and all programming of computers has ultimately to be done in machine-code.

Each CPU design requires different machine code instructions, and writing directly in machine code is a very tedious business, mainly because of the amount of detail that is needed. There may, for example, be several hundred steps of machine code needed for an action like extracting a square root of a number. Because this type of programming is so tedious, *high-level* languages have been devised. A high level language consists of a set of instruction words and a syntax. The instruction words are like ordinary English words, words such as PRINT, DO, REPEAT, UNTIL and so on. The 'syntax' means the way in which the words are used. A computer which can use a high-level language contains a program, written in machine code, which can interpret the meaning of the instruction words and convert a program written in this way into machine code. This can be done in two ways. One way is called 'interpretation', and this is the method that is used by most of the very small microcomputers. In an interpreter, each instruction word, and the things it operates on (numbers, letters, words) will be converted into a set of machine-code numbers, and the machine code is run. This is done immediately after an instruction is interpreted, and the next instruction is then converted into machine code and run. The conversion process takes time – it consists of reading the instruction word, and looking for a matching word in the part of the memory that holds the interpreter program. When a match is found, this leads to an address for the machine-code that carries out the action. All of this code is already stored as part of the interpreter, but finding it takes time. This means that an interpreted program runs comparatively slowly, because the looking-up process for each instruction takes time, and the instruction cannot be carried out until the looking-up is completed.

The alternative is called 'compiling'. The instructions are read, looked up, and translated into machine code as before. This time, though, each section of machine code is recorded, to form a complete machine-code program. Once this compiling action has been done, the recorded machine code program can be put into the memory and run at any time. This program will run fast, because it does not have to wait for any looking up processes to be completed. All of the looking up actions have been done once during compiling, and do not have to be done again.

The microprocessor

The microprocessor evolved from the CPU of the older types of computer. Since the CPU consists entirely of gates and registers, and since gates and registers can be fabricated in IC form, it was theoretically possible to make the whole CPU of a computer in one chip. This was feasible only when the techniques of IC construction permitted very large scale integration (VLSI), because the number of devices that need to be fabricated on one chip to form a CPU is daunting. This was first achieved by Intel (USA) in 1969. The first commercially produced microprocessor, the Intel 4004, was soon after followed by the 8008, and then the Intel 8080 which became the basis of many of the early microcomputers.

A microprocessor, then, is a one-chip CPU, which consists of gates and registers under the control of an instruction register. Like the traditional CPU, the microprocessor uses a program counter register (PC) which stores memory addresses, and which is used in the action of reading a machine-code program. Also like the traditional CPU, the microprocessor contains a 'microprogram' which decides what actions can be carried out in response to each command code in the instruction register. The important features of the microprocessor consist of how many bits of data can be processed in each operation, how many units of memory can be addressed, and what actions can be carried out by bits of data. Different microprocessor types have evolved because of the very different ranges of applications, with entirely different requirements. For example, where a microprocessor is to be used for machine or process control, it is regarded as a programmable logic circuit. For such applications, it may not need to work with many bits at a time, may not need to be able to carry out arithmetic, may not need to use a large range of memory addresses. It may, however, require to operate at high clock speeds. For computer use, on the other hand, the microprocessor needs to use several bits of data at a time, eight at least. It must be able to work with large amounts of memory, because computers are traditionally used for handling large amounts of information. The speed of operation is not quite so important, though it should be as high as is consistent with the other factors. From the days of the 8080, then, the design of microprocessors has evolved in two quite different directions, resulting in specialized microprocessors for control applications, and more general types which are used mainly in computing applications. The basic ideas of both types, however, are the same. Whatever type of problem can be solved by hardware, meaning a fixed circuit of gates and registers designed for the purpose, can also be solved by software.

Software means a set of instructions to a microprocessor, and the benefit of using software is the ease with which it can be changed. An alteration to the specification of a circuit board means that a batch of boards has to be scrapped and a new design built. An alteration to a software specification means only that different code numbers have to be stored in memory. Since programs can be recorded on magnetic disks and loaded into RAM memory in a matter of seconds, altering software at the design stage is very easy. Even when the program is to be produced in quantity, it can be put on to a EPROM. This is a form of ROM on which storage is permanent for as long as the chip is shielded from ultra-violet radiation. If faults (or 'bugs') are found on the program, new EPROMS can be made, and old ones re-programmed. Only when the software is known to be almost faultfree does it have to be manufactured in quantity in ROM form. It is worth noting that large elaborate programs are never entirely fault-free, but it is reasonable to expect any faults to be minor and to affect only unusual actions or applications.

Practical details

Almost every type of microprocessor will make use of data lines and address lines. A few types of microprocessors are 'dedicated'. This means that they have been designed for one purpose only, and some are made with all the RAM and ROM memory that they need provided and connected internally. Microprocessors of this type may not need data or memory connections, and have only output control lines to send signals to whatever machines they control. These chips are so specialized that we will not consider them further here. The more important types from the point of view of the designer or service engineer are the general-purpose microprocessors. These may be intended for machine control or for computing purposes, but they are designed in similar ways, and can be dealt with here as being of the same type.

The number of data pins on the microprocessor decides the size of number that can be dealt with in each step. Early processors used four data lines, allowing a four-bit number to be manipulated. Four binary bits, however, allow the representation of numbers only in the range 0 to 15, and if larger numbers have to be used, they have to be split into four-bit units. This makes any processing actions rather slow, and four-bit machines, as these processors are called are not used in computing. They do still have applications to machine control, however. Another use of four-bit microprocessors is as 'bit-slice machines'. A bit slice microprocessor is one which is designed to work in parallel with others, so that a set of such microprocessors can behave like one single large unit. A set of eight 4-bit slice microprocessors, for example, can give the effect of one 32-bit microprocessor. Bit-slice machines are being used less, now that genuine 32-bit microprocessors are becoming available.

The most common provision of data pins is eight, and a group of eight bits is known as a byte. Using an 8-byte data set allows a range of numbers from 0 to 255 (denary) to be manipulated, and the importance of this is that it includes all the numbers of the ASCII code set. The microprocessors which were used in the first microcomputers were all of this class, and many of the second generation of these processors, notably the Z-80 and the 6502, are still in production. Once again, larger numbers are handled by using several steps, commonly two bytes for an integer, and five bytes for a real number. For actions like word-processing, however, in which only ASCII codes are manipulated, the use of eight bits is perfectly adequate. More modern designs have concentrated on 16-bit machines, which use sixteen data pins. This allows numbers in the range 0 to 65 536 (2 to the power 16) to be manipulated in one step. Several processors of this type allow two sixteen-bit numbers to be assembled into a 32-bit number inside the microprocessor and then manipulated in this form. This technique, the 16/32 bit machine, has greatly extended the potential power of micro-computers. When the operating system programs catch up with the technical advances in chip-manufacture, a completely new set of prospects for small computers will be opened up.

The data lines that connect to the data pins of the microprocessor are used to transfer data signals in either direction. This allows any number that is stored in the memory to be copied into a register within the microprocessor. Similarly, any number that is stored in a register can be copied into the memory. The same lines are used to transfer signals in either direction, because all microprocessor actions are in sequence. The memory chips each use a read/write pin whose signal voltage (0 or 1) determines whether the memory can be read or written. This read/write pin is controlled by the microprocessor. Data pins are normally 'three-state' controlled. This means that the pins can be isolated from the internal connections when the microprocessor is neither reading nor writing. This avoids any possibility of induced signals on the data lines causing errors.

The address pins of the microprocessor determine how many units of memory can be used. If we take the example of a microprocessor with eight data lines, each unit of memory will be 8-bits, one byte. Such a microprocessor would normally have 16 address lines. This allows 2 to the power 16 addresses, 65 536 in all, to be used. In other words, 65 536 bytes of memory can be addressed separately, with random access to any one address. This amount is often known as 64K, with K having the rather specialized meaning of 1024 units (two to the power 10). This amount may seem reasonable, but it is in fact rather a limitation by modern standards. In the early days of microcomputing, it was enough to provide for signals in from a keyboard, and signals out to a cathode-ray tube monitor, and little else. It was also acceptable to make use of a high level language with only a limited number of instruction words. All of this could mean that a ROM containing the essential machine code for operating the computer and implementing the high-level language needed only 8K of memory, leaving a possible 56K of RAM. Many machines were sold with much less than this, 8K was considered quite a respectable amount of RAM free for the user, some machines used 4K and one even used only 1K. Nowadays, computer users expect a much larger operating system which controls the action of the machine, and even for an 8-bit microprocessor, an operating system of 32K is not uncommon. Another modern practice is to use some of the RAM for holding information for the screen display (shape and colour codes for each part of the screen), and this can take another 20K or so. All this leaves only about 12K of memory, which is quite inadequate for any serious programming.

The limitations on eight-bit machines have been dealt with in two ways. One is to use the technique of 'bank-switching'. Since the computer carries out every action in one-at-a-time sequence, it never needs to gain access to both ROM and RAM at the same time. It is possible, then, to switch the memories. All of the memory chips are connected to the same set of sixteen lines of the address bus. The chip-enable pins, which activate the memory chips, are connected to a switching unit which is controlled by the microprocessor. When ROM is needed, the RAM is switched out and the ROM is switched in. When RAM is needed, the opposite switching connections are used. Since the switching can be carried out in nanoseconds, the delay is not obvious to the user. The other possibility is to make use of specially-designed 8-bit microprocessors which have additional address pins. Adding another four address pins, for example, allows a microprocessor to address one megabyte of memory (1048K). The substitution of one of these microprocessor chips for an older type allows a microcomputer design to be rehashed for greatly enhanced memory with the least possible change in the general design. The modern 16-bit microprocessor chips, however, use additional address lines. The use of 24 address lines allows 16 megabytes of memory to be addressed, and this is well beyond the needs of current programs. A few designs use multiplexed addressing, with the address pins from the microprocessor leading to a demultiplexer chip so as to place the correct address signals on to the address lines.

As well as the data and address bus lines, the microprocessor will also be connected to a control bus. This consists of the set of lines that must be used to switch items in and out of use. One of the control lines, for example will be the read/write control for memory. Another will be concerned with ports, which we will come to later. There will also be input lines in this bus, most concerned with the interrupt system. A signal on one of these lines will allow the action of the microprocessor to be interrupted after completing the action. The design of the chip allows for such interrupt signals to place some pre-arranged address into the program counter so that a routine can be run each time the interrupt pulse arrives. One common use for this system is in responding to a keyboard.

Circuit organization

Figure 5.24 shows the simplest possible block diagram for a microprocessor system, which could represent a machine-control application or a computer. In this diagram, the address bus has not been shown, and only the data bus and a control bus appear. The system is one in which signals will be fed in, processed, and fed out again. A machine

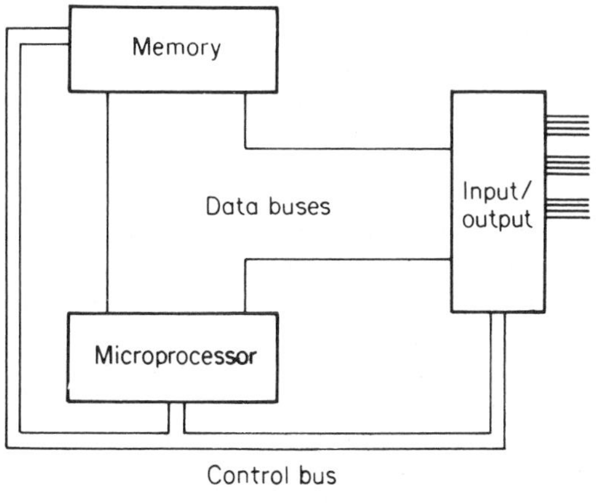
Memory
Data buses
Input/
output
Microprocessor
Control bus

Figure 5.24. A block diagram of connections between a microprocessor, memory and input/output port. Each bus consists of a set of connecting lines

control system, for example, would use input signals from measuring devices, and its outputs would be the signals that controlled the machine. For a computer, the input signals would be from a keyboard, and the output to a screen. The important point is that the lines of the data bus must connect the microprocessor chip to the memory and also to the block that is marked input/output. One control signal will then determine whether the memory or the input/output is to be used, and another control signal will determine whether data is being read by the microprocessor or written from the microprocessor. These control signals will come from the microprocessor, and what ultimately decides on what signals are sent out is the program that is used to control the microprocessor. If, for example, the control program contains a read-from-outside command, then the microprocessor will send out the signal that selects reading, in other words, data passing from the input/output to the microprocessor.

The block which has been marked as input/output in this diagram is more often known as a port. A port is a form of gate, usually with register action combined. It should be possible to isolate the output lines of the port completely from the microprocessor data bus. If this were not done, then external signals could interfere with the signals on the data bus, causing damage to the microprocessor. Any connection between the data bus and the port lines must then be indirect, and normally employs a bi-directional register. This also has advantages in connection with timing. For example, because the microprocessor operates at a high clock rate, it would be very difficult to ensure that an external signal was present on the port lines at the correct time. If the external signals are stored in a register, however, they can be read by the microprocessor whenever the correct piece of program is executed. Similarly, a byte of data which is written out from the microprocessor to the port can remain stored (latched) at the port until it is read by external circuits. This type of action requires additional control pins on the port. There must at least be one pin on which a signal can indicate that the port has data to send out, and another which signifies that there is data for the microprocessor to read in. Most port chips go very much further than this, and provide for just about every conceivable input or output action, synchronization, and indication. This makes port chips rather complicated; they are, in fact, after the microprocessor itself, the most complicated chip in a microprocessor system.

The simple block diagram of *Figure 5.24* provides for all the elementary actions of a microprocessor system. The microprocessor can read an instruction byte from the memory and then carry out the commands of that byte. This can mean, for example, reading a byte from the port and storing it in memory. It can also imply the opposite, reading a byte from the memory and sending it out through the port. An astonishing amount of computing and machine-control applications use just these actions. For other applications, the microprocessor can act on bytes, using its logic and arithmetic sections as needed. A simple example will probably illustrate this better. Suppose that the microprocessor unit is part of a machine controller, and the machine is drilling a hole which is to be 12 mm deep. One of the output lines of the port controls the drill feed, another controls the motor. A set of input lines brings signals from a measuring micrometer which measures the depth of the hole. How will signals pass between the machine and the microprocessor system in order to maintain control?

In outline, the program consists of a loop. This means that a set of commands will be repeated until a condition is satisfied. The condition in this example is that the signal from the micrometer corresponds to the 12 mm measurement. Until this signal is at its correct value, the output line must continue to make the drill bit move into the work, and the drill motor must be activated. The loop will start by reading the signals from the micrometer. These will be compared with another byte – the byte that corresponds to 12.00 mm. At the start of drilling, these will not be equal. It is this comparison step that decides whether the drilling goes on or not. If the two quantities are not equal, the microprocessor loads from memory the bits that will be sent out to keep the drill motor moving, and also to advance the drill by another fraction of a millimeter. The next program command is a 'loop jump', meaning that the microprocessor has to return to the first step again. This reads the micrometer again, compares with 12.00 mm, and once again sends out the drill control signals. Eventually, the signal from the micrometer will be equal to the 12.00 mm setting, and this time things change. The program jumps to a new address, and a different byte is loaded into the microprocessor, one which will pull the drill out. This is sent, and the program goes through another loop, measuring the depth and comparing it with zero. When zero depth is reached, the drill has been pulled all the way out of the hole. The program now leaves this loop, and loads a byte that will stop the drill motor. Even for this very simple example, a lot of program steps have been needed.

Connecting to buses

The way in which a microprocessor is connected to the other chips on a board can be very baffling at first sight. The problem of working out what is connected to where is made more difficult by the almost universal use of double-sided PCBs, in which tracks suddenly burrow through the board and reappear on the other side. This is particularly difficult to follow when the track goes through at a point where the pin of an IC is soldered in, or when it dives through at a place which is covered by the body of an IC. On a circuit diagram, however, it is usually possible to follow where the lines go, and the pattern on the board will then make more sense.

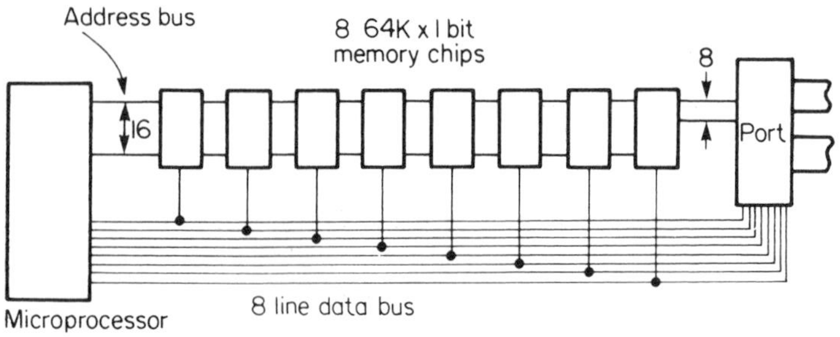
Address bus
8 64K x I bit
memory chips
8
16
Port
Microprocessor
8 line data bus

Figure 5.25. A typical connection pattern of memory chips, port and microprocessor for a small computer or control system. Each memory chip contributes one bit to the data bus, which has been shown in more detail. The lines of the address bus connect to each chip, and eight of them connect to the port

A typical pattern of connecting memory and a port is shown in *Figure 5.25.* For a small computer, each memory chip would be a 64K type, meaning that each chip allows the addressing of 64K bits, and any one of these bits can be connected to the input/output data pin. The read/write pin voltage will then determine whether the selected bit is to be written or read. Since each memory chip contributes one bit, eight chips will be needed in a 64K RAM system to provide the full set of eight lines in the data bus. All sixteen address lines of the address bus will connect to the address pins of each memory chip. Only eight lines of the address bus are shown connected to the port chip, however. This is a fairly common way of using a port chip, and it still provides for 256 possible port addresses, more than is ever needed. All of the data lines are connected to the port. With a scheme like this, it is possible to use more than one port, with different address lines connected so that specific addresses would activate the port. Another common system for a single port is to have one address used for port writing and another for port reading.

Other parts of the system

The microprocessor, ports and memory chips form the main parts of any microprocessor system, but a number of other chips will be found in a lot of examples. One indispensible part of the system is the clock pulse generator. Several microprocessor chips contain built-in clocks, so that all that is visible of the clock circuit is a timing circuit (usually a quartz crystal, or a resistor-capacitor network) attached to two or sometimes three terminals of the microprocessor IC. For other IC types, a separate clock is needed. This is usually provided either by a specialized chip or by a general-purpose CMOS inverter circuit connected as an

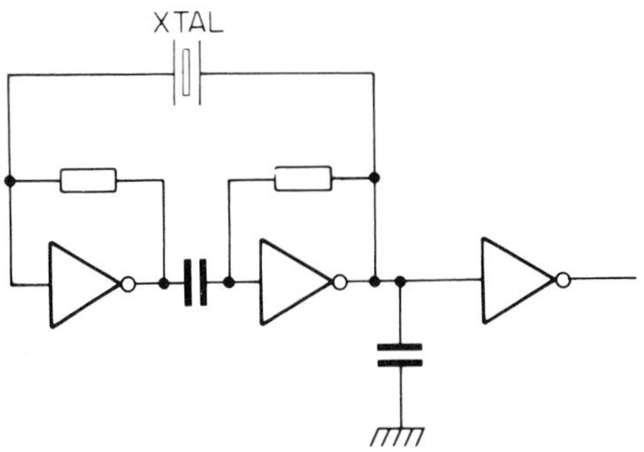
XTAL

Figure 5.26. The type of oscillator circuit which is used when a microprocessor chip has no built-in clock circuits. The inverters used will be CMOS or TTL, and the capacitor values depend on the crystal frequency. Resistor values of around 1K are used

oscillator (*Figure 5.26*). For a computer, or for any circuit in which timing is important, the clock *must* be crystal-controlled.

Many circuits make use of A-D converters. If a microprocessor system, for example, is to use inputs from analogue instruments, the input signals will be in the form of varying voltages. If these are to be used in a program, they must be converted into digital form, and a chip which carries out this action is used. This chip makes use of the clock pulses to sample the analogue voltage and convert it into a number. This number will often be a two-byte number, although for some simple applications, a single byte conversion is often more than adequate. A single byte conversion will allow a range of voltage to be represented by 256 steps. If this covers a range of 0 to 10 V, for example, then the discrimination of the unit will be about 1/25 V, or 40 mV. This is considerably better than can be read by most analogue voltmeters. A few computers, notably the BBC micro, possess a built-in analogue to digital converter. The BBC micro system uses four channels of A-D conversion, so allowing four sets of inputs to be worked with. The converter uses 10 bits, so that the number 0 corresponds to zero voltage, and 1123 (denary) to maximum voltage. The A-D converter can also be programmed to use 8-bits if this is more convenient. Since an eight-bit conversion needs only 4 ms to carry out, and a ten-bit conversion needs 10 ms, the 8-bit method is better for many purposes. The presence of this built-in A-D conversion makes the BBC micro very useful for all forms of process control work. In some cases, it is much cheaper to buy and adapt the BBC micro than it would be to buy purpose-built control systems, and it is certainly much cheaper than custom-made control equipment. The main limitation is the time of conversion, which limits sampling to a maximum of 250 per second for the 8-bit conversion on one channel. Any scheme which required, for example, ten-bit conversion on all four channels would run fairly slowly.

Pinouts

The following pages illustrate labelled pinout diagrams for the two most common eight-bit microcomputers, the 6502 and the Z80, and a pinout diagram for the sixteen-bit 68010. Since all microprocessor systems are clocked, the purity of the clock signal is very important, and all of the timings assume that the clock signal will be within specification limits. Servicing work on microprocessor equipment is made much easier if specialized logic analyzers are available. A good storage oscilloscope is also very useful in order to show the time relationships of signals which may not occur frequently enough to trigger a conventional display.

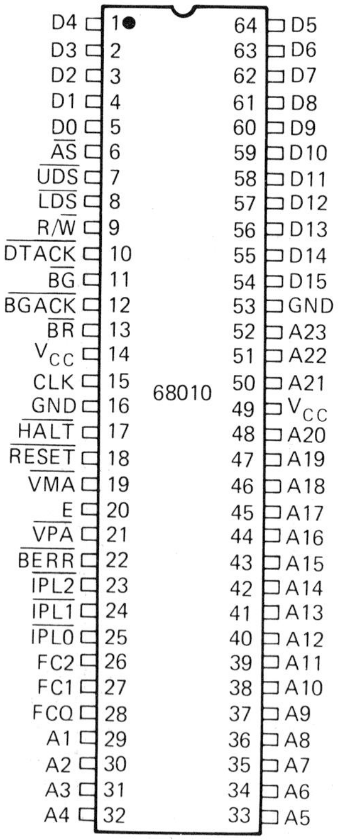
68010
D4 1
D3 2
D2 3
D1 4
D0 5
AS 6
UDS 7
LDS 8
R/W 9
DTACK 10
BG 11
BGACK 12
BR 13
VCC 14
CLK 15
GND 16
HALT 17
RESET 18
VMA 19
E 20
VPA 21
BERR 22
IPL2 23
IPL1 24
IPL0 25
FC2 26
FC1 27
FC0 28
A1 29
A2 30
A3 31
A4 32
64 D5
63 D6
62 D7
61 D8
60 D9
59 D10
58 D11
57 D12
56 D13
55 D14
54 D15
53 GND
52 A23
51 A22
50 A21
49 VCC
48 A20
47 A19
46 A18
45 A17
44 A16
43 A15
42 A14
41 A13
40 A12
39 A11
38 A10
37 A9
36 A8
35 A7
34 A6
33 A5

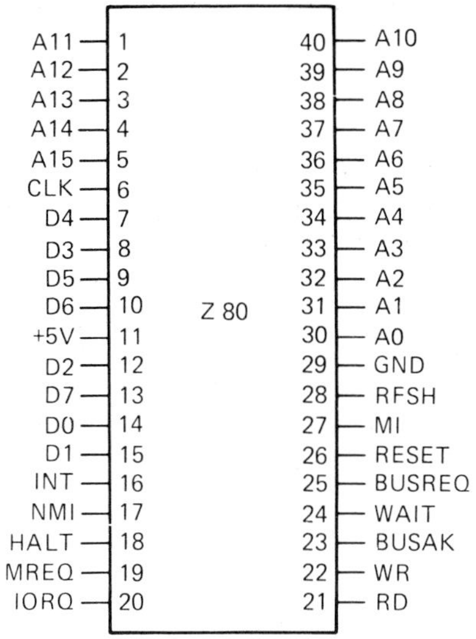
Z 80
A11 1
A12 2
A13 3
A14 4
A15 5
CLK 6
D4 7
D3 8
D5 9
D6 10
+5V 11
D2 12
D7 13
D0 14
D1 15
INT 16
NMI 17
HALT 18
MREQ 19
IORQ 20
RD 21
WR 22
BUSAK 23
WAIT 24
BUSREQ 25
RESET 26
MI 27
RFSH 28
GND 29
A0 30
A1 31
A2 32
A3 33
A4 34
A5 35
A6 36
A7 37
A8 38
A9 39
A10 40

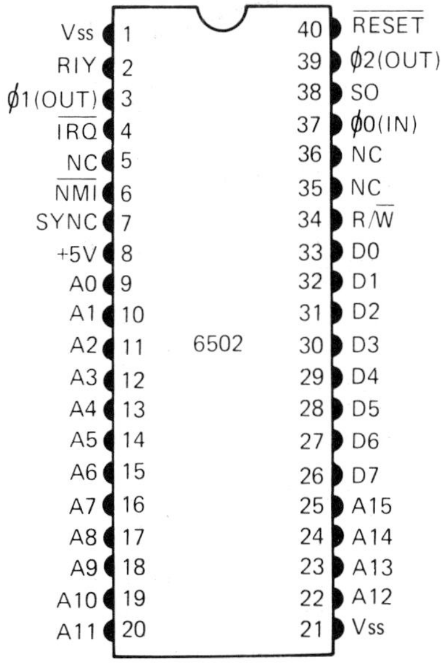
6502
Vss 1
RIY 2
ϕ1(OUT) 3
IRQ 4
NC 5
NMI 6
SYNC 7
+5V 8
A0 9
A1 10
A2 11
A3 12
A4 13
A5 14
A6 15
A7 16
A8 17
A9 18
A10 19
A11 20
Vss 21
A12 22
A13 23
A14 24
A15 25
D7 26
D6 27
D5 28
D4 29
D3 30
D2 31
D1 32
D0 33
R/W 34
NC 35
NC 36
ϕ0(IN) 37
SO 38
ϕ2(OUT) 39
RESET 40

Table 5.4. TTL PINOUTS

7400
QUADRUPLE 2-INPUT POSITIVE-NAND GATES

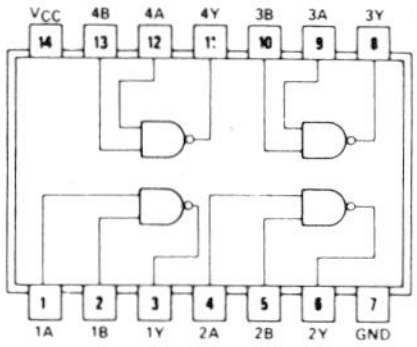
VCC 4B 4A 4Y 3B 3A 3Y
1A 1B 1Y 2A 2B 2Y GND

7401
QUADRUPLE 2-INPUT POSITIVE-NAND GATES WITH OPEN-COLLECTOR OUTPUTS

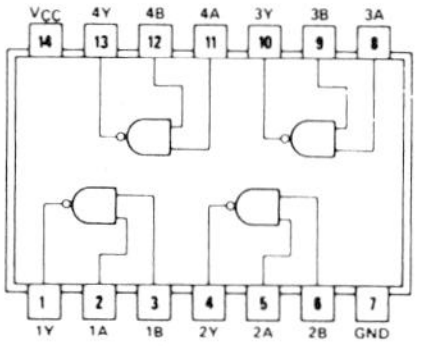
VCC 4Y 4B 4A 3Y 3B 3A
1Y 1A 1B 2Y 2A 2B GND

74H01

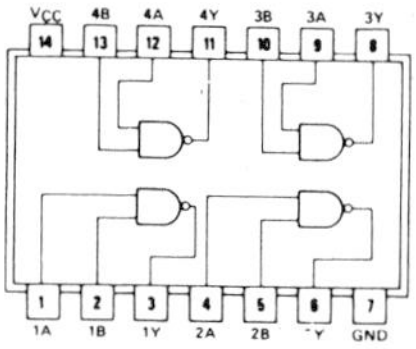
VCC 4B 4A 4Y 3B 3A 3Y
1A 1B 1Y 2A 2B GND

7402
QUADRUPLE 2-INPUT POSITIVE-NOR GATES

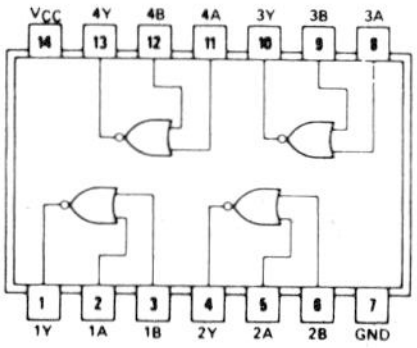
VCC 4Y 4B 4A 3Y 3B 3A
1Y 1A 1B 2Y 2A 2B GND

7403
QUADRUPLE 2-INPUT POSITIVE-NAND GATES WITH OPEN-COLLECTOR OUTPUTS

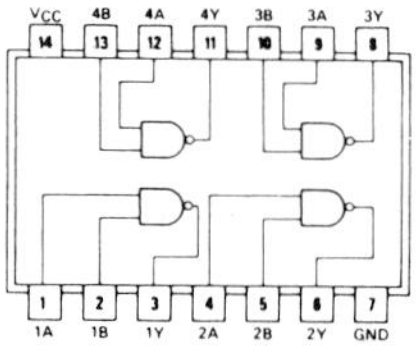
VCC 4B 4A 4Y 3B 3A 3Y
1A 1B 1Y 2A 2B 2Y GND

7404
HEX INVERTERS

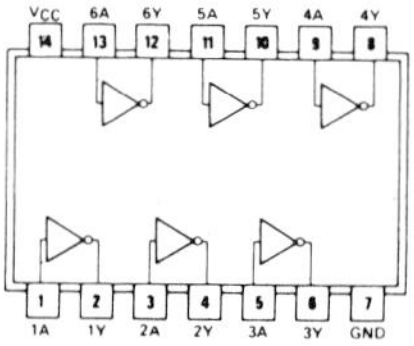
VCC 6A 6Y 5A 5Y 4A 4Y
1A 1Y 2A 2Y 3A 3Y GND

7405
HEX INVERTERS WITH OPEN-COLLECTOR OUTPUTS

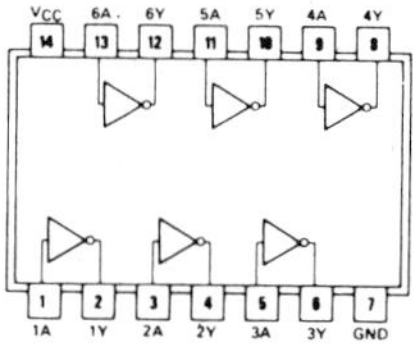
VCC 6A 6Y 5A 5Y 4A 4Y
1A 1Y 2A 2Y 3A 3Y GND

7406
HEX INVERTER BUFFERS/DRIVERS WITH OPEN-COLLECTOR HIGH-VOLTAGE OUTPUTS

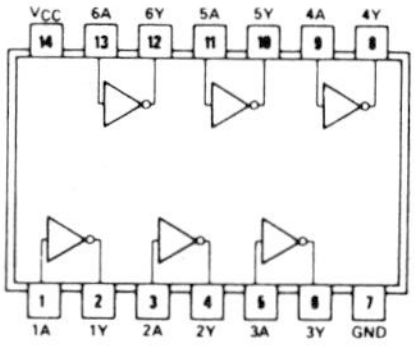
VCC 6A 6Y 5A 5Y 4A 4Y
1A 1Y 2A 2Y 3A 3Y GND

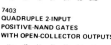

7407
HEX BUFFERS/DRIVERS WITH OPEN-COLLECTOR HIGH-VOLTAGE OUTPUTS

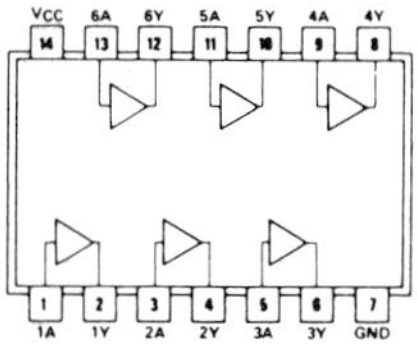
VCC 6A 6Y 5A 5Y 4A 4Y
1A 1Y 2A 2Y 3A 3Y GND

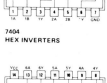

7408
QUADRUPLE 2-INPUT POSITIVE-AND GATES

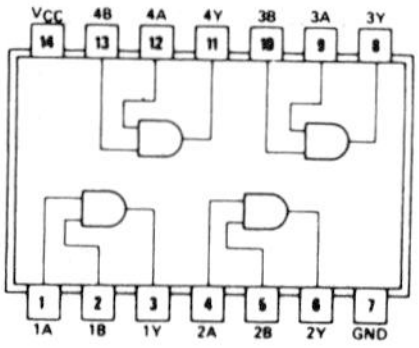
VCC 4B 4A 4Y 3B 3A 3Y
1A 1B 1Y 2A 2B 2Y GND

7409
QUADRUPLE 2-INPUT POSITIVE-AND GATES WITH OPEN-COLLECTOR OUTPUTS

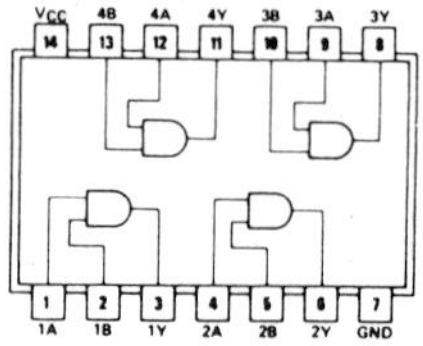
VCC 4B 4A 4Y 3B 3A 3Y
1A 1B 1Y 2A 2B 2Y GND

7410
TRIPLE 3-INPUT POSITIVE-NAND GATES

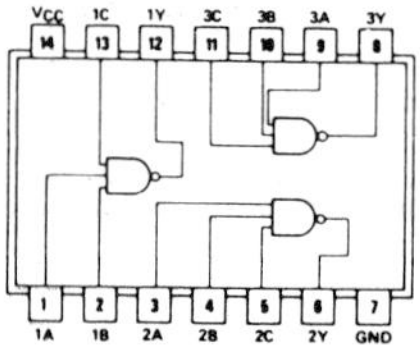
VCC 1C 1Y 3C 3B 3A 3Y
1A 1B 2A 2B 2C 2Y GND

74H11
TRIPLE 3-INPUT POSITIVE-AND GATES

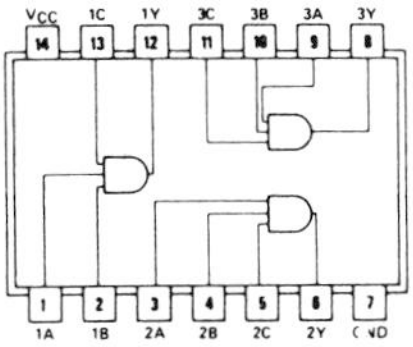
VCC 1C 1Y 3C 3B 3A 3Y
14 13 12 11 10 9 8
1 2 3 4 5 6 7
1A 1B 2A 2B 2C 2Y GND

7412
TRIPLE 3-INPUT POSITIVE-NAND GATES WITH OPEN-COLLECTOR OUTPUTS

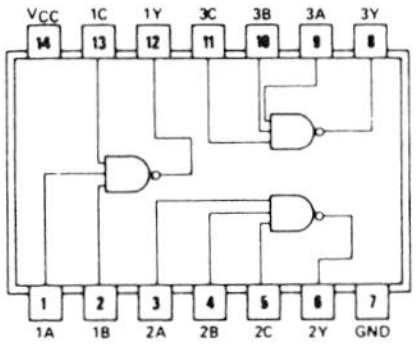
VCC 1C 1Y 3C 3B 3A 3Y
14 13 12 11 10 9 8
1 2 3 4 5 6 7
1A 1B 2A 2B 2C 2Y GND

7413
DUAL 4-INPUT POSITIVE-NAND SCHMITT TRIGGERS

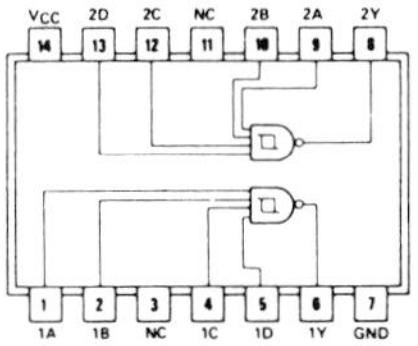
VCC 2D 2C NC 2B 2A 2Y
14 13 12 11 10 9 8
1 2 3 4 5 6 7
1A 1B NC 1C 1D 1Y GND

7414
HEX SCHMITT-TRIGGER INVERTERS

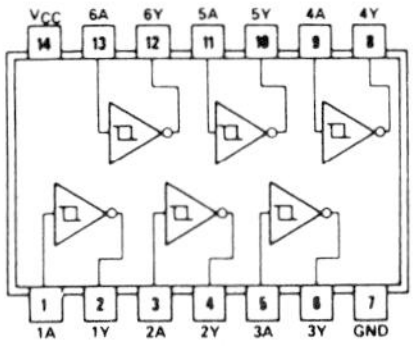
VCC 6A 6Y 5A 5Y 4A 4Y
14 13 12 11 10 9 8
1 2 3 4 5 6 7
1A 1Y 2A 2Y 3A 3Y GND

74H15
TRIPLE 3-INPUT POSITIVE-AND GATES WITH OPEN-COLLECTOR OUTPUTS

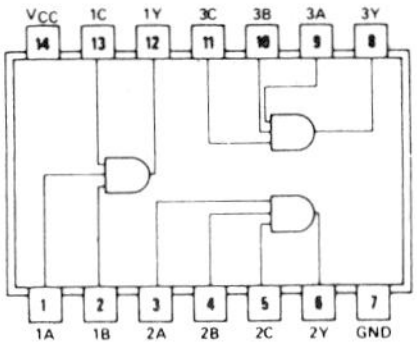
VCC 1C 1Y 3C 3B 3A 3Y
14 13 12 11 10 9 8
1 2 3 4 5 6 7
1A 1B 2A 2B 2C 2Y GND

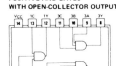

7416
HEX INVERTER BUFFERS/DRIVERS WITH OPEN-COLLECTOR HIGH-VOLTAGE OUTPUTS

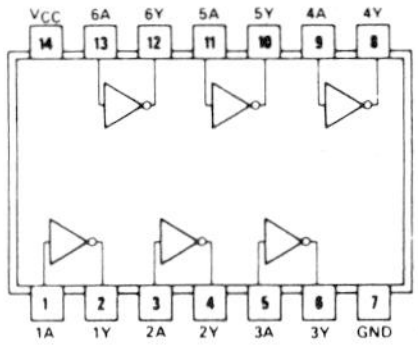
VCC 6A 6Y 5A 5Y 4A 4Y
14 13 12 11 10 9 8
1 2 3 4 5 6 7
1A 1Y 2A 2Y 3A 3Y GND

7417
HEX BUFFERS/DRIVERS WITH OPEN-COLLECTOR HIGH-VOLTAGE OUTPUTS

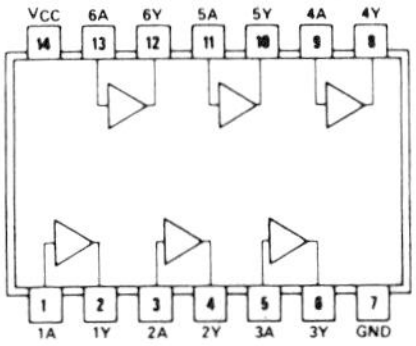
VCC 6A 6Y 5A 5Y 4A 4Y
14 13 12 11 10 9 8
1 2 3 4 5 6 7
1A 1Y 2A 2Y 3A 3Y GND

7420
DUAL 4-INPUT POSITIVE-NAND GATES

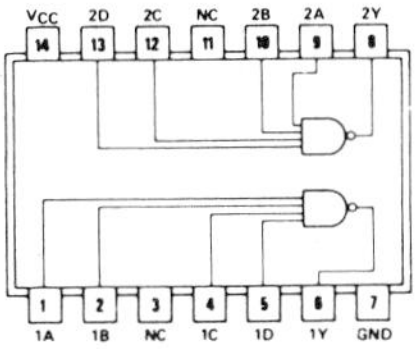
VCC 2D 2C NC 2B 2A 2Y
14 13 12 11 10 9 8
1 2 3 4 5 6 7
1A 1B NC 1C 1D 1Y GND

74H21
DUAL 4-INPUT POSITIVE-AND GATES

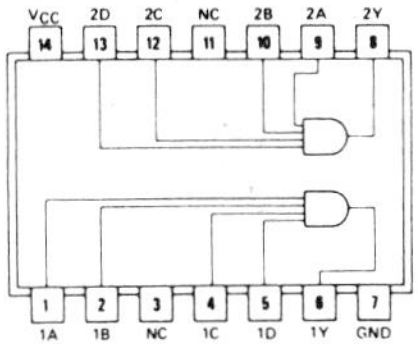
VCC 2D 2C NC 2B 2A 2Y
14 13 12 11 10 9 8
1 2 3 4 5 6 7
1A 1B NC 1C 1D 1Y GND

7422
DUAL 4-INPUT POSITIVE-NAND GATES WITH OPEN-COLLECTOR OUTPUTS

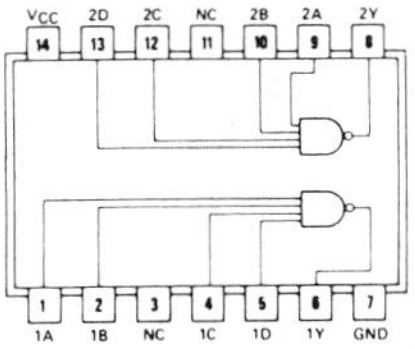
VCC 2D 2C NC 2B 2A 2Y
14 13 12 11 10 9 8
1 2 3 4 5 6 7
1A 1B NC 1C 1D 1Y GND

7423
EXPANDABLE DUAL 4-INPUT POSITIVE-NOR GATES WITH STROBE

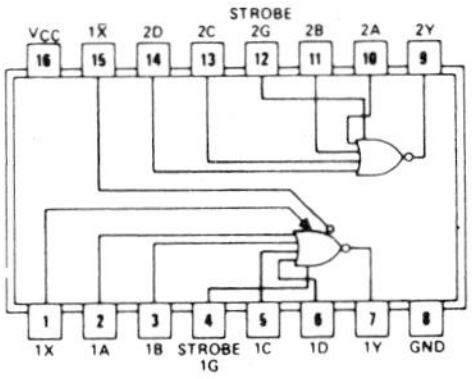
VCC 1X̄ 2D 2C STROBE 2G 2B 2A 2Y
16 15 14 13 12 11 10 9
1 2 3 4 5 6 7 8
1X 1A 1B STROBE 1G 1C 1D 1Y GND

7425
DUAL 4-INPUT POSITIVE-NOR GATES WITH STROBE

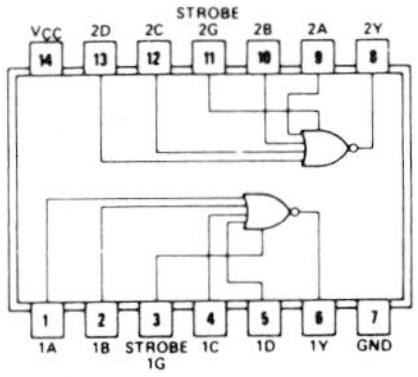
VCC 2D 2C STROBE 2G 2B 2A 2Y
14 13 12 11 10 9 8
1 2 3 4 5 6 7
1A 1B STROBE 1G 1C 1D 1Y GND

7426
QUADRUPLE 2-INPUT HIGH-VOLTAGE INTERFACE POSITIVE-NAND GATES

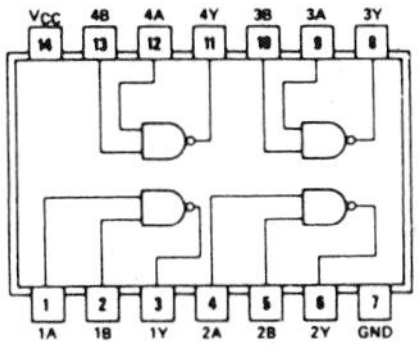
VCC 4B 4A 4Y 3B 3A 3Y
14 13 12 11 10 9 8
1 2 3 4 5 6 7
1A 1B 1Y 2A 2B 2Y GND

7427
TRIPLE 3-INPUT POSITIVE-NOR GATES

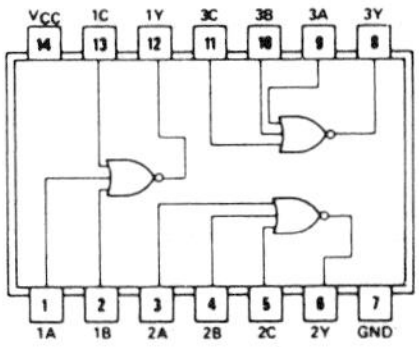
VCC 1C 1Y 3C 3B 3A 3Y
14 13 12 11 10 9 8
1 2 3 4 5 6 7
1A 1B 2A 2B 2C 2Y GND

7428
QUADRUPLE 2-INPUT POSITIVE-NOR BUFFERS

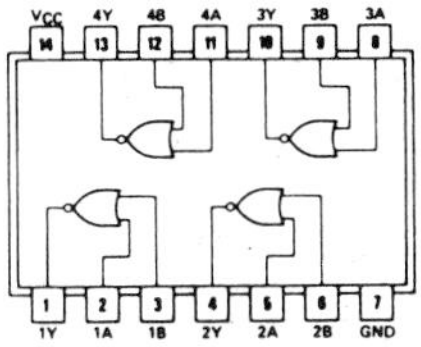
VCC 4Y 4B 4A 3Y 3B 3A
14 13 12 11 10 9 8
1 2 3 4 5 6 7
1Y 1A 1B 2Y 2A 2B GND

7430
8-INPUT
POSITIVE-NAND GATES

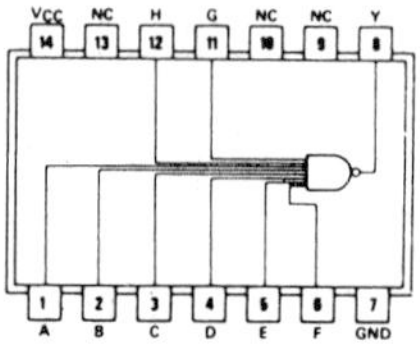
VCC NC H G NC NC Y
A B C D E F GND

7432
QUADRUPLE 2-INPUT
POSITIVE-OR GATES

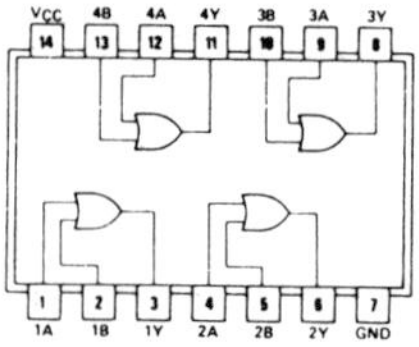
VCC 4B 4A 4Y 3B 3A 3Y
1A 1B 1Y 2A 2B 2Y GND

7433
QUADRUPLE 2-INPUT
POSITIVE-NOR BUFFERS
WITH OPEN-COLLECTOR OUTPUTS

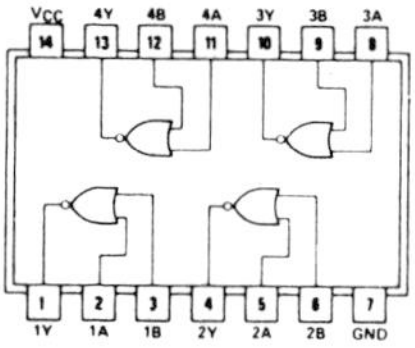
VCC 4Y 4B 4A 3Y 3B 3A
1Y 1A 1B 2Y 2A 2B GND

7437
QUADRUPLE 2-INPUT
POSITIVE-NAND BUFFERS

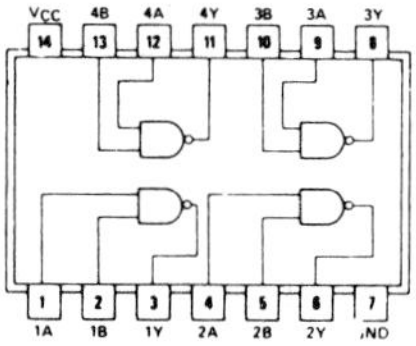
VCC 4B 4A 4Y 3B 3A 3Y
1A 1B 1Y 2A 2B 2Y GND

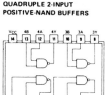

7438
QUADRUPLE 2-INPUT
POSITIVE-NAND BUFFERS
WITH OPEN-COLLECTOR OUTPUTS

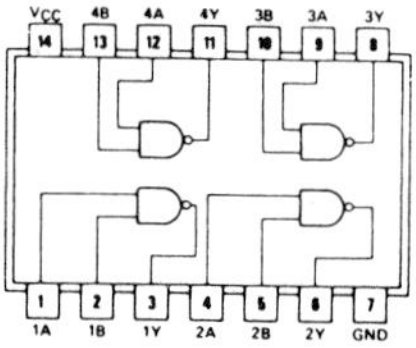
VCC 4B 4A 4Y 3B 3A 3Y
1A 1B 1Y 2A 2B 2Y GND

7440
DUAL 4-INPUT
POSITIVE-NAND BUFFERS

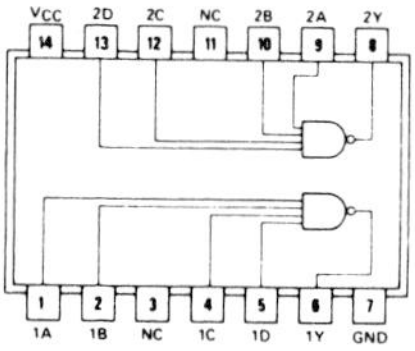
VCC 2D 2C NC 2B 2A 2Y
1A 1B NC 1C 1D 1Y GND

7442A
4 LINE-TO-10-LINE DECODERS

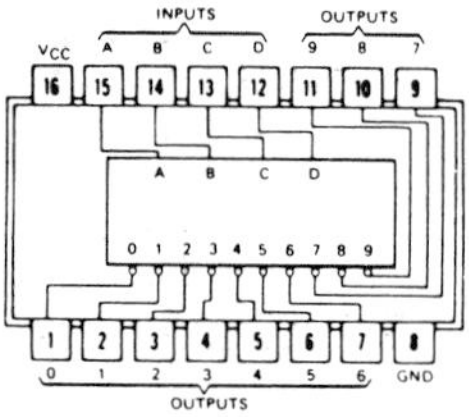
INPUTS
OUTPUTS
VCC A B C D 9 8 7
0 1 2 3 4 5 6 GND
OUTPUTS

7445
BCD-TO-DECIMAL DECODER/DRIVER

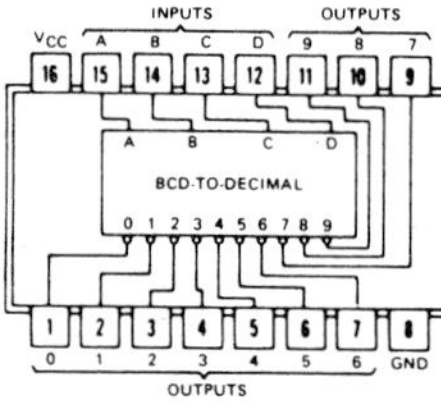
INPUTS
OUTPUTS
VCC A B C D 9 8 7
BCD-TO-DECIMAL
0 1 2 3 4 5 6 GND
OUTPUTS

7446
BCD-TO-SEVEN-SEGMENT
DECODERS/DRIVERS

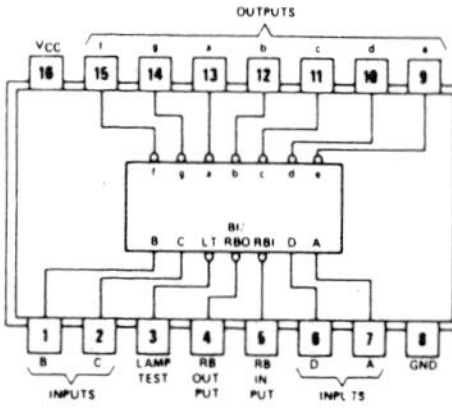
OUTPUTS
VCC
B C LAMP TEST RB OUT PUT RB IN PUT D A GND
INPUTS
INPUTS

7448
BCD-TO-SEVEN-SEGMENT
DECODERS/DRIVERS

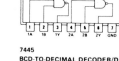

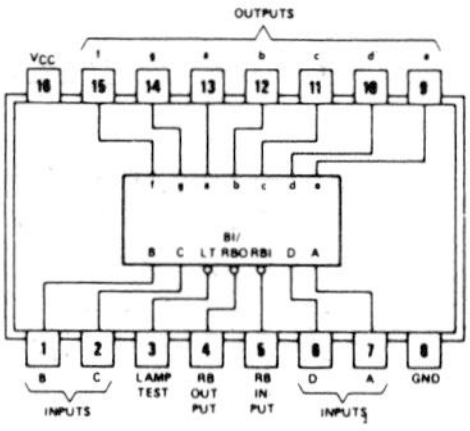
OUTPUTS
VCC
B C LAMP TEST RB OUT PUT RB IN PUT D A GND
INPUTS
INPUTS

74LS49
BCD-TO-SEVEN-SEGMENT
DECODERS/DRIVERS

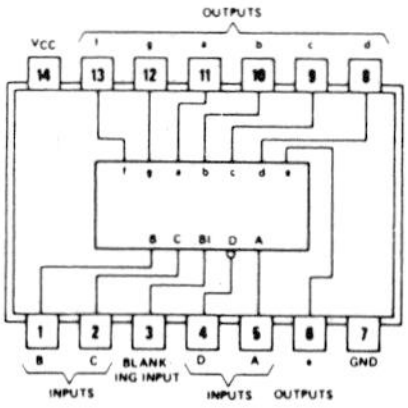
OUTPUTS
VCC
B C BLANKING INPUT D A GND
INPUTS
INPUTS OUTPUTS

7450
DUAL 2-WIDE 2-INPUT
AND-OR-INVERT GATES
(ONE GATE EXPANDABLE)

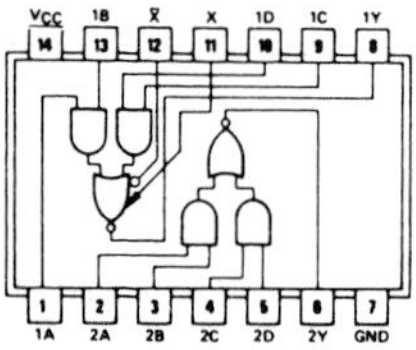
VCC 1B X̄ X 1D 1C 1Y
1A 2A 2B 2C 2D 2Y GND

7451
AND-OR-INVERT GATES

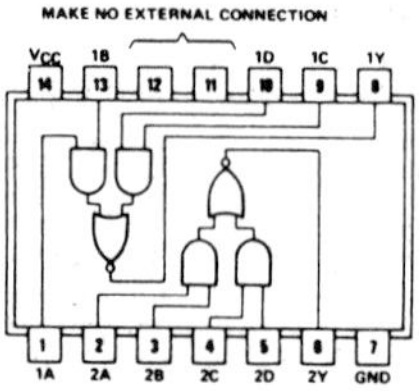
MAKE NO EXTERNAL CONNECTION
VCC 1B 1D 1C 1Y
1A 2A 2B 2C 2D 2Y GND

74L51

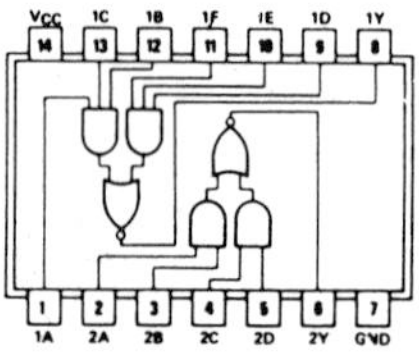
VCC 1C 1B 1F IE 1D 1Y
1A 2A 2B 2C 2D 2Y GND

74H52
EXPANDABLE 4-WIDE
AND-OR GATES

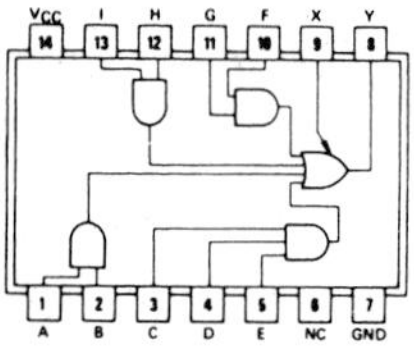
VCC I H G F X Y
A B C D E NC GND

7453
EXPANDABLE 4-WIDE AND-OR-INVERT GATES

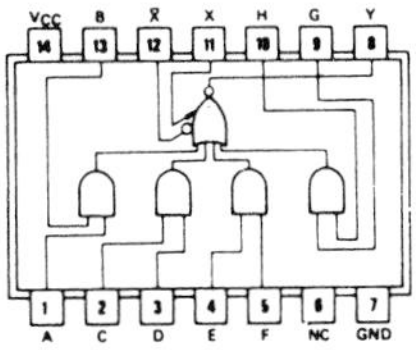
VCC B X̄ X H G Y
14 13 12 11 10 9 8
1 2 3 4 5 6 7
A C D E F NC GND

74H53

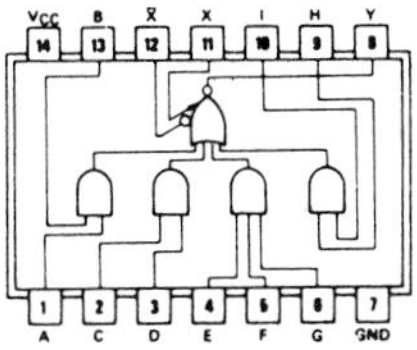
VCC B X̄ X I H Y
14 13 12 11 10 9 8
1 2 3 4 5 6 7
A C D E F G GND

7454
4-WIDE AND-OR-INVERT GATES

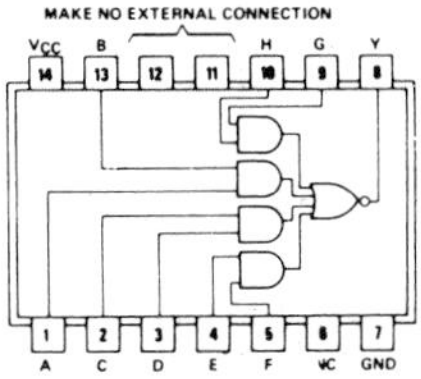
MAKE NO EXTERNAL CONNECTION
VCC B H G Y
14 13 12 11 10 9 8
1 2 3 4 5 6 7
A C D E F NC GND

74H54

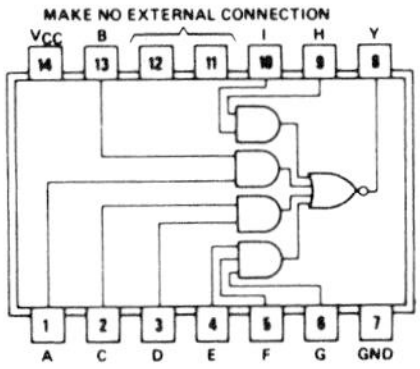
MAKE NO EXTERNAL CONNECTION
VCC B I H Y
14 13 12 11 10 9 8
1 2 3 4 5 6 7
A C D E F G GND

74L54

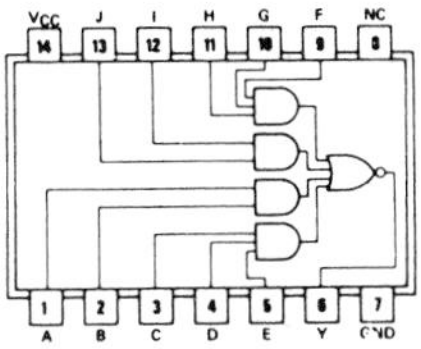
VCC J I H G F NC
14 13 12 11 10 9 8
1 2 3 4 5 6 7
A B C D E Y GND

74H55
2-WIDE 4-INPUT AND-OR-INVERT GATES

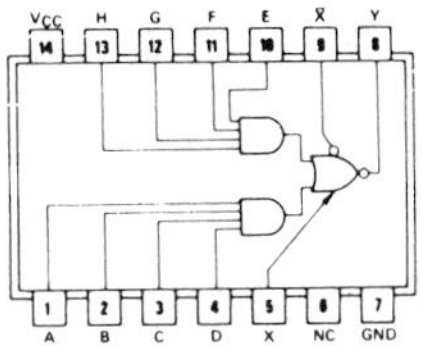
VCC H G F E X̄ Y
14 13 12 11 10 9 8
1 2 3 4 5 6 7
A B C D X NC GND

74L55

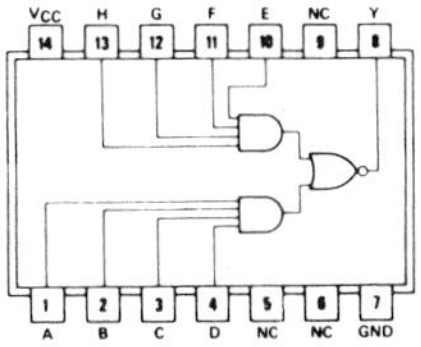
VCC H G F E NC Y
14 13 12 11 10 9 8
1 2 3 4 5 6 7
A B C D NC NC GND

7460
DUAL 4-INPUT EXPANDERS

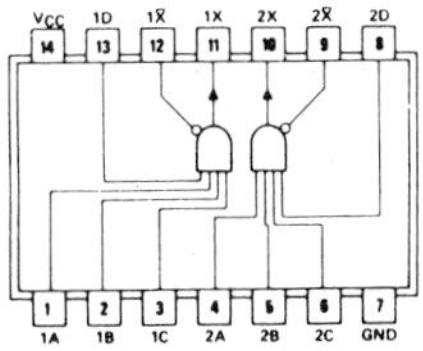
VCC 1D 1X̄ 1X 2X 2X̄ 2D
14 13 12 11 10 9 8
1 2 3 4 5 6 7
1A 1B 1C 2A 2B 2C GND

74H61
TRIPLE 3-INPUT EXPANDERS

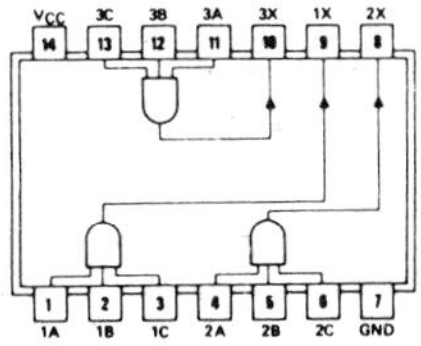
VCC 3C 3B 3A 3X 1X 2X
14 13 12 11 10 9 8
1 2 3 4 5 6 7
1A 1B 1C 2A 2B 2C GND

74H62
4-WIDE AND-OR EXPANDERS

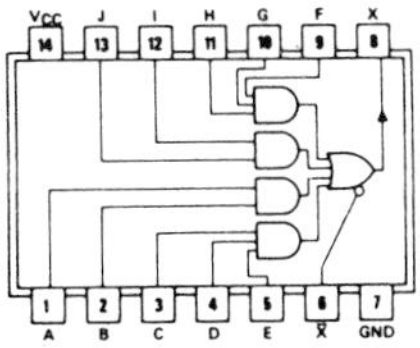
VCC J I H G F X
14 13 12 11 10 9 8
1 2 3 4 5 6 7
A B C D E X̄ GND

74LS63
HEX CURRENT-SENSING INTERFACE GATES

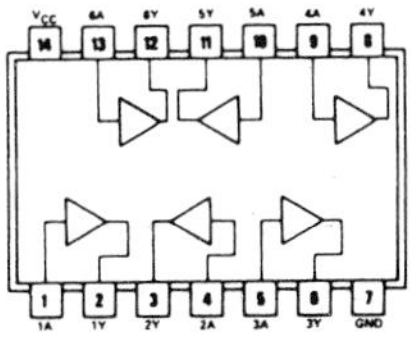
14 13 12 11 10 9 8
1 2 3 4 5 6 7

74S64
4-2-3-2 INPUT AND-OR-INVERT GATES

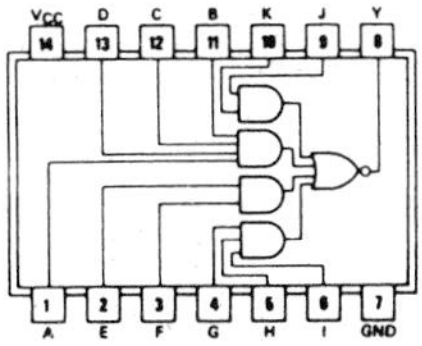
VCC D C B K J Y
14 13 12 11 10 9 8
1 2 3 4 5 6 7
A E F G H I GND

7470
AND-GATED J-K POSITIVE-EDGE-TRIGGERED FLIP-FLOPS WITH PRESET AND CLEAR

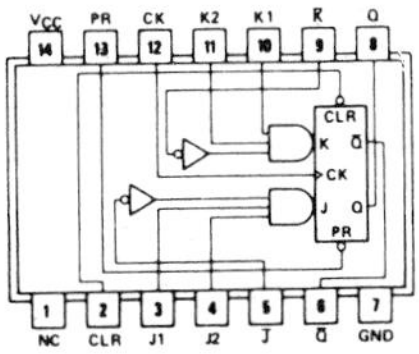
VCC PR CK K2 K1 K̄ Q
14 13 12 11 10 9 8
CLR K Q̄ CK J Q PR
1 2 3 4 5 6 7
NC CLR J1 J2 J̄ Q̄ GND

74H71
AND-OR-GATED J-K MASTER-SLAVE FLIP-FLOPS WITH PRESET

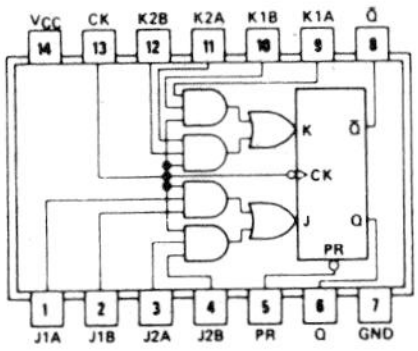
VCC CK K2B K2A K1B K1A Q̄
14 13 12 11 10 9 8
K Q̄ CK J Q PR
1 2 3 4 5 6 7
J1A J1B J2A J2B PR Q GND

74L71
AND-GATED R-S MASTER-SLAVE FLIP-FLOPS WITH PRESET AND CLEAR

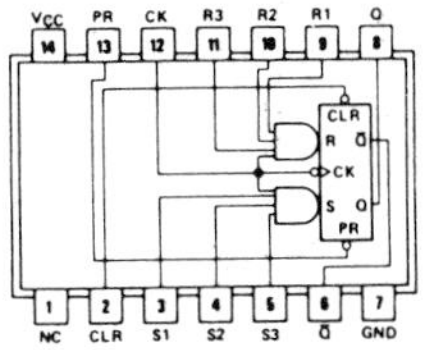
VCC PR CK R3 R2 R1 Q
14 13 12 11 10 9 8
CLR R Q̄ CK S Q PR
1 2 3 4 5 6 7
NC CLR S1 S2 S3 Q̄ GND

7472
AND-GATED J-K MASTER-SLAVE FLIP-FLOPS WITH PRESET AND CLEAR

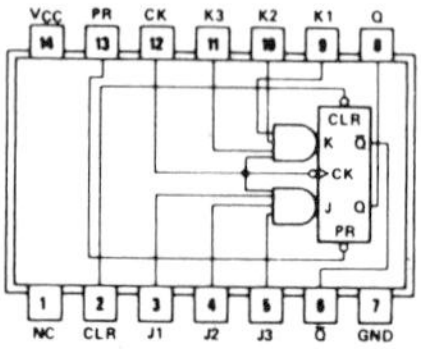
VCC PR CK K3 K2 K1 Q
NC CLR J1 J2 J3 Q̄ GND

7473
DUAL J-K FLIP-FLOPS WITH CLEAR

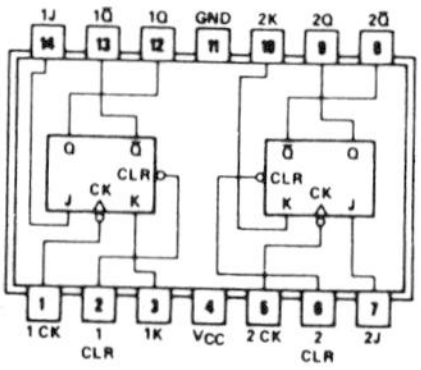
1J 1Q̄ 1Q GND 2K 2Q 2Q̄
1CK 1 CLR 1K VCC 2CK 2 CLR 2J

7474
DUAL D-TYPE POSITIVE-EDGE-TRIGGERED FLIP-FLOPS WITH PRESET AND CLEAR

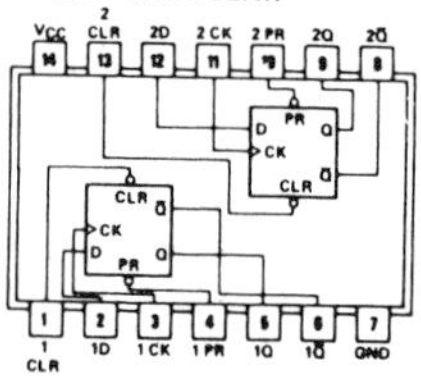
VCC 2 CLR 2D 2CK 2PR 2Q 2Q̄
1 CLR 1D 1CK 1PR 1Q 1Q̄ GND

7475
4-BIT BISTABLE LATCHES

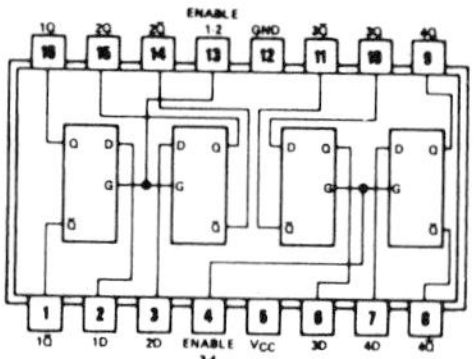
1Q 2Q 2Q̄ ENABLE 1-2 GND 3Q̄ 3Q 4Q
1Q̄ 1D 2D ENABLE 3-4 VCC 3D 4D 4Q̄

7476
DUAL J-K FLIP-FLOPS WITH PRESET AND CLEAR

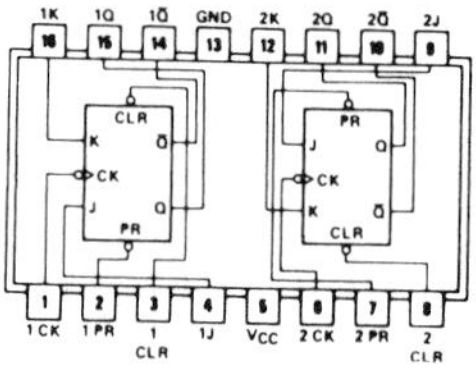
1K 1Q 1Q̄ GND 2K 2Q 2Q̄ 2J
1CK 1PR 1 CLR 1J VCC 2CK 2PR 2 CLR

74H78
DUAL J-K FLIP-FLOPS WITH PRESET, COMMON CLEAR, AND COMMON CLOCK

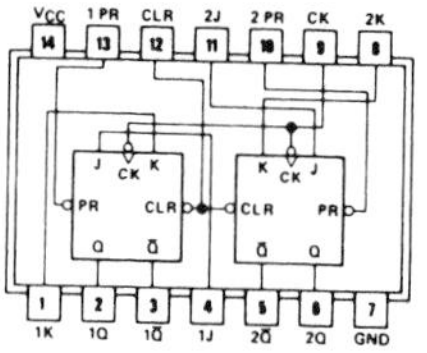
VCC 1PR CLR 2J 2PR CK 2K
1K 1Q 1Q̄ 1J 2Q̄ 2Q GND

74L78

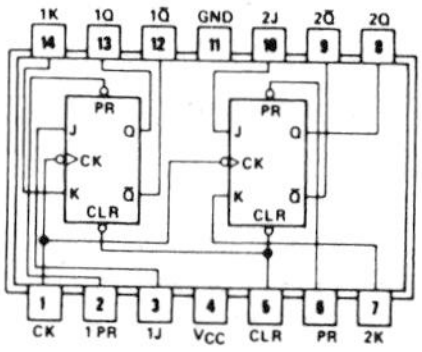
1K 1Q 1Q̄ GND 2J 2Q̄ 2Q
CK 1PR 1J VCC CLR PR 2K

7480
GATED FULL ADDERS

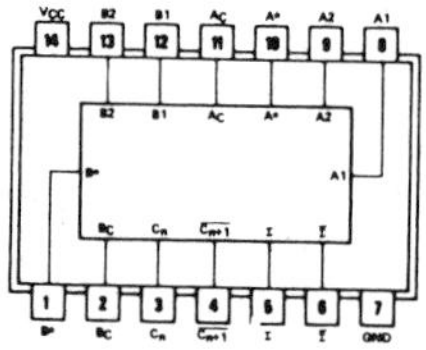

7481
16-BIT RANDOM-ACCESS MEMORIES

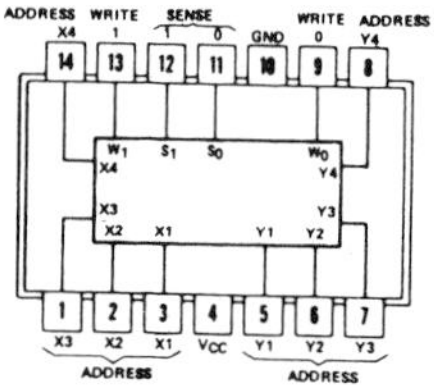
ADDRESS X4
WRITE 1
SENSE 1 0
GND
WRITE 0
ADDRESS Y4
X3 X2 X1 VCC Y1 Y2 Y3
ADDRESS
ADDRESS

7482
2-BIT BINARY FULL ADDERS

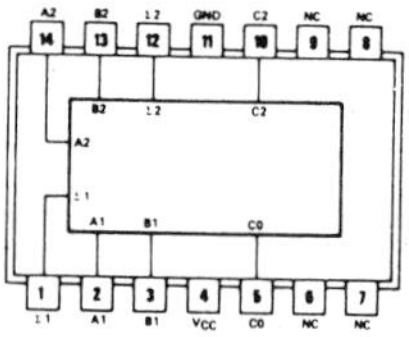

7483
4-BIT BINARY FULL ADDERS WITH FAST CARRY

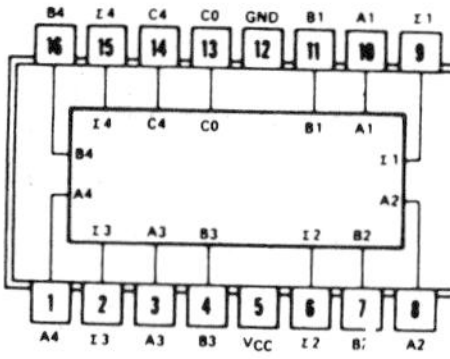
B4 Σ4 C4 C0 GND B1 A1 Σ1
A4 Σ3 A3 B3 VCC Σ2 B2 A2

7484
16-BIT RANDOM-ACCESS MEMORIES

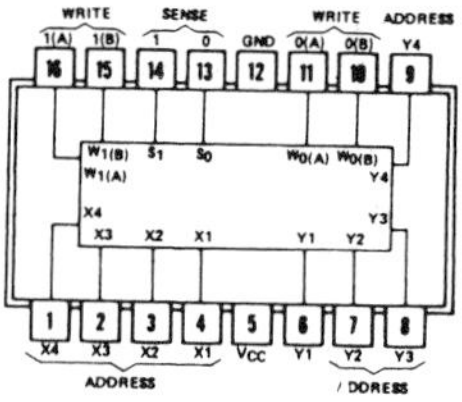
WRITE
SENSE
GND
WRITE
ADDRESS Y4
X4 X3 X2 X1 VCC Y1 Y2 Y3
ADDRESS

7485
4-BIT MAGNITUDE COMPARATORS

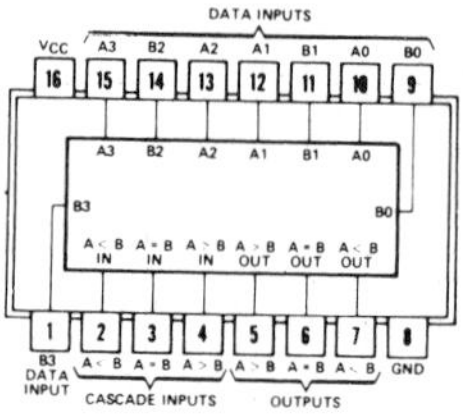
DATA INPUTS
VCC A3 B2 A2 A1 B1 A0 B0
B3 DATA INPUT
CASCADE INPUTS
OUTPUTS
GND

7486
QUADRUPLE 2-INPUT EXCLUSIVE-OR GATES

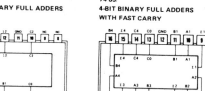

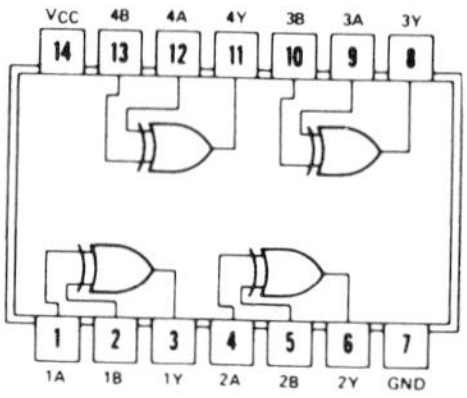
VCC 4B 4A 4Y 3B 3A 3Y
1A 1B 1Y 2A 2B 2Y GND

74L86

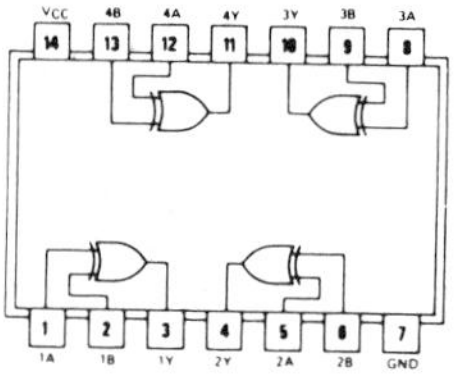
VCC 4B 4A 4Y 3Y 3B 3A
1A 1B 1Y 2Y 2A 2B GND

74H87
4-BIT TRUE/COMPLEMENT, ZERO/ONE ELEMENTS

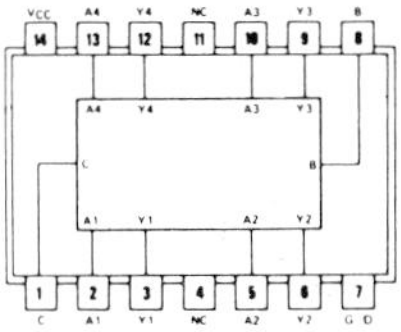
VCC A4 Y4 NC A3 Y3 B
14 13 12 11 10 9 8
1 2 3 4 5 6 7
C A1 Y1 NC A2 Y2 GND

7488A
256-BIT READ-ONLY MEMORIES

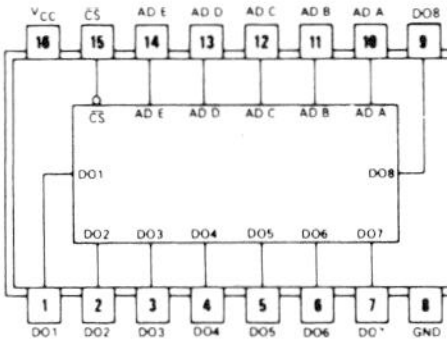
VCC CS AD E AD D AD C AD B AD A DO8
16 15 14 13 12 11 10 9
1 2 3 4 5 6 7 8
DO1 DO2 DO3 DO4 DO5 DO6 DO7 GND

7489
64-BIT READ/WRITE MEMORIES

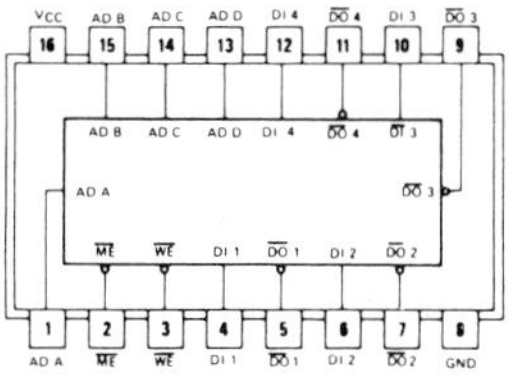
VCC AD B AD C AD D DI 4 DO 4 DI 3 DO 3
16 15 14 13 12 11 10 9
1 2 3 4 5 6 7 8
AD A ME WE DI 1 DO 1 DI 2 DO 2 GND

7490A
DECADE COUNTERS

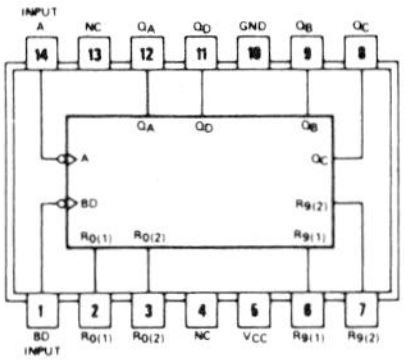
INPUT A NC QA QD GND QB QC
14 13 12 11 10 9 8
1 2 3 4 5 6 7
BD INPUT R0(1) R0(2) NC VCC R9(1) R9(2)

7491A
8-BIT SHIFT REGISTERS

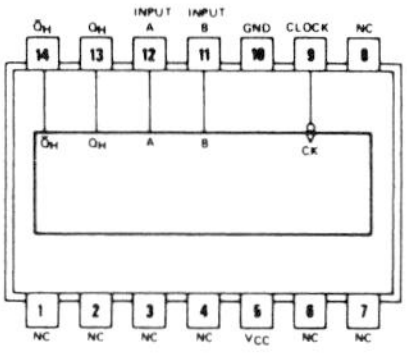
QH QH INPUT A INPUT B GND CLOCK NC
14 13 12 11 10 9 8
1 2 3 4 5 6 7
NC NC NC NC VCC NC NC

7492
DIVIDE-BY-TWELVE COUNTERS

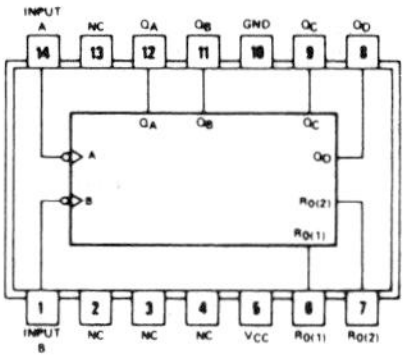
INPUT A NC QA QB GND QC QD
14 13 12 11 10 9 8
1 2 3 4 5 6 7
INPUT B NC NC NC VCC R0(1) R0(2)

7493A
4-BIT BINARY COUNTERS

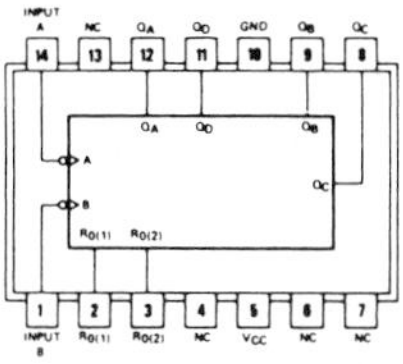
INPUT A NC QA QD GND QB QC
14 13 12 11 10 9 8
1 2 3 4 5 6 7
INPUT B R0(1) R0(2) NC VCC NC NC

7494
4-BIT SHIFT REGISTERS

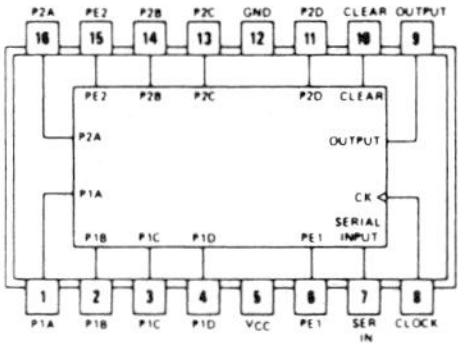
P2A PE2 P2B P2C GND P2D CLEAR OUTPUT
16 15 14 13 12 11 10 9
1 2 3 4 5 6 7 8
P1A P1B P1C P1D VCC PE1 SER IN CLOCK

7495A
4-BIT SHIFT REGISTERS

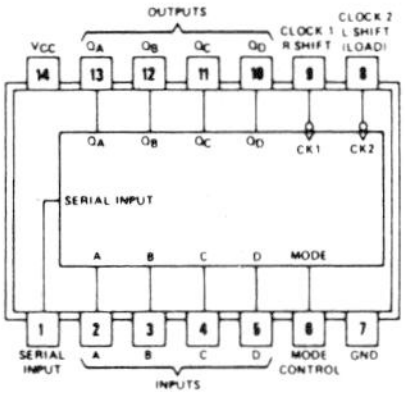
OUTPUTS
VCC QA QB QC QD CLOCK 1 R SHIFT CLOCK 2 L SHIFT (LOAD)
14 13 12 11 10 9 8
1 2 3 4 5 6 7
SERIAL INPUT A B C D MODE CONTROL GND
INPUTS

7496
5-BIT SHIFT REGISTERS

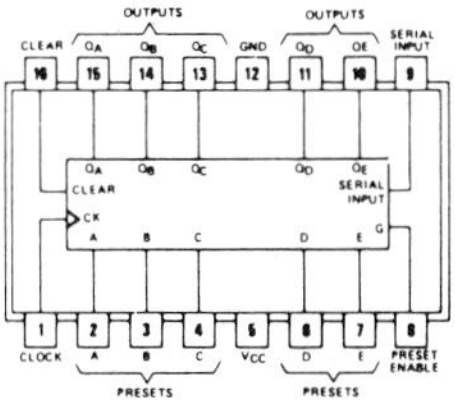
OUTPUTS OUTPUTS
CLEAR QA QB QC GND QD QE SERIAL INPUT
16 15 14 13 12 11 10 9
1 2 3 4 5 6 7 8
CLOCK A B C VCC D E PRESET ENABLE
PRESETS PRESETS

7497
SYNCHRONOUS 6-BIT BINARY RATE MULTIPLIERS

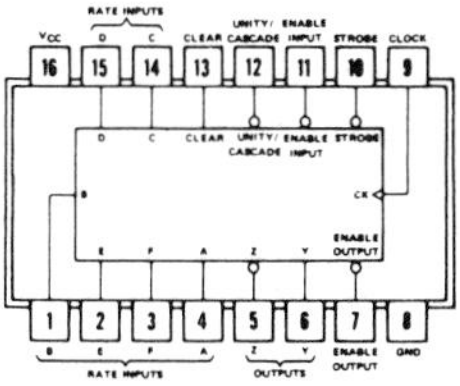
RATE INPUTS
VCC D C CLEAR UNITY/CASCADE ENABLE INPUT STROBE CLOCK
16 15 14 13 12 11 10 9
1 2 3 4 5 6 7 8
B E F A Z Y ENABLE OUTPUT GND
RATE INPUTS OUTPUTS

74L98
4-BIT DATA SELECTOR/STORAGE REGISTERS

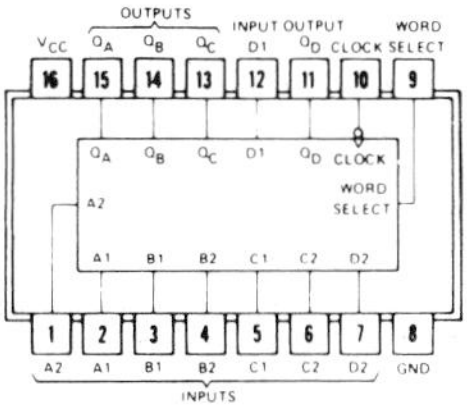
OUTPUTS
VCC QA QB QC INPUT D1 OUTPUT QD CLOCK WORD SELECT
16 15 14 13 12 11 10 9
1 2 3 4 5 6 7 8
A2 A1 B1 B2 C1 C2 D2 GND
INPUTS

74L99
4-BIT BIDIRECTIONAL UNIVERSAL SHIFT REGISTERS

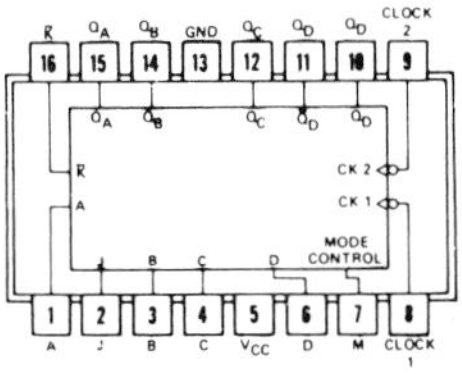
R QA QB GND QC QC QD CLOCK 2
16 15 14 13 12 11 10 9
1 2 3 4 5 6 7 8
A J B C VCC D M CLOCK 1

74100
8-BIT BISTABLE LATCHES

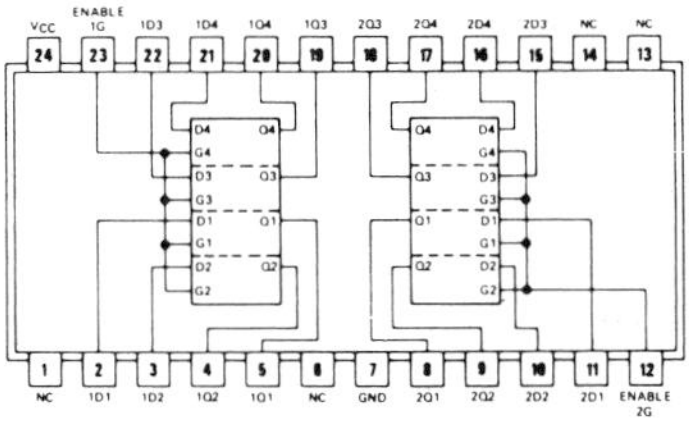
VCC ENABLE 1G 1D3 1D4 1Q4 1Q3 2Q3 2Q4 2D4 2D3 NC NC
24 23 22 21 20 19 18 17 16 15 14 13
1 2 3 4 5 6 7 8 9 10 11 12
NC 1D1 1D2 1Q2 1Q1 NC GND 2Q1 2Q2 2D2 2D1 ENABLE 2G

74H101
AND-OR-GATED J-K NEGATIVE-EDGE-TRIGGERED FLIP-FLOPS WITH PRESET

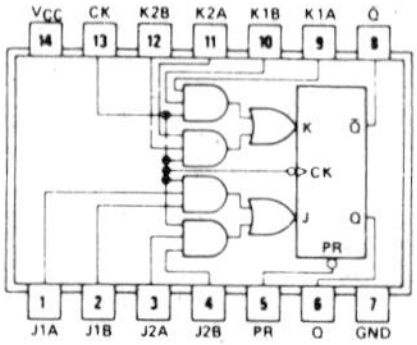
VCC CK K2B K2A K1B K1A Q̄
14 13 12 11 10 9 8
1 2 3 4 5 6 7
J1A J1B J2A J2B PR Q GND

74H102
AND-GATED J-K NEGATIVE-EDGE-TRIGGERED FLIP-FLOPS WITH PRESET AND CLEAR

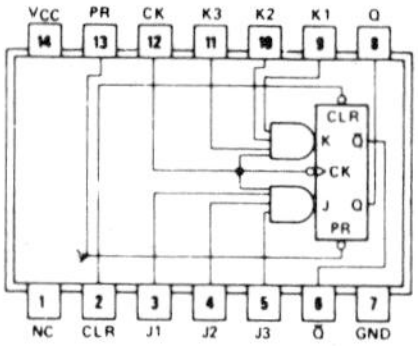
VCC PR CK K3 K2 K1 Q
14 13 12 11 10 9 8
1 2 3 4 5 6 7
NC CLR J1 J2 J3 Q̄ GND

74H103
DUAL J-K NEGATIVE-EDGE-TRIGGERED FLOPS WITH CLEAR

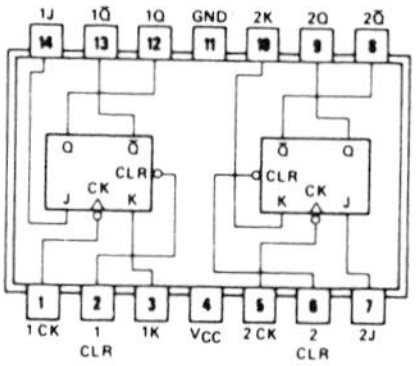
1J 1Q̄ 1Q GND 2K 2Q 2Q̄
14 13 12 11 10 9 8
1 2 3 4 5 6 7
1CK 1 CLR 1K VCC 2CK 2 CLR 2J

74H106
DUAL J-K NEGATIVE-EDGE-TRIGGERED FLIP-FLOPS WITH PRESET AND CLEAR

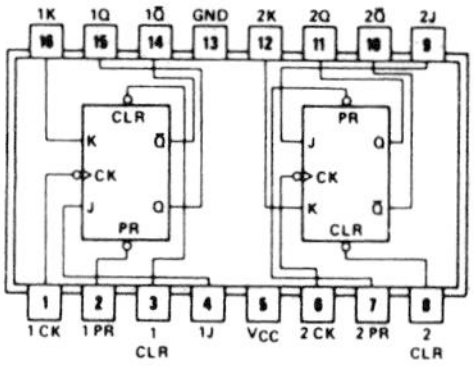
1K 1Q 1Q̄ GND 2K 2Q 2Q̄ 2J
16 15 14 13 12 11 10 9
1 2 3 4 5 6 7 8
1CK 1PR 1 CLR 1J VCC 2CK 2PR 2 CLR

74107
DUAL J-K FLIP-FLOPS WITH CLEAR

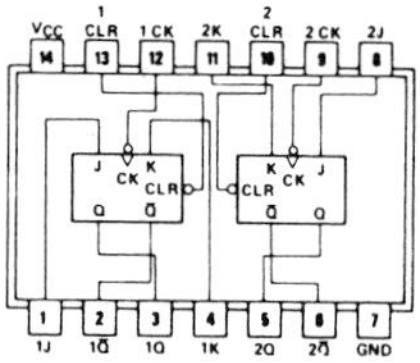
VCC 1 CLR 1CK 2K 2 CLR 2CK 2J
14 13 12 11 10 9 8
1 2 3 4 5 6 7
1J 1Q̄ 1Q 1K 2Q 2Q̄ GND

74H108
DUAL J-K NEGATIVE-EDGE-TRIGGERED FLIP-FLOPS WITH PRESET, COMMON CLEAR, AND COMMON CLOCK

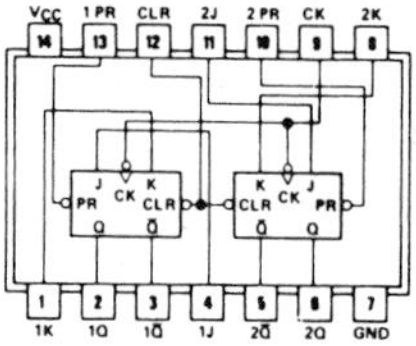
VCC 1PR CLR 2J 2PR CK 2K
14 13 12 11 10 9 8
1 2 3 4 5 6 7
1K 1Q 1Q̄ 1J 2Q̄ 2Q GND

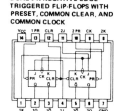

74109
DUAL J-K̄ POSITIVE-EDGE-TRIGGERED FLIP-FLOPS WITH PRESET AND CLEAR

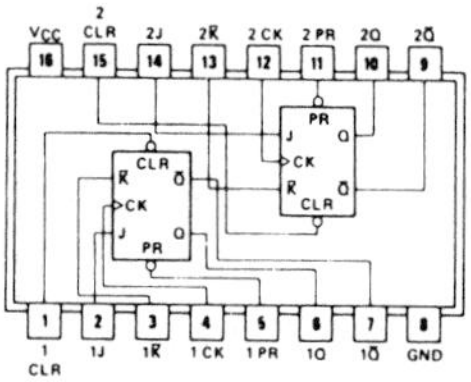
VCC 2 CLR 2J 2K̄ 2CK 2PR 2Q 2Q̄
16 15 14 13 12 11 10 9
1 2 3 4 5 6 7 8
1 CLR 1J 1K̄ 1CK 1PR 1Q 1Q̄ GND

74110
AND-GATED J-K MASTER-SLAVE FLIP-FLOPS WITH DATA LOCKOUT

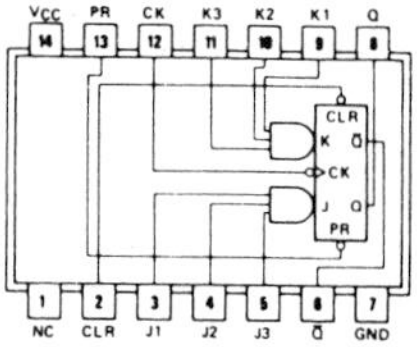
VCC PR CK K3 K2 K1 Q
14 13 12 11 10 9 8
1 2 3 4 5 6 7
NC CLR J1 J2 J3 Q̄ GND

74111
DUAL J-K MASTER-SLAVE FLIP-FLOPS WITH DATA LOCKOUT

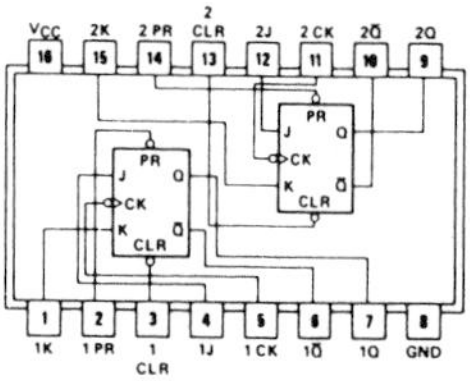
VCC 2K 2PR 2 CLR 2J 2CK 2Q̄ 2Q
16 15 14 13 12 11 10 9
1 2 3 4 5 6 7 8
1K 1PR 1 CLR 1J 1CK 1Q̄ 1Q GND

74LS112A
DUAL J-K NEGATIVE-EDGE-TRIGGERED FLIP-FLOPS WITH PRESET AND CLEAR

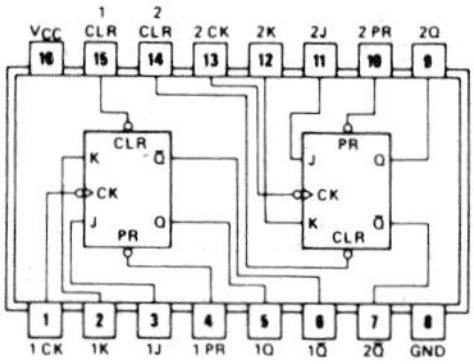
VCC 1 CLR 2 CLR 2CK 2K 2J 2PR 2Q
16 15 14 13 12 11 10 9
1 2 3 4 5 6 7 8
1CK 1K 1J 1PR 1Q 1Q̄ 2Q̄ GND

74LS113A
DUAL J-K NEGATIVE-EDGE-TRIGGERED FLIP-FLOPS WITH PRESET

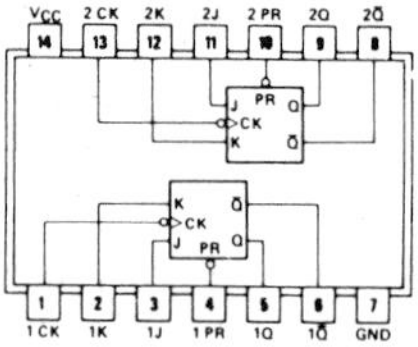
VCC 2CK 2K 2J 2PR 2Q 2Q̄
14 13 12 11 10 9 8
1 2 3 4 5 6 7
1CK 1K 1J 1PR 1Q 1Q̄ GND

74LS114A
DUAL J-K NEGATIVE-EDGE-TRIGGERED FLIP-FLOPS WITH PRESET, COMMON CLEAR, AND COMMON CLOCK

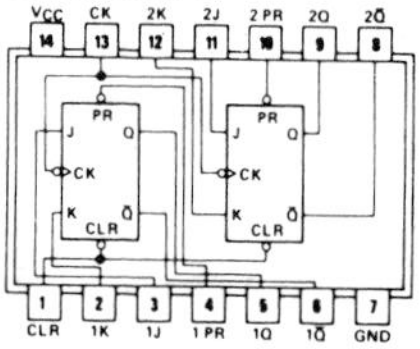
VCC CK 2K 2J 2PR 2Q 2Q̄
14 13 12 11 10 9 8
1 2 3 4 5 6 7
CLR 1K 1J 1PR 1Q 1Q̄ GND

74116
DUAL 4-BIT LATCHES

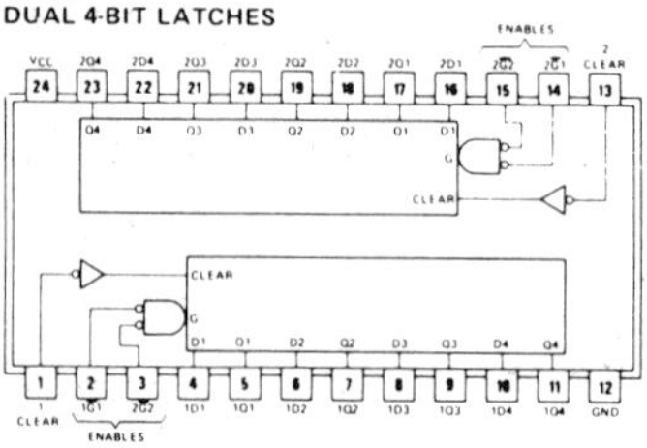
ENABLES
VCC 2D4 2Q4 2Q3 2D3 2Q2 2D2 2Q1 2D1 2Ḡ2 2Ḡ1 2 CLEAR
24 23 22 21 20 19 18 17 16 15 14 13
1 2 3 4 5 6 7 8 9 10 11 12
1 CLEAR 1Ḡ1 1Ḡ2 1D1 1Q1 1D2 1Q2 1D3 1Q3 1D4 1Q4 GND
ENABLES

74120
DUAL PULSE SYNCHRONIZERS/DRIVERS

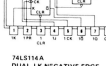

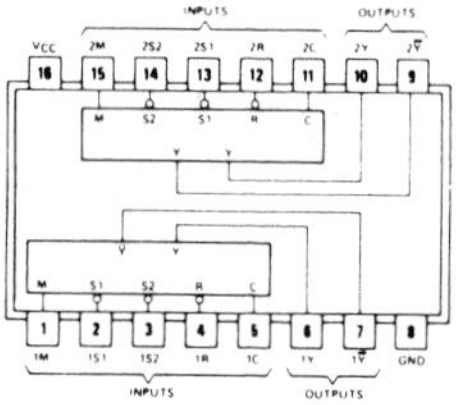
INPUTS OUTPUTS
VCC 2M 2S2 2S1 2R 2C 2Y 2Ȳ
16 15 14 13 12 11 10 9
1 2 3 4 5 6 7 8
1M 1S1 1S2 1R 1C 1Y 1Ȳ GND
INPUTS OUTPUTS

74121
MONOSTABLE MULTIVIBRATORS

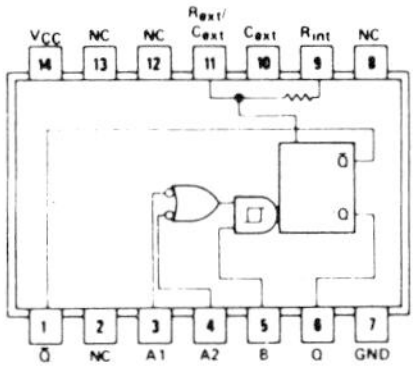
VCC NC NC Rext/Cext Cext Rint NC
14 13 12 11 10 9 8
1 2 3 4 5 6 7
Q̄ NC A1 A2 B Q GND

74122
RETRIGGERABLE MONOSTABLE MULTIVIBRATORS WITH CLEAR

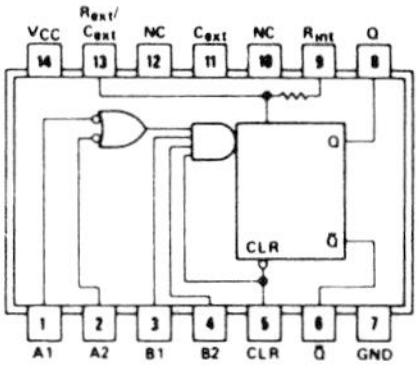
VCC Rext/Cext NC Cext NC Rint Q
14 13 12 11 10 9 8
CLR
1 2 3 4 5 6 7
A1 A2 B1 B2 CLR Q̄ GND

74123
DUAL RETRIGGERABLE MONOSTABLE MULTIVIBRATORS WITH CLEAR

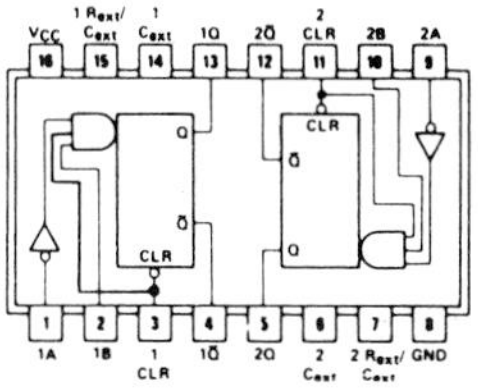
VCC 1 Rext/Cext 1 Cext 1Q 2Q̄ 2 CLR 2B 2A
16 15 14 13 12 11 10 9
CLR
1 2 3 4 5 6 7 8
1A 1B 1 CLR 1Q̄ 2Q 2 Cext 2 Rext/Cext GND

74LS124
DUAL VOLTAGE-CONTROLLED OSCILLATORS

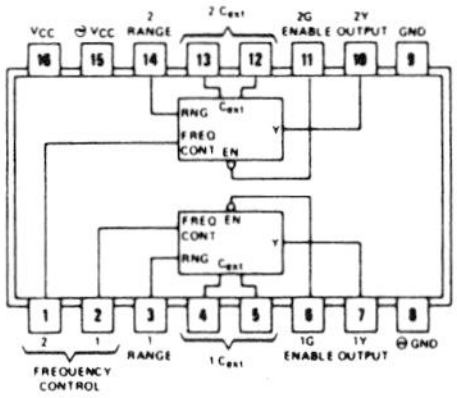
VCC VCC 2 RANGE 2 Cext 2G ENABLE 2Y OUTPUT GND
16 15 14 13 12 11 10 9
RNG Cext
FREQ CONT
EN
1 2 3 4 5 6 7 8
FREQUENCY CONTROL
1 RANGE
1 Cext
1G ENABLE 1Y OUTPUT GND

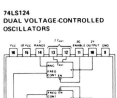

74125
QUADRUPLE BUS BUFFER GATES WITH THREE-STATE OUTPUTS

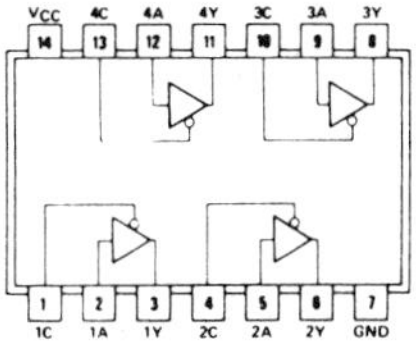
VCC 4C 4A 4Y 3C 3A 3Y
14 13 12 11 10 9 8
1 2 3 4 5 6 7
1C 1A 1Y 2C 2A 2Y GND

74126
QUADRUPLE BUS BUFFER GATES WITH THREE-STATE OUTPUTS

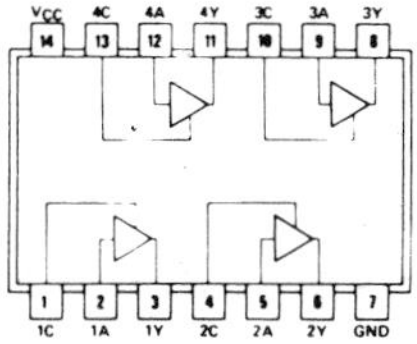
VCC 4C 4A 4Y 3C 3A 3Y
14 13 12 11 10 9 8
1 2 3 4 5 6 7
1C 1A 1Y 2C 2A 2Y GND

74128
SN74128 . . . 50-OHM LINE DRIVER

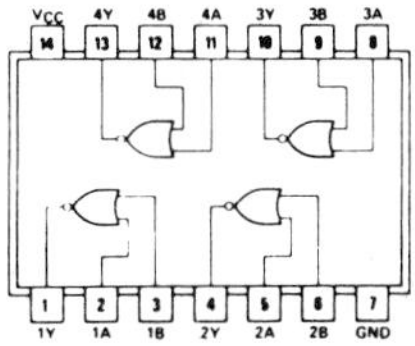
VCC 4Y 4B 4A 3Y 3B 3A
14 13 12 11 10 9 8
1 2 3 4 5 6 7
1Y 1A 1B 2Y 2A 2B GND

74132
QUADRUPLE 2-INPUT POSITIVE-NAND SCHMITT TRIGGERS

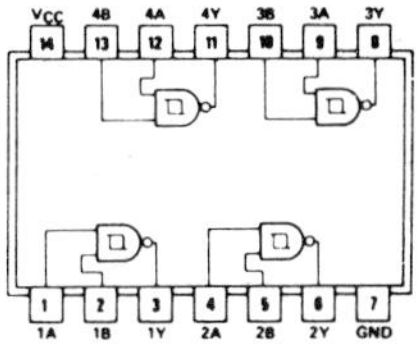
VCC 4B 4A 4Y 3B 3A 3Y
14 13 12 11 10 9 8
1 2 3 4 5 6 7
1A 1B 1Y 2A 2B 2Y GND

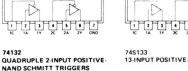

74S133
13-INPUT POSITIVE-NAND GATES

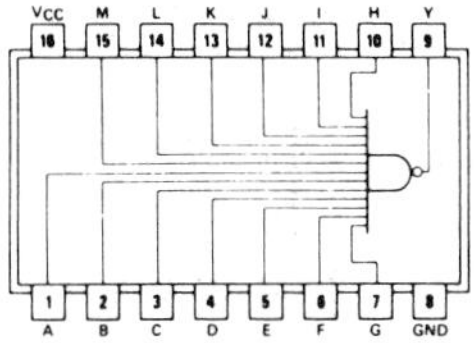
VCC M L K J I H Y
16 15 14 13 12 11 10 9
1 2 3 4 5 6 7 8
A B C D E F G GND

74S134
12-INPUT POSITIVE-NAND GATES WITH THREE-STATE OUTPUTS

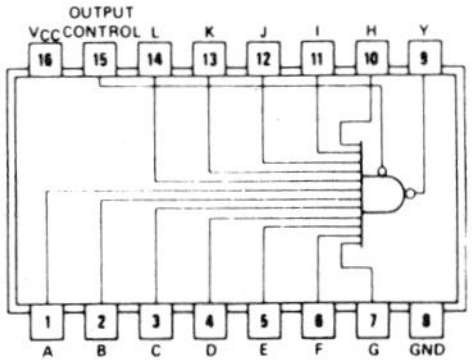
VCC OUTPUT CONTROL L K J I H Y
16 15 14 13 12 11 10 9
1 2 3 4 5 6 7 8
A B C D E F G GND

74S135
QUAD EXCLUSIVE-OR/NOR GATES

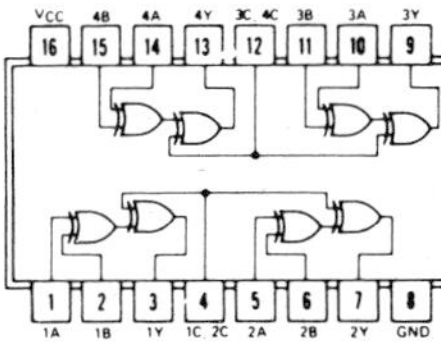
VCC 4B 4A 4Y 3C 4C 3B 3A 3Y
16 15 14 13 12 11 10 9
1 2 3 4 5 6 7 8
1A 1B 1Y 1C 2C 2A 2B 2Y GND

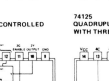

74136
QUAD EXCLUSIVE-OR GATES

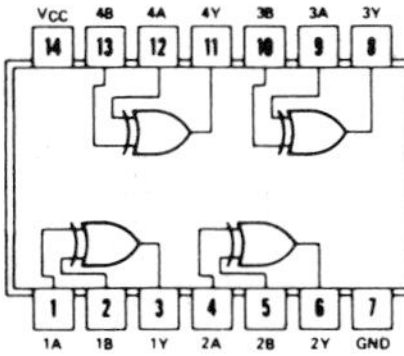
VCC 4B 4A 4Y 3B 3A 3Y
14 13 12 11 10 9 8
1 2 3 4 5 6 7
1A 1B 1Y 2A 2B 2Y GND

74LS138
3-TO-8 LINE DECODERS/ MULTIPLEXERS

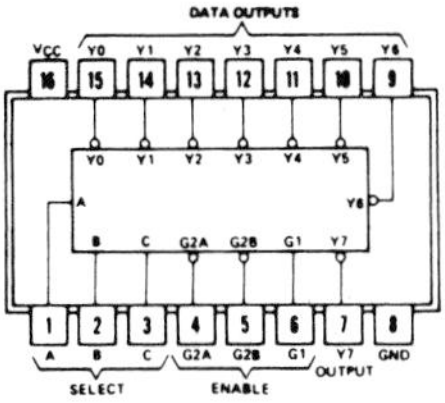
DATA OUTPUTS
VCC Y0 Y1 Y2 Y3 Y4 Y5 Y6
16 15 14 13 12 11 10 9
Y0 Y1 Y2 Y3 Y4 Y5
A Y6
B C G2A G2B G1 Y7
1 2 3 4 5 6 7 8
A B C G2A G2B G1 Y7 OUTPUT GND
SELECT ENABLE

74LS139
DUAL 2-TO-4 LINE DECODERS/ MULTIPLEXERS

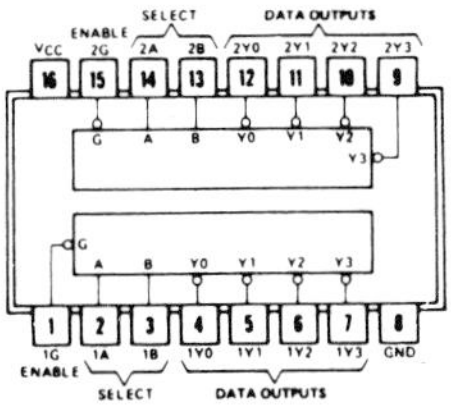
ENABLE SELECT DATA OUTPUTS
VCC 2G 2A 2B 2Y0 2Y1 2Y2 2Y3
16 15 14 13 12 11 10 9
G A B Y0 Y1 Y2 Y3
G A B Y0 Y1 Y2 Y3
1 2 3 4 5 6 7 8
1G ENABLE 1A 1B 1Y0 1Y1 1Y2 1Y3 GND
SELECT DATA OUTPUTS

74S140
DUAL 4-INPUT POSITIVE-NAND 50-OHM LINE DRIVERS

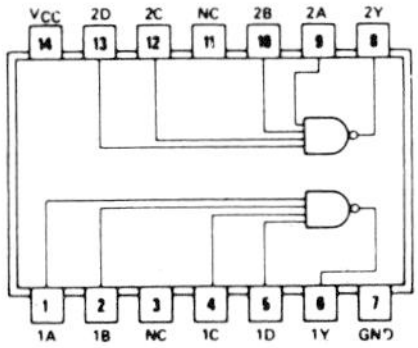
VCC 2D 2C NC 2B 2A 2Y
14 13 12 11 10 9 8
1 2 3 4 5 6 7
1A 1B NC 1C 1D 1Y GND

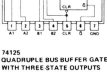

74141
BCD-TO-DECIMAL DECODER/ DRIVER

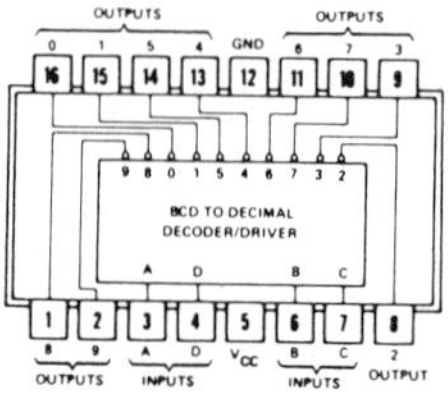
OUTPUTS
OUTPUTS
GND
BCD TO DECIMAL
DECODER/DRIVER
OUTPUTS
INPUTS
VCC
INPUTS
OUTPUT

74142
COUNTER/LATCH/DECODER/ DRIVER

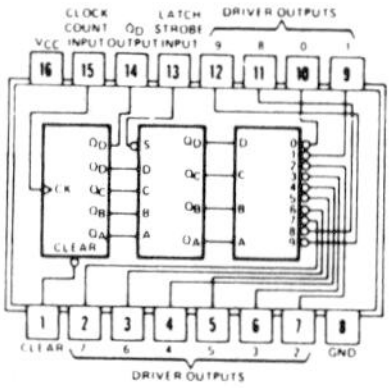
CLOCK
COUNT INPUT
LATCH STROBE INPUT
DRIVER OUTPUTS
VCC
CLEAR
GND
DRIVER OUTPUTS

74143 74144
COUNTERS/LATCHES/DECODERS/ DRIVERS

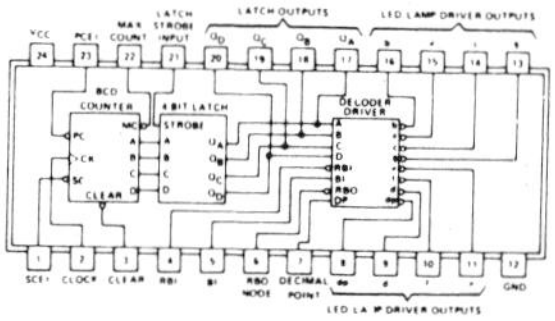
VCC
LATCH OUTPUTS
LED LAMP DRIVER OUTPUTS
BCD COUNTER
4 BIT LATCH
DECODER DRIVER
CLEAR
CLOCK
RBI
BI
RBO NODE
DECIMAL POINT
GND
LED LAMP DRIVER OUTPUTS

74145
BCD-TO-DECIMAL DECODERS/DRIVERS FOR LAMPS, RELAYS, MOS

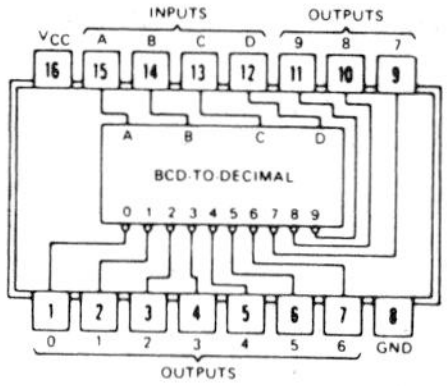
INPUTS
OUTPUTS
VCC
BCD-TO-DECIMAL
OUTPUTS
GND

74147
10-LINE DECIMAL TO 4-LINE BCD PRIORITY ENCODERS

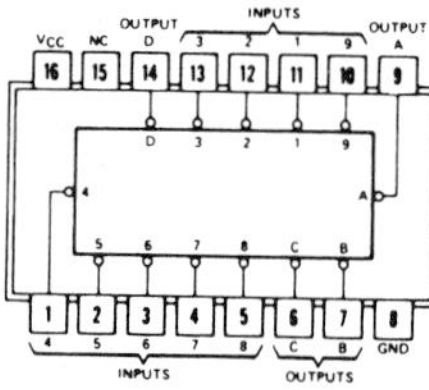
OUTPUT
INPUTS
OUTPUT
VCC
NC
INPUTS
OUTPUTS
GND

74148
8-LINE-TO-3-LINE OCTAL PRIORITY ENCODERS

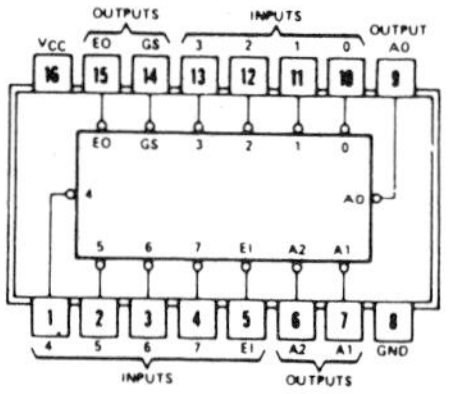
OUTPUTS
INPUTS
OUTPUT
VCC
EO
GS
A0
INPUTS
OUTPUTS
GND

74150
1-OF-16 DATA SELECTORS/ MULTIPLEXERS

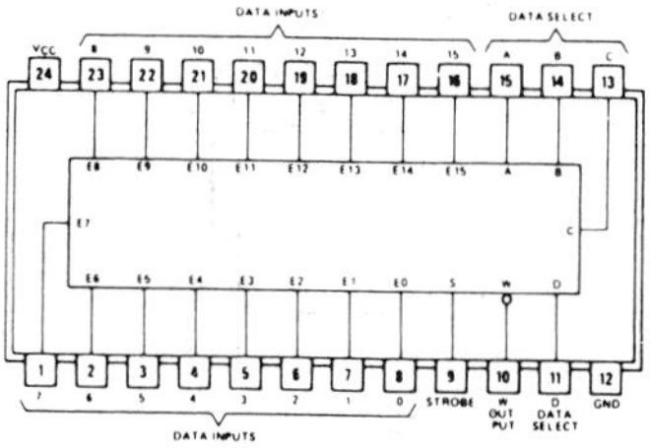
DATA INPUTS
DATA SELECT
VCC
STROBE
W OUTPUT
D DATA SELECT
GND
DATA INPUTS

74151A
1-OF-8 DATA SELECTORS/MULTIPLEXERS

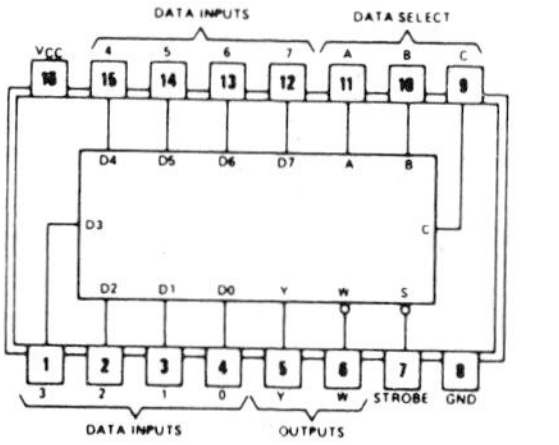
DATA INPUTS
DATA SELECT
VCC
STROBE
GND
DATA INPUTS
OUTPUTS

74153
DUAL 4-LINE TO 1-LINE DATA SELECTORS/MULTIPLEXERS

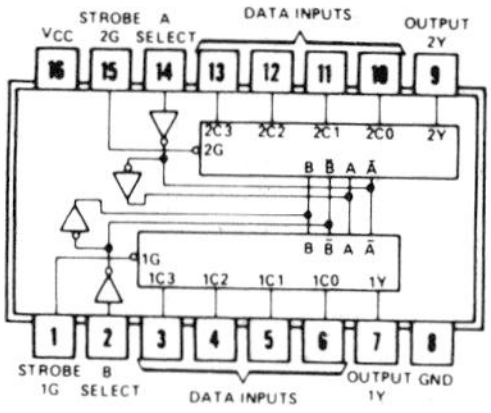
STROBE 2G
A SELECT
DATA INPUTS
OUTPUT 2Y
VCC
STROBE 1G
B SELECT
DATA INPUTS
OUTPUT 1Y
GND

74154
4-LINE TO 16-LINE DECODERS/ DEMULTIPLEXERS

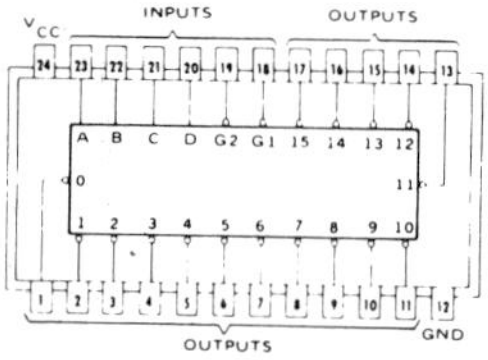
VCC
INPUTS
OUTPUTS
OUTPUTS
GND

74155
DECODERS/DEMULTIPLEXERS

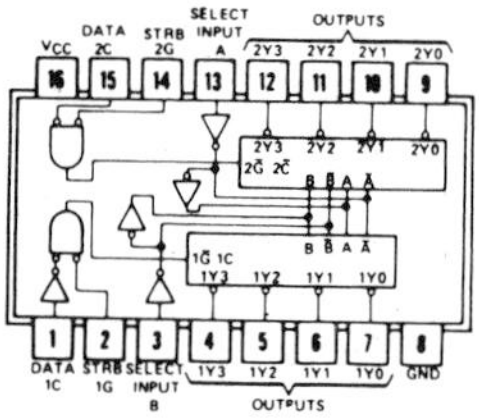
DATA 2C
STRB 2G
SELECT INPUT A
OUTPUTS
VCC
DATA 1C
STRB 1G
SELECT INPUT B
OUTPUTS
GND

74157
QUAD 2- TO 1-LINE DATA SELECTORS/MULTIPLEXERS

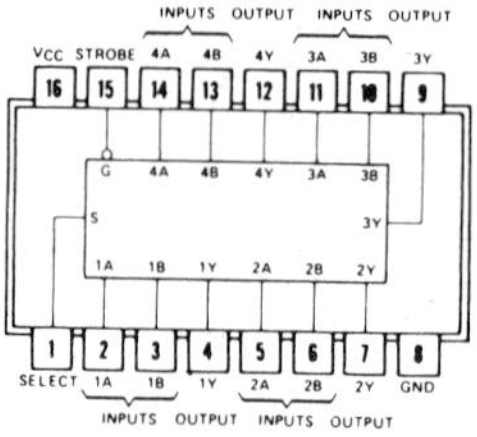
INPUTS
OUTPUT
INPUTS
OUTPUT
VCC
STROBE
SELECT
GND
INPUTS
OUTPUT
INPUTS
OUTPUT

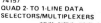

74159
4- TO 16-LINE DECODERS/ DEMULTIPLEXERS

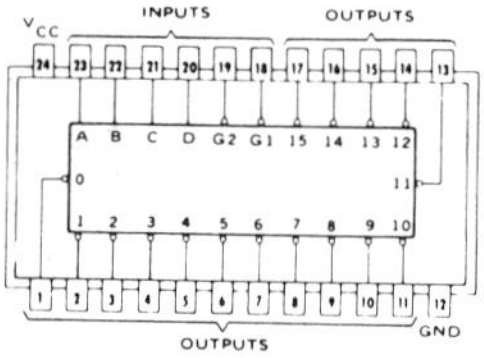
VCC
INPUTS
OUTPUTS
OUTPUTS
GND

74160 74161 74162 74163
SYNCHRONOUS 4-BIT COUNTERS

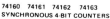

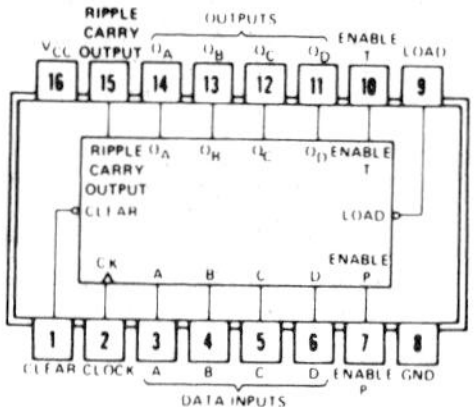
RIPPLE CARRY OUTPUT
OUTPUTS
ENABLE T
LOAD
VCC
CLEAR
CLOCK
ENABLE P
GND
DATA INPUTS

74164
8-BIT PARALLEL OUTPUT SERIAL SHIFT REGISTERS

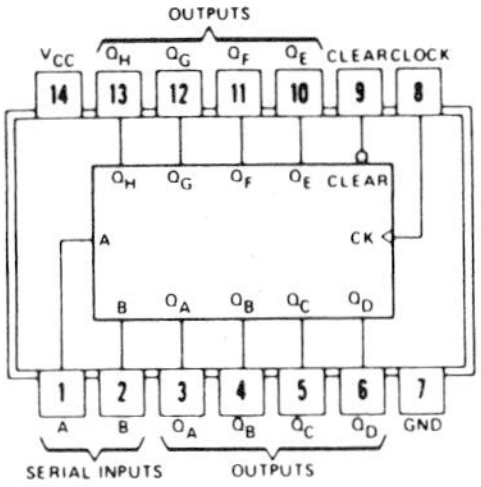
OUTPUTS
VCC QH QG QF QE CLEAR CLOCK
14 13 12 11 10 9 8
QH QG QF QE CLEAR
A CK
B QA QB QC QD
1 2 3 4 5 6 7
A B QA QB QC QD GND
SERIAL INPUTS OUTPUTS

74165
PARALLEL-LOAD 8-BIT SHIFT REGISTERS WITH COMPLEMENTARY OUTPUTS

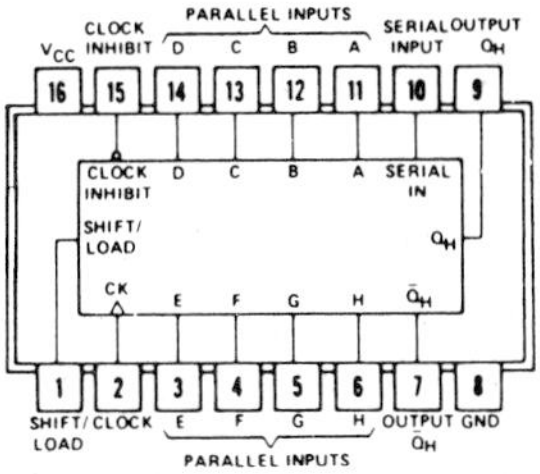
CLOCK INHIBIT PARALLEL INPUTS SERIAL INPUT OUTPUT QH
VCC D C B A
16 15 14 13 12 11 10 9
CLOCK INHIBIT D C B A SERIAL IN
SHIFT/ LOAD QH
CK E F G H Q̄H
1 2 3 4 5 6 7 8
SHIFT/ LOAD CLOCK E F G H OUTPUT Q̄H GND
PARALLEL INPUTS

74166
8-BIT SHIFT REGISTERS

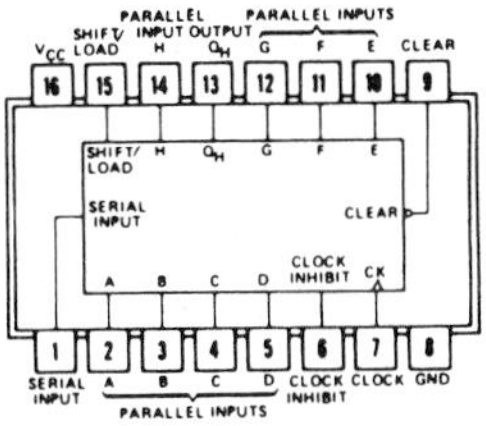
SHIFT/ LOAD PARALLEL INPUT H OUTPUT QH PARALLEL INPUTS G F E CLEAR
VCC
16 15 14 13 12 11 10 9
SHIFT/ LOAD H QH G F E
SERIAL INPUT CLEAR
CLOCK INHIBIT CK
A B C D
1 2 3 4 5 6 7 8
SERIAL INPUT A B C D CLOCK INHIBIT CLOCK GND
PARALLEL INPUTS

74167
SYNCHRONOUS DECADE RATE MULTIPLIERS

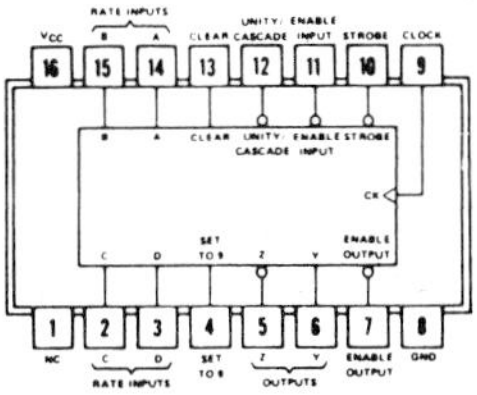
16 15 14 13 12 11 10 9
1 2 3 4 5 6 7 8

74S168 74S169
4-BIT UP/DOWN SYNCHRONOUS COUNTERS

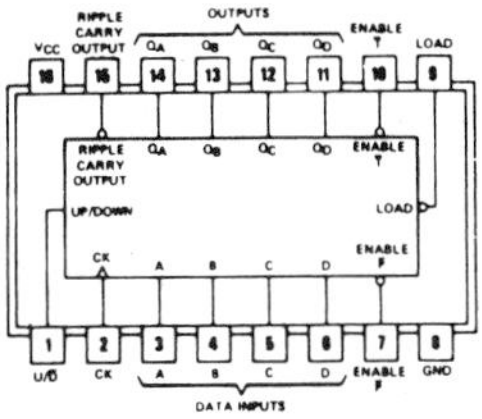
RIPPLE CARRY OUTPUT
OUTPUTS
VCC QA QB QC QD ENABLE T LOAD
16 15 14 13 12 11 10 9
RIPPLE CARRY OUTPUT QA QB QC QD ENABLE T
UP/DOWN LOAD
ENABLE P
CK A B C D
1 2 3 4 5 6 7 8
U/D CK A B C D ENABLE P GND
DATA INPUTS

74170
4-BY-4 REGISTER FILES

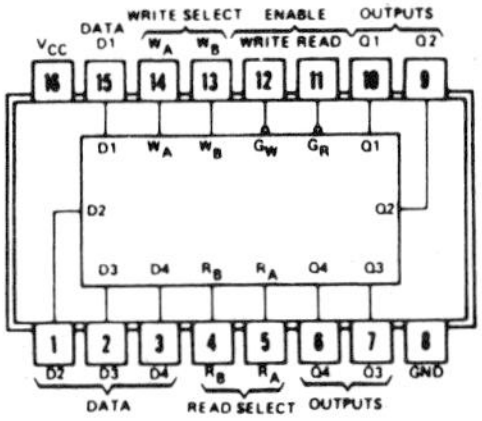
DATA D1 WRITE SELECT WA WB ENABLE WRITE READ OUTPUTS Q1 Q2
VCC
16 15 14 13 12 11 10 9
D1 WA WB GW GR Q1
D2 Q2
D3 D4 RB RA Q4 Q3
1 2 3 4 5 6 7 8
D2 D3 D4 RB RA Q4 Q3 GND
DATA READ SELECT OUTPUTS

74172
16-BIT REGISTER FILE

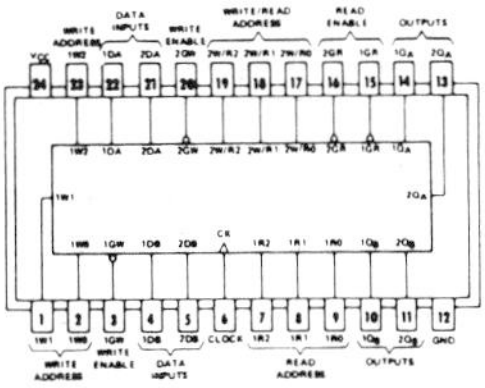
24 23 22 21 20 19 18 17 16 15 14 13
1 2 3 4 5 6 7 8 9 10 11 12

74173
4-BIT D-TYPE REGISTERS

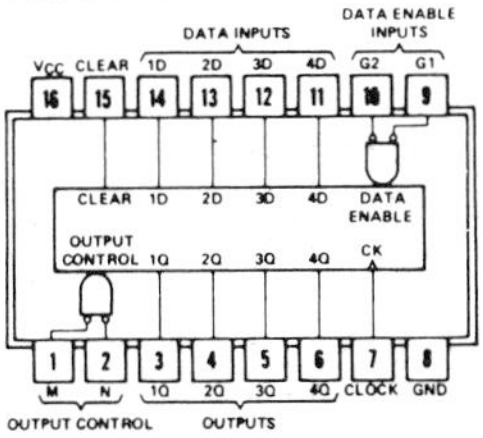
DATA ENABLE INPUTS
DATA INPUTS
VCC CLEAR 1D 2D 3D 4D G2 G1
16 15 14 13 12 11 10 9
CLEAR 1D 2D 3D 4D DATA ENABLE
OUTPUT CONTROL 1Q 2Q 3Q 4Q CK
1 2 3 4 5 6 7 8
M N 1Q 2Q 3Q 4Q CLOCK GND
OUTPUT CONTROL OUTPUTS

74174
HEX D-TYPE FLIP-FLOPS

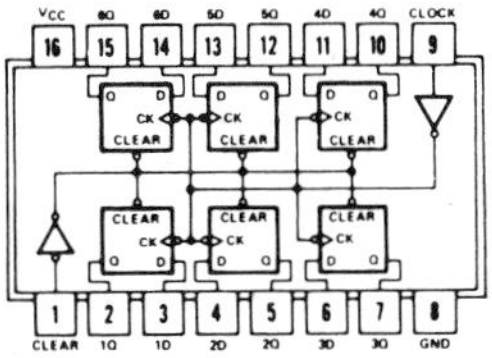
VCC 6Q 6D 5D 5Q 4D 4Q CLOCK
16 15 14 13 12 11 10 9
1 2 3 4 5 6 7 8
CLEAR 1Q 1D 2D 2Q 3D 3Q GND

74175
QUAD D-TYPE FLIP-FLOPS

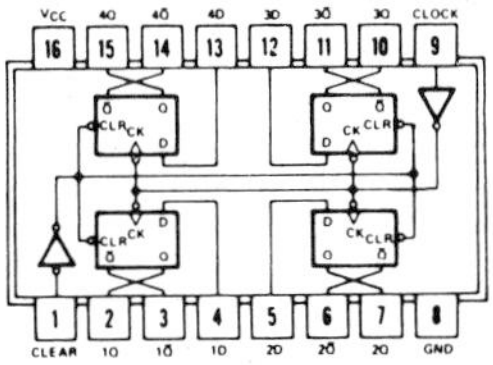
VCC 4Q 4Q̄ 4D 3D 3Q̄ 3Q CLOCK
16 15 14 13 12 11 10 9
1 2 3 4 5 6 7 8
CLEAR 1Q 1Q̄ 1D 2D 2Q̄ 2Q GND

74176 74177
PRESETABLE COUNTERS/LATCHES

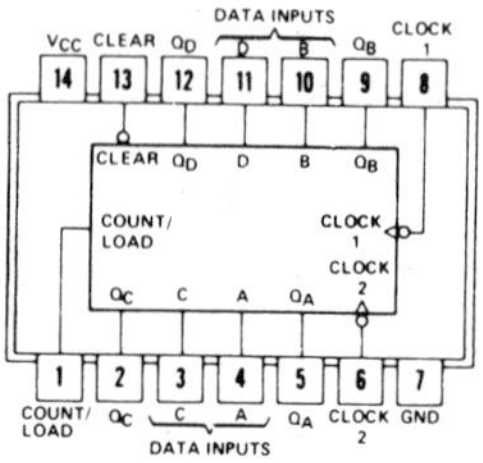
DATA INPUTS
VCC CLEAR QD D B QB CLOCK 1
14 13 12 11 10 9 8
CLEAR QD D B QB
COUNT/ LOAD
CLOCK 1
CLOCK 2
QC C A QA
1 2 3 4 5 6 7
COUNT/ LOAD QC C A QA CLOCK 2 GND
DATA INPUTS

74178
4-BIT UNIVERSAL SHIFT REGISTERS

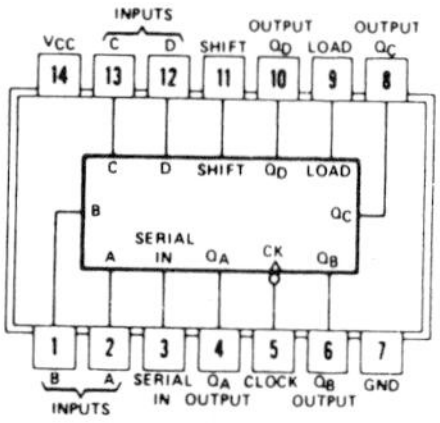
INPUTS
OUTPUT OUTPUT
VCC C D SHIFT QD LOAD QC
14 13 12 11 10 9 8
C D SHIFT QD LOAD
B QC
SERIAL IN
A QA CK QB
1 2 3 4 5 6 7
B A SERIAL IN QA OUTPUT CLOCK QB OUTPUT GND
INPUTS

74179
4-BIT UNIVERSAL SHIFT REGISTERS

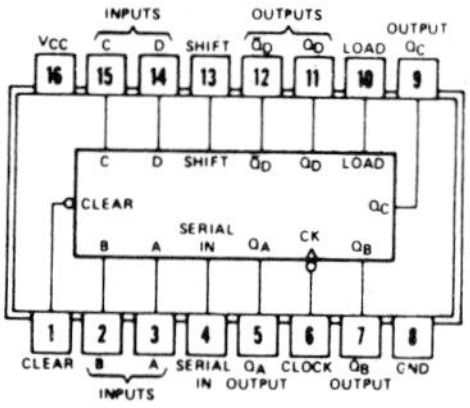
INPUTS
OUTPUTS
OUTPUT
VCC C D SHIFT QD QD LOAD QC
16 15 14 13 12 11 10 9
C D SHIFT QD QD LOAD
CLEAR QC
SERIAL IN
B A QA CK QB
1 2 3 4 5 6 7 8
CLEAR B A SERIAL IN QA OUTPUT CLOCK QB OUTPUT GND
INPUTS

74180
9-BIT ODD/EVEN PARITY GENERATORS/CHECKERS

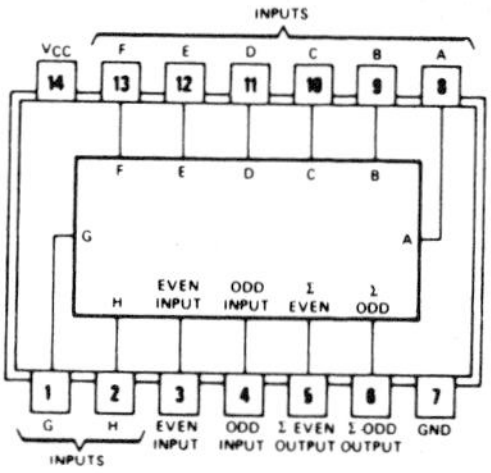
INPUTS
VCC F E D C B A
14 13 12 11 10 9 8
F E D C B
G A
H EVEN INPUT ODD INPUT Σ EVEN Σ ODD
1 2 3 4 5 6 7
G H EVEN INPUT ODD INPUT Σ EVEN OUTPUT Σ ODD OUTPUT GND
INPUTS

74181
ARITHMETIC LOGIC UNITS/FUNCTION GENERATORS

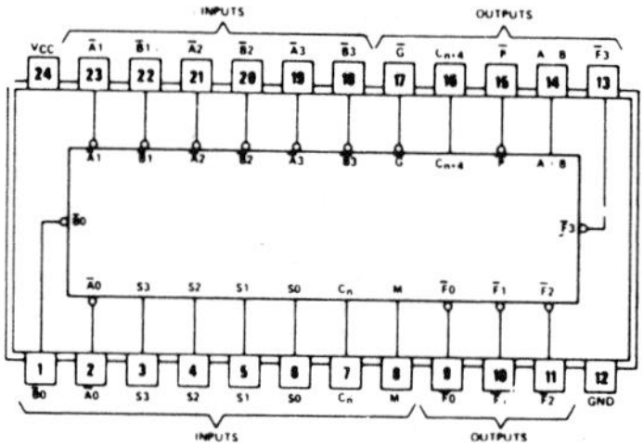
INPUTS
OUTPUTS
VCC
24 23 22 21 20 19 18 17 16 15 14 13
1 2 3 4 5 6 7 8 9 10 11 12
A0 S3 S2 S1 S0 Cn M
GND
INPUTS
OUTPUTS

74182
LOOK-AHEAD CARRY GENERATORS

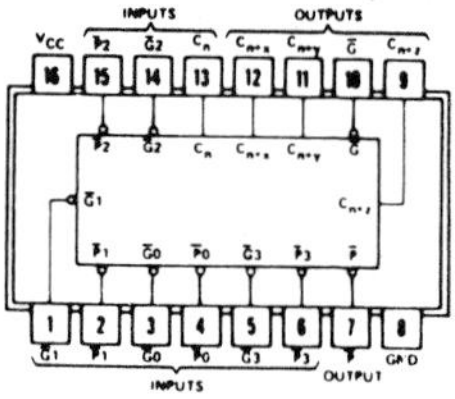
INPUTS
OUTPUTS
VCC
16 15 14 13 12 11 10 9
1 2 3 4 5 6 7 8
GND
INPUTS
OUTPUT

74LS183
DUAL CARRY-SAVE FULL ADDERS

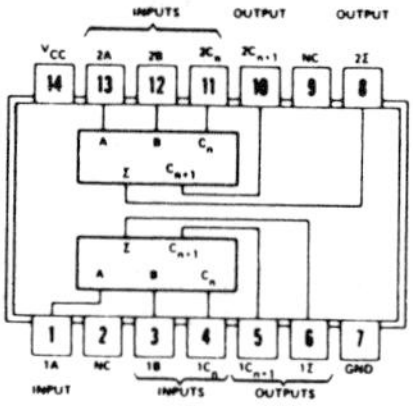
INPUTS
OUTPUT
OUTPUT
VCC
14 13 12 11 10 9 8
1 2 3 4 5 6 7
1A NC 1B GND
INPUT
INPUTS
OUTPUTS

74184 74185A
CODE CONVERTERS

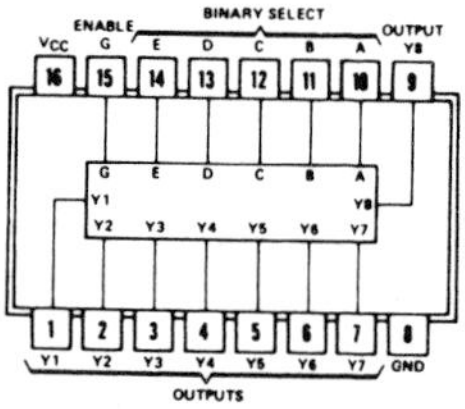
ENABLE
BINARY SELECT
OUTPUT
VCC G E D C B A Y8
16 15 14 13 12 11 10 9
G E D C B A
Y1 Y8
Y2 Y3 Y4 Y5 Y6 Y7
1 2 3 4 5 6 7 8
Y1 Y2 Y3 Y4 Y5 Y6 Y7 GND
OUTPUTS

74186
512-BIT PROGRAMABLE READ-ONLY MEMORIES

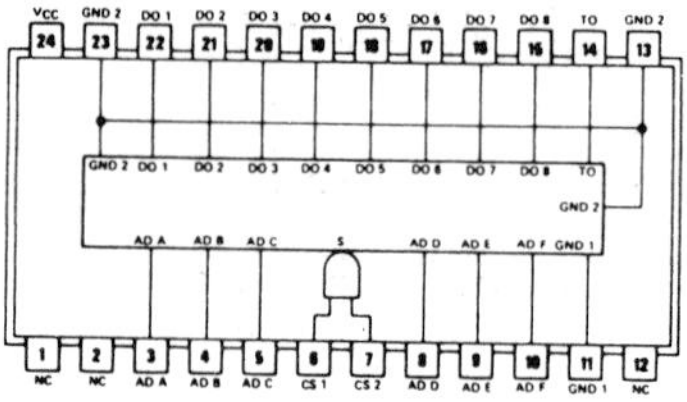
VCC GND 2 DO 1 DO 2 DO 3 DO 4 DO 5 DO 6 DO 7 DO 8 TO GND 2
24 23 22 21 20 19 18 17 16 15 14 13
1 2 3 4 5 6 7 8 9 10 11 12
NC NC AD A AD B AD C CS 1 CS 2 AD D AD E AD F GND 1 NC

74187
1024-BIT READ-ONLY MEMORIES

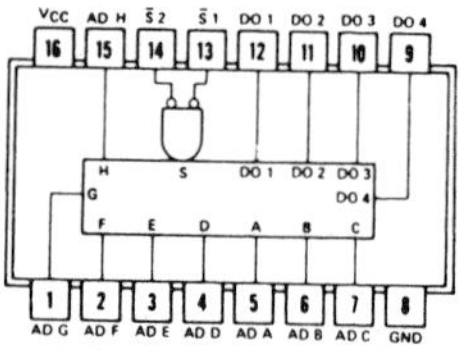
VCC AD H S2 S1 DO 1 DO 2 DO 3 DO 4
16 15 14 13 12 11 10 9
H S DO 1 DO 2 DO 3
G DO 4
F E D A B C
1 2 3 4 5 6 7 8
AD G AD F AD E AD D AD A AD B AD C GND

74188A
256-BIT PROGRAMMABLE READ-ONLY MEMORIES

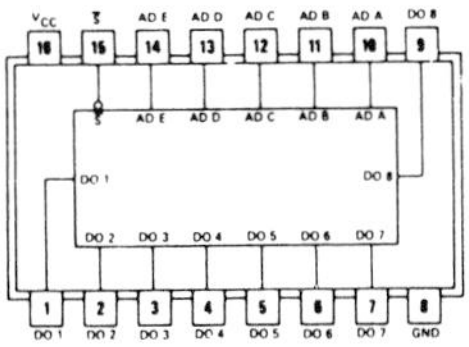
VCC
AD E
AD D
AD C
AD B
AD A
DO 8
DO 1
DO 2
DO 3
DO 4
DO 5
DO 6
DO 7
GND

74S189
64-BIT RANDOM-ACCESS MEMORIES

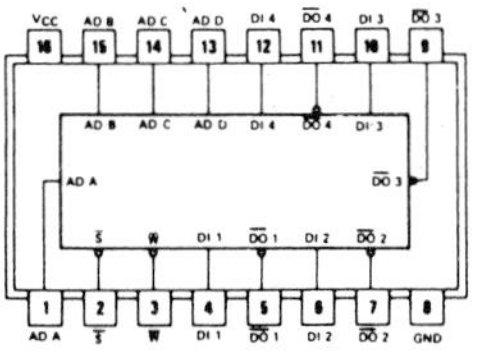
VCC
AD B
AD C
AD D
DI 4
DO 4
DI 3
DO 3
AD A
S
W
DI 1
DO 1
DI 2
DO 2
GND

74190 74191
SYNCHRONOUS UP/DOWN COUNTERS

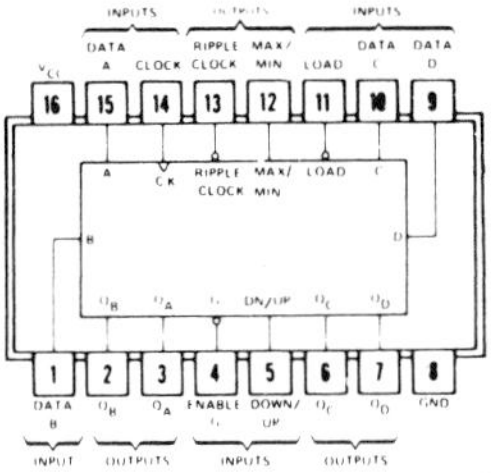
INPUTS
OUTPUTS
INPUTS
DATA A
CLOCK
RIPPLE CLOCK
MAX/MIN
LOAD
DATA C
DATA D
DATA B
ENABLE G
DOWN/UP
GND
INPUT
OUTPUTS
INPUTS
OUTPUTS

74192 74193
SYNCHRONOUS UP/DOWN DUAL CLOCK COUNTERS

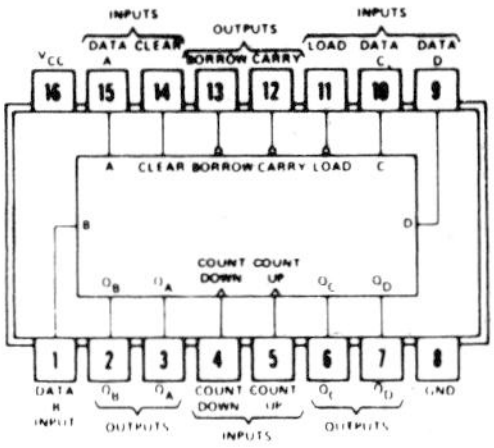
INPUTS
OUTPUTS
INPUTS
DATA A
CLEAR
BORROW
CARRY
LOAD
DATA C
DATA D
DATA B
COUNT DOWN
COUNT UP
GND
INPUT
OUTPUTS
INPUTS
OUTPUTS

74194
4-BIT BIDIRECTIONAL UNIVERSAL SHIFT REGISTERS

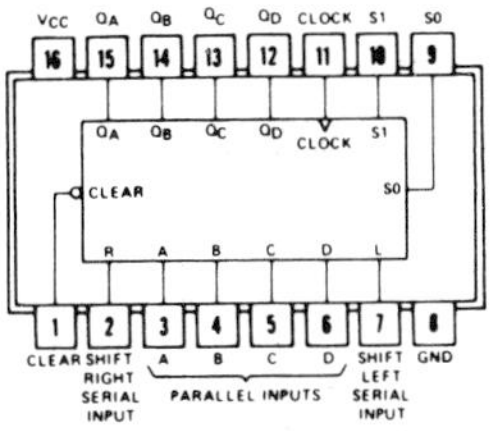
VCC
CLOCK
S1
S0
CLEAR
SHIFT RIGHT SERIAL INPUT
PARALLEL INPUTS
SHIFT LEFT SERIAL INPUT
GND

74195
4-BIT PARALLEL-ACCESS SHIFT REGISTERS

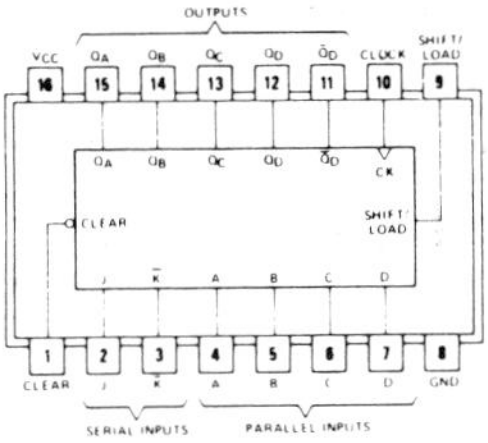
OUTPUTS
VCC
CLOCK
SHIFT/LOAD
CLEAR
SERIAL INPUTS
PARALLEL INPUTS
GND

74196 74197
PRESETABLE COUNTERS/LATCHES

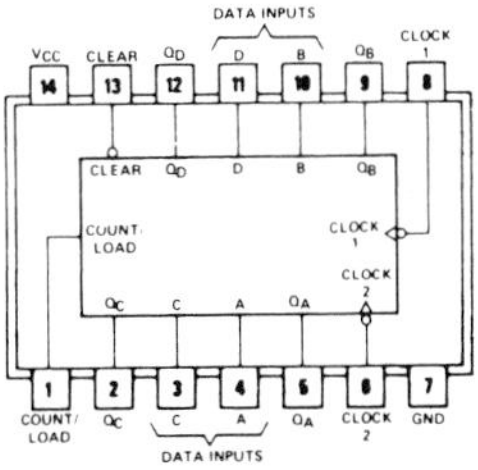
DATA INPUTS
VCC
CLEAR
CLOCK 1
COUNT/LOAD
CLOCK 2
GND
DATA INPUTS

74198
8-BIT BIDIRECTIONAL UNIVERSAL SHIFT REGISTERS

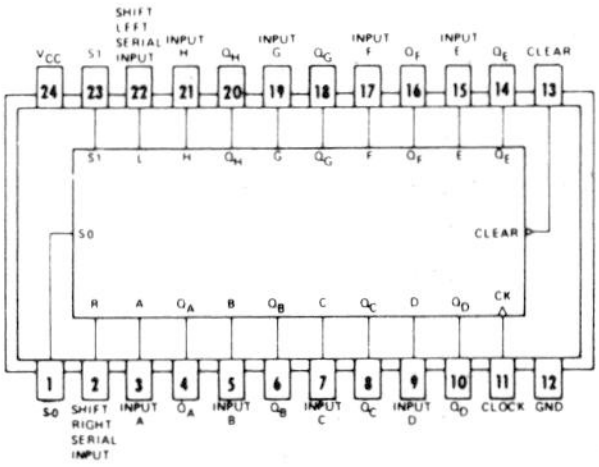
VCC
S1
SHIFT LEFT SERIAL INPUT
INPUT H
INPUT G
INPUT F
INPUT E
CLEAR
S0
SHIFT RIGHT SERIAL INPUT
INPUT A
INPUT B
INPUT C
INPUT D
CLOCK
GND

74199
8-BIT BIDIRECTIONAL UNIVERSAL SHIFT REGISTERS

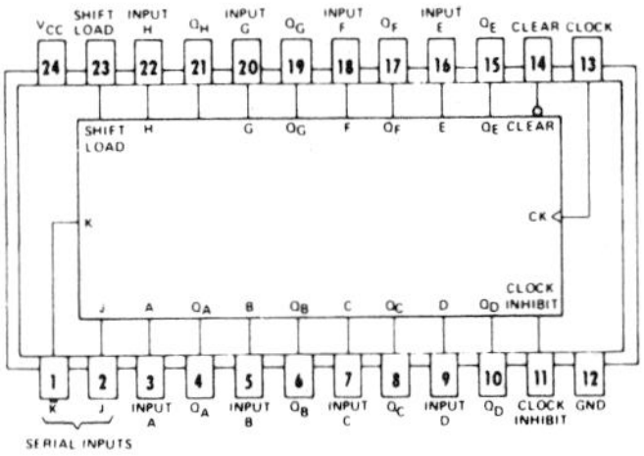
VCC
SHIFT LOAD
INPUT H
INPUT G
INPUT F
INPUT E
CLEAR
CLOCK
INPUT A
INPUT B
INPUT C
INPUT D
CLOCK INHIBIT
GND
SERIAL INPUTS

74LS200A
256-BIT RANDOM-ACCESS MEMORIES

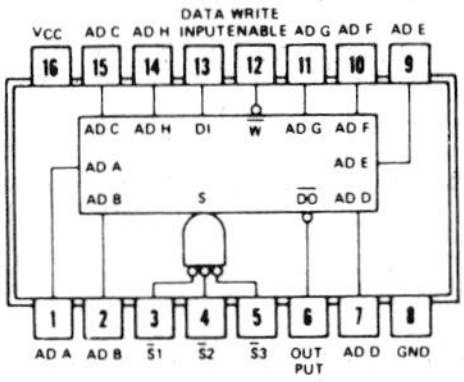
VCC
AD C
AD H
DATA INPUT
WRITE ENABLE
AD G
AD F
AD E
AD A
AD B
S1
S2
S3
OUT PUT
AD D
GND

74S201
256-BIT RANDOM-ACCESS MEMORIES

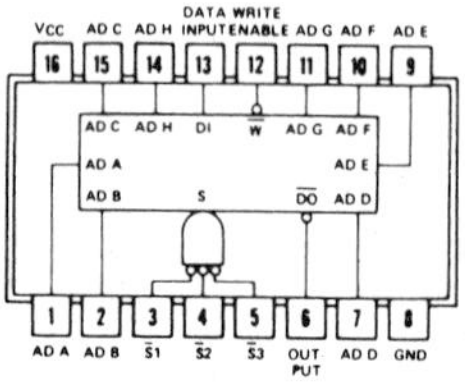
VCC AD C AD H DATA INPUT WRITE ENABLE AD G AD F AD E
AD A AD B S̄1 S̄2 S̄3 OUT PUT AD D GND

74LS202
256-BIT READ/WRITE MEMORIES WITH POWER DOWN

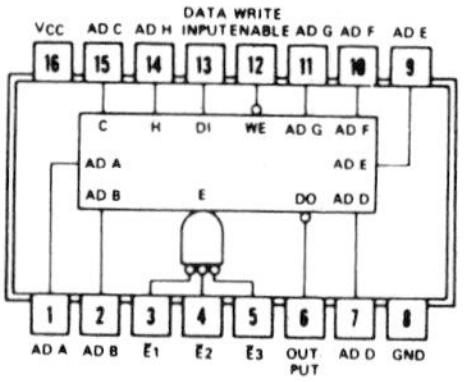
VCC AD C AD H DATA INPUT WRITE ENABLE AD G AD F AD E
AD A AD B Ē1 Ē2 Ē3 OUT PUT AD D GND

74LS207
RANDOM-ACCESS MEMORIES

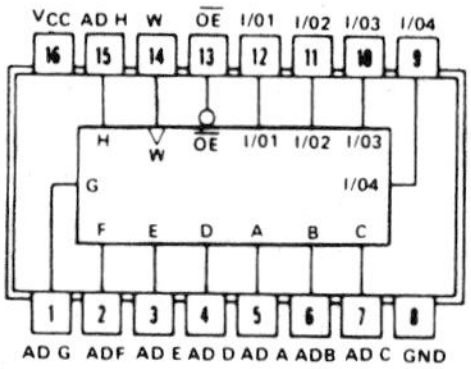
VCC AD H W OE I/01 I/02 I/03 I/04
AD G ADF AD E AD D AD A ADB AD C GND

74LS208
RANDOM-ACCESS MEMORIES

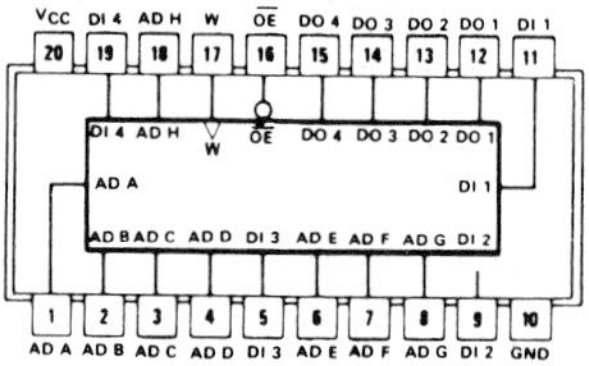
VCC DI 4 AD H W OE DO 4 DO 3 DO 2 DO 1 DI 1
AD A AD B AD C AD D DI 3 AD E AD F AD G DI 2 GND

74LS214 74LS215
RANDOM-ACCESS MEMORIES

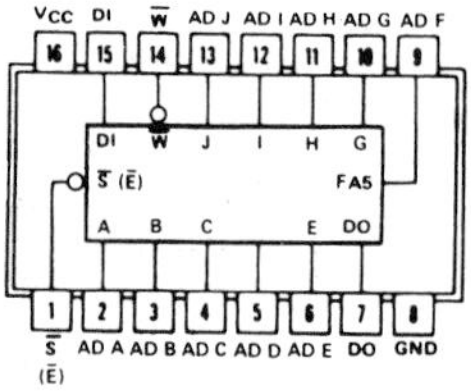
VCC DI W AD J AD I AD H AD G AD F
S (Ē) AD A AD B AD C AD D AD E DO GND

74221
DUAL MONOSTABLE MULTIVIBRATORS

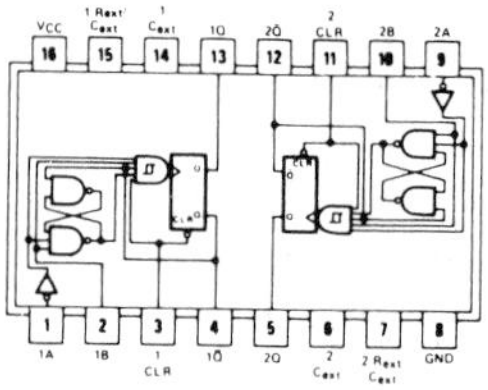

74S225
ASYNCHRONOUS FIRST IN, FIRST OUT MEMORIES

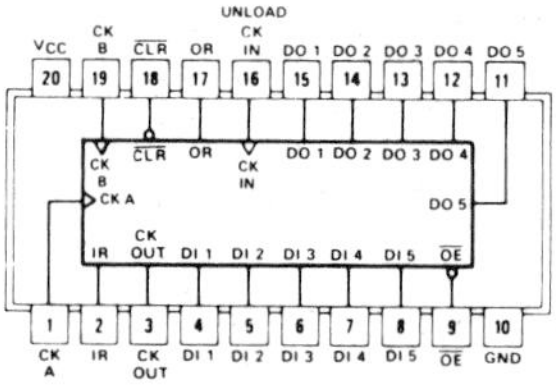
VCC CK B CLR OR UNLOAD CK IN DO 1 DO 2 DO 3 DO 4 DO 5
CK A IR CK OUT DI 1 DI 2 DI 3 DI 4 DI 5 OE GND

74S226
4-BIT PARALLEL LATCHED BUS TRANSCEIVERS

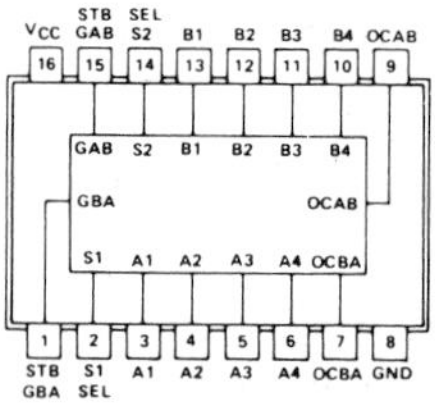
VCC STB GAB SEL S2 B1 B2 B3 B4 OCAB
STB GBA S1 SEL A1 A2 A3 A4 OCBA GND

74LS240
OCTAL BUFFERS/LINE DRIVERS/LINE RECEIVERS

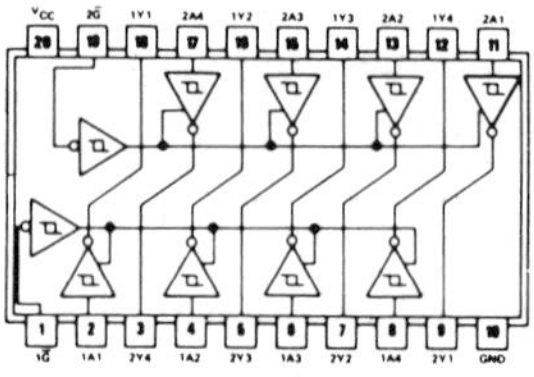

74LS241
OCTAL BUFFERS/LINE DRIVERS/LINE RECEIVERS

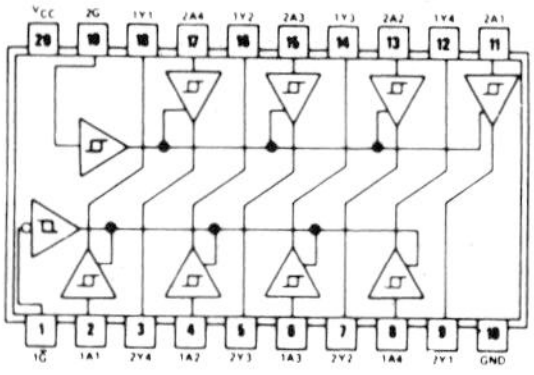

74LS242
QUADRUPLE BUS TRANSCEIVERS

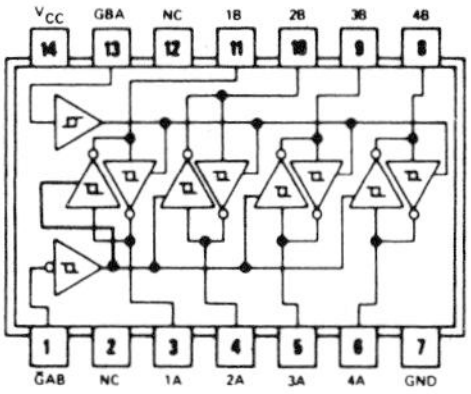
VCC GBA NC 1B 2B 3B 4B
GAB NC 1A 2A 3A 4A GND

74LS243
QUADRUPLE BUS TRANCEIVERS

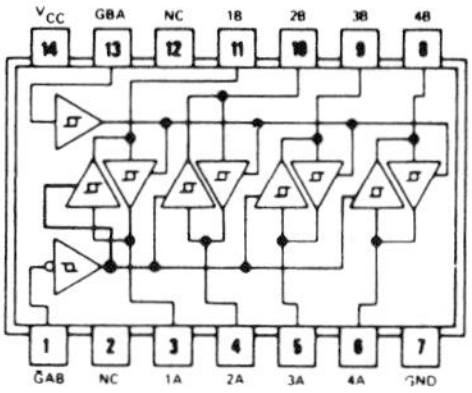
VCC GBA NC 1B 2B 3B 4B
GAB NC 1A 2A 3A 4A GND

74LS244
OCTAL BUFFERS/LINE DRIVERS/LINE RECEIVERS

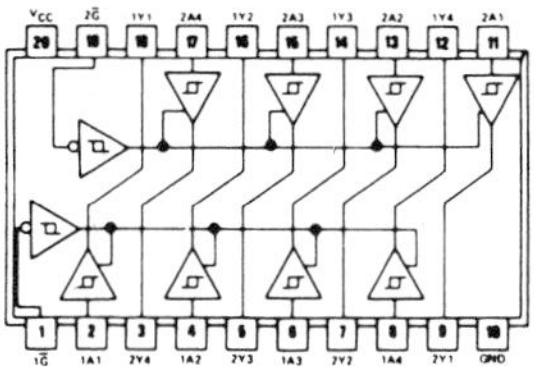

74LS245
OCTAL BUS TRANCEIVERS

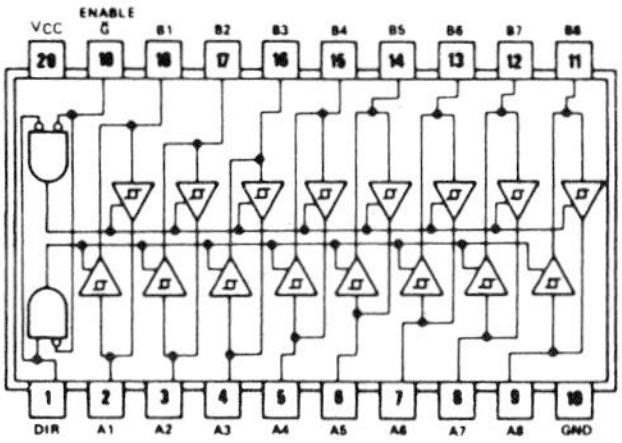
VCC
ENABLE
G
B1 B2 B3 B4 B5 B6 B7 B8
DIR A1 A2 A3 A4 A5 A6 A7 A8 GND

74246 74247
BCD-TO-SEVEN-SEGMENT DECODERS/DRIVERS

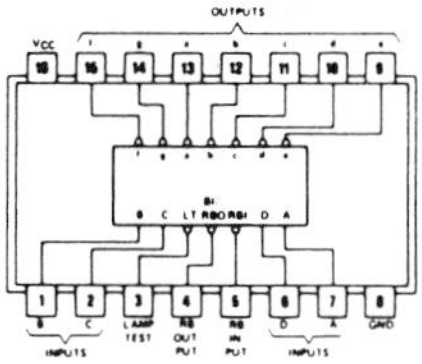

74248 74249
BCD-TO-SEVEN-SEGMENT DECODERS/DRIVERS

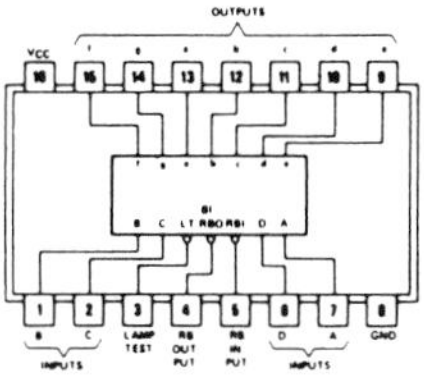

74251
DATA SELECTORS/ MULTIPLEXERS

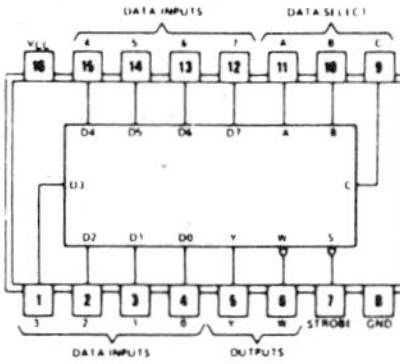

74LS253
DUAL DATA SELECTORS/MULTIPLEXERS

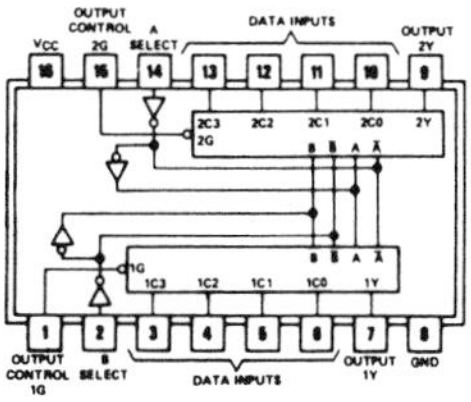
OUTPUT CONTROL 2G
A SELECT
DATA INPUTS
OUTPUT 2Y
VCC
OUTPUT CONTROL 1G
B SELECT
DATA INPUTS
OUTPUT 1Y
GND

74LS257A
QUAD DATA SELECTORS/MULTIPLEXERS

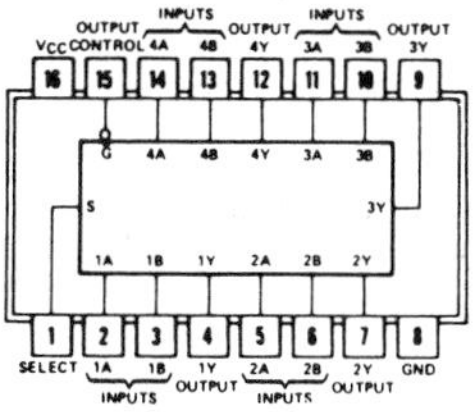
VCC
OUTPUT CONTROL
INPUTS
4A 4B
OUTPUT 4Y
INPUTS
3A 3B
OUTPUT 3Y
SELECT
1A 1B INPUTS
1Y OUTPUT
2A 2B INPUTS
2Y OUTPUT
GND

74LS258A
QUAD DATA SELECTORS/MULTIPLEXERS

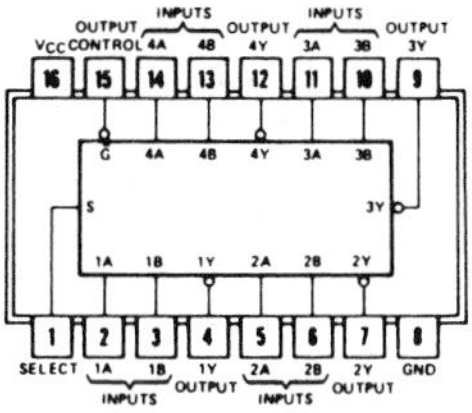
VCC
OUTPUT CONTROL
INPUTS
4A 4B
OUTPUT 4Y
INPUTS
3A 3B
OUTPUT 3Y
SELECT
1A 1B INPUTS
1Y OUTPUT
2A 2B INPUTS
2Y OUTPUT
GND

74259
EIGHT-BIT ADDRESSABLE LATCHES

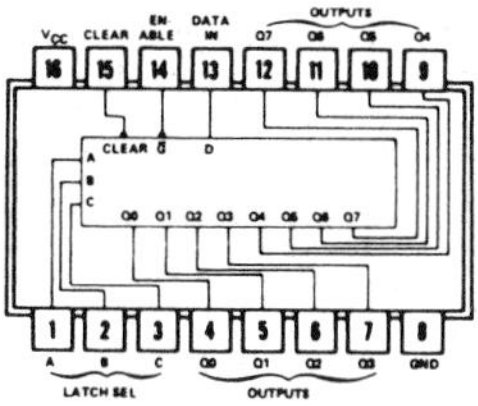
VCC CLEAR
EN ABLE
DATA IN
OUTPUTS
A B C
LATCH SEL
OUTPUTS
GND

74S260
DUAL 5-INPUT POSITIVE NOR GATES

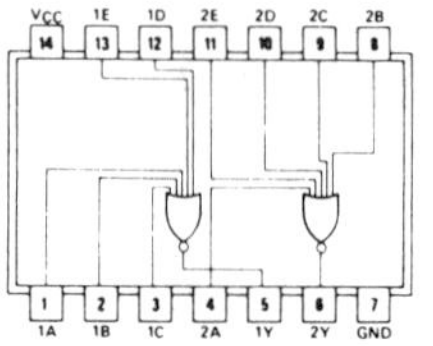
VCC 1E 1D 2E 2D 2C 2B
1A 1B 1C 2A 1Y 2Y GND

74LS261
2-BIT BY 4-BIT PARALLEL BINARY MULTIPLIERS

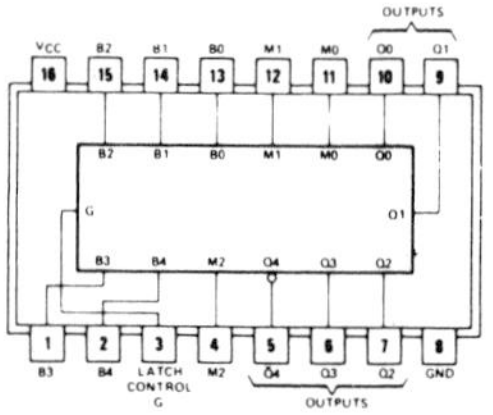
OUTPUTS
VCC B2 B1 B0 M1 M0 Q0 Q1
B3 B4 LATCH CONTROL G M2 Q4 Q3 Q2 GND
OUTPUTS

74265
QUAD COMPLEMENTARY-OUTPUT ELEMENTS

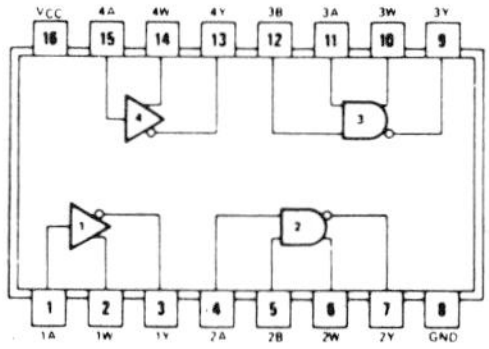
VCC 4A 4W 4Y 3B 3A 3W 3Y
1A 1W 1Y 2A 2B 2W 2Y GND

74LS266
QUAD 2-INPUT EXCLUSIVE-NOR GATES WITH OPEN-COLLECTOR OUTPUTS

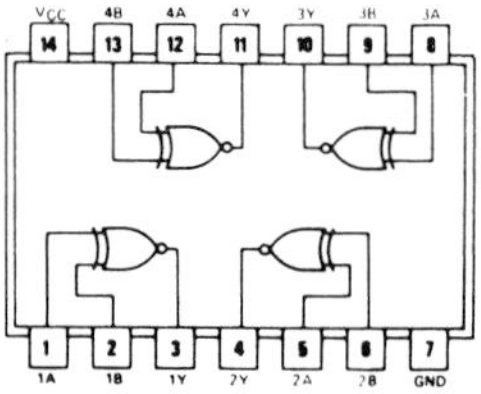
VCC 4B 4A 4Y 3Y 3B 3A
1A 1B 1Y 2Y 2A 2B GND

74S270
2048-BIT READ-ONLY MEMORIES

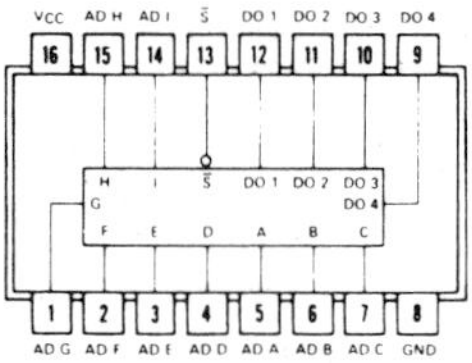
VCC AD H AD I S̄ DO 1 DO 2 DO 3 DO 4
AD G AD F AD E AD D AD A AD B AD C GND

74S271
2048-BIT READ-ONLY MEMORIES

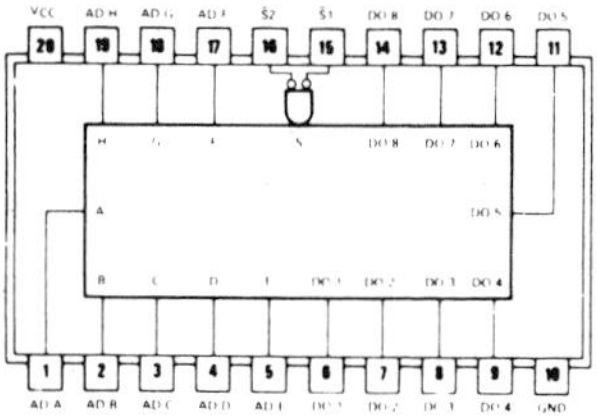

74273
OCTAL D-TYPE FLIP-FLOPS

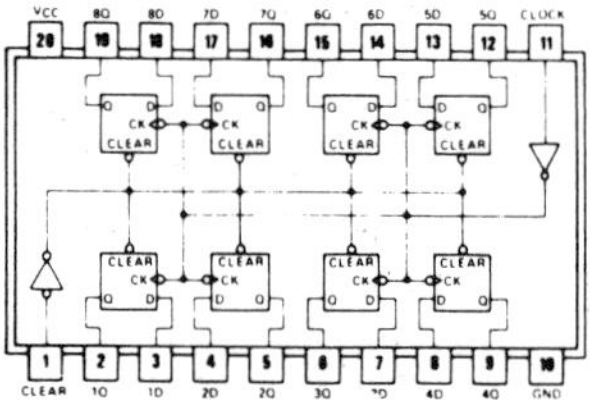
VCC 8Q 8D 7D 7Q 6Q 6D 5D 5Q CLOCK
CLEAR 1Q 1D 2D 2Q 3Q 3D 4D 4Q GND

74S274
4-BIT BY 4-BIT BINARY MULTIPLIERS

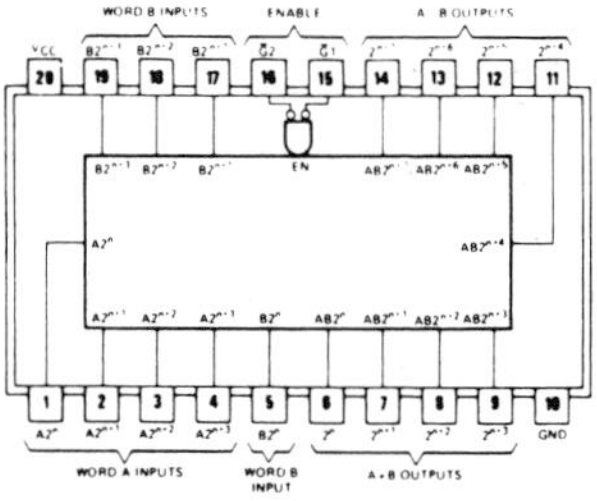
WORD B INPUTS
ENABLE
A × B OUTPUTS
WORD A INPUTS
WORD B INPUT
A × B OUTPUTS
GND

74LS275
7-BIT SLICE WALLACE TREES

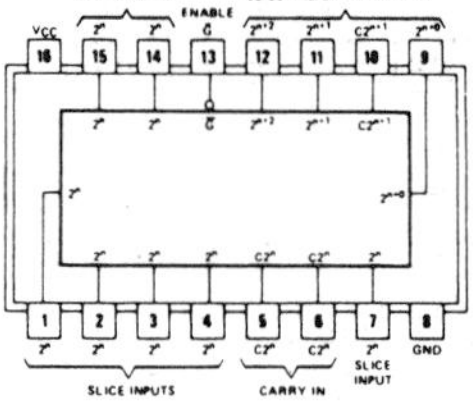
SLICE INPUTS
ENABLE
SLICE AND CARRY OUTPUTS
VCC
GND
SLICE INPUT
SLICE INPUTS
CARRY IN

74276
QUAD J-K̄ FLIP-FLOPS

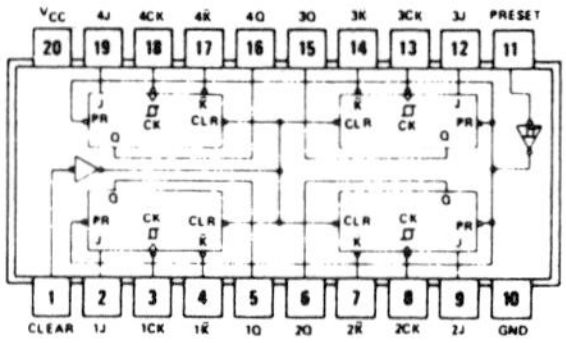
VCC 4J 4CK 4K̄ 4Q 3Q 3K̄ 3CK 3J PRESET
CLEAR 1J 1CK 1K̄ 1Q 2Q 2K̄ 2CK 2J GND

4-BIT CASCADEABLE PRIORITY REGISTERS

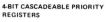

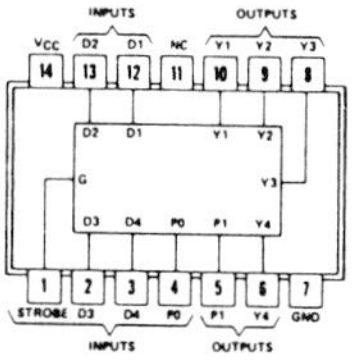
INPUTS
OUTPUTS
VCC D2 D1 NC Y1 Y2 Y3
STROBE D3 D4 P0 P1 Y4 GND
INPUTS
OUTPUTS

74279
QUAD $\overline{S}$-$\overline{R}$ LATCHES

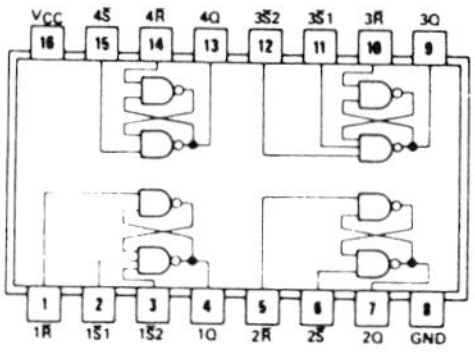
VCC 4S 4R 4Q 3S2 3S1 3R 3Q
1R 1S1 1S2 1Q 2R 2S 2Q GND

74LS280
9-BIT ODD/EVEN PARITY GENERATORS/CHECKERS

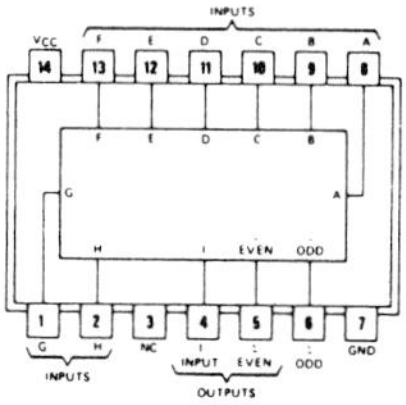
INPUTS
VCC F E D C B A
G H NC I EVEN ODD GND
INPUTS
OUTPUTS

74S281
4-BIT PARALLEL BINARY ACCUMULATORS

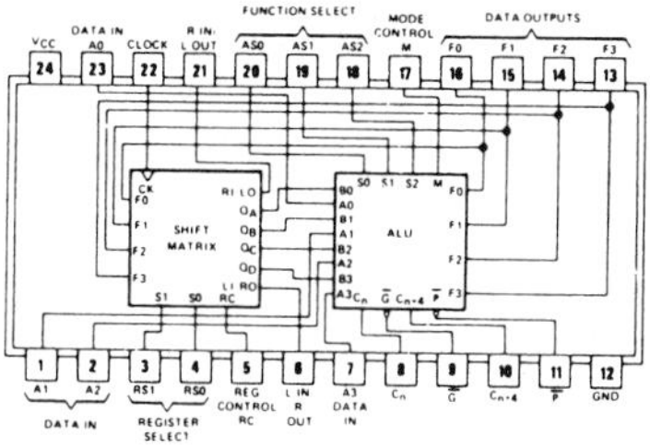
FUNCTION SELECT
MODE CONTROL
DATA OUTPUTS
VCC
DATA IN A0
CLOCK
R IN L OUT
AS0 AS1 AS2 M F0 F1 F2 F3
SHIFT MATRIX
ALU
A1 A2 RS1 RS0
REG CONTROL RC
L IN R OUT
A3 DATA IN
Cn G Cn+4 P GND
DATA IN
REGISTER SELECT

74283
4-BIT BINARY FULL ADDERS

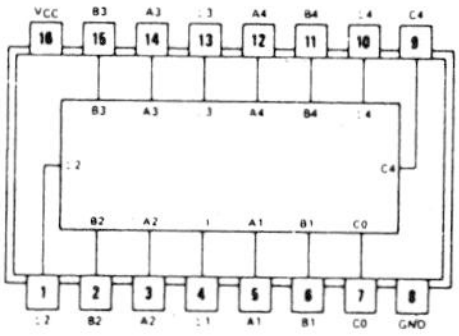
VCC B3 A3 Σ3 A4 B4 Σ4 C4
Σ2 B2 A2 Σ1 A1 B1 C0 GND

74284
4-BIT-BY-4-BIT PARALLEL BINARY MULTIPLIERS USED WITH '285

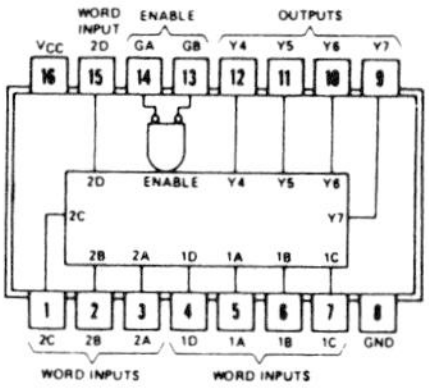
WORD INPUT
ENABLE
OUTPUTS
VCC 2D GA GB Y4 Y5 Y6 Y7
2C 2B 2A 1D 1A 1B 1C GND
WORD INPUTS
WORD INPUTS

74285
4-BIT-BY-4-BIT PARALLEL BINARY MULTIPLIERS USED WITH '284

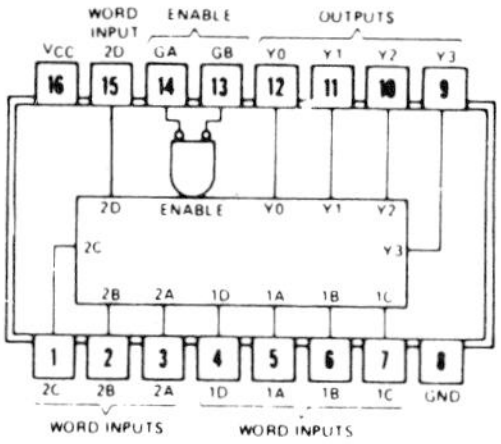
WORD INPUT
ENABLE
OUTPUTS
VCC 2D GA GB Y0 Y1 Y2 Y3
2C 2B 2A 1D 1A 1B 1C GND
WORD INPUTS
WORD INPUTS

74S287
1024-BIT PROGRAMMABLE READ-ONLY MEMORIES

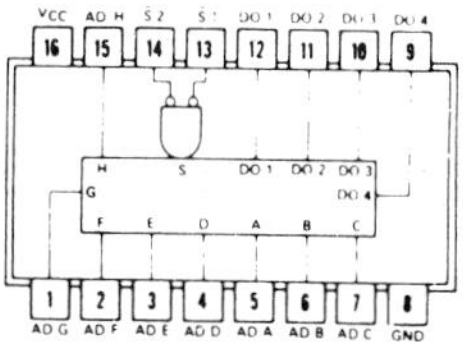
VCC AD H S2 S1 DO 1 DO 2 DO 3 DO 4
AD G AD F AD E AD D AD A AD B AD C GND

74S288
256-BIT PROGRAMMABLE READ-ONLY MEMORIES

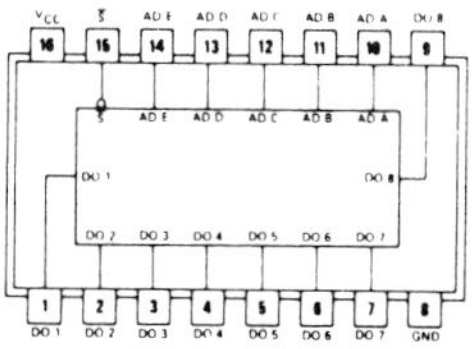
VCC S AD E AD D AD C AD B AD A DO 8
DO 1 DO 2 DO 3 DO 4 DO 5 DO 6 DO 7 GND

74S289
64-BIT RANDOM-ACCESS MEMORIES

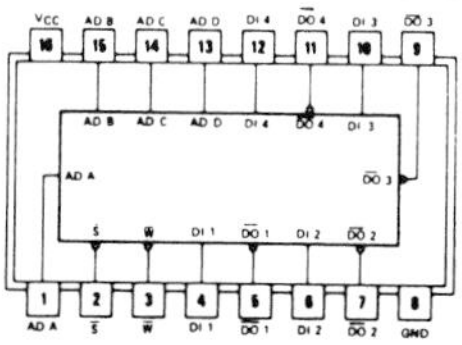
VCC AD B AD C AD D DI 4 DO 4 DI 3 DO 3
AD A S W DI 1 DO 1 DI 2 DO 2 GND

74290
DECADE COUNTERS

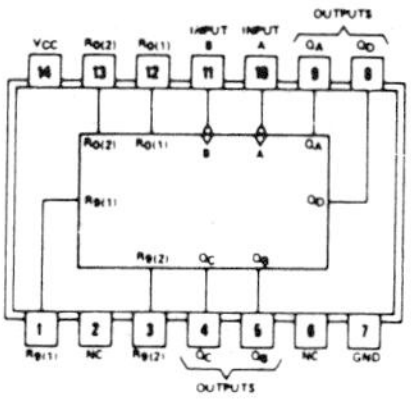
OUTPUTS
VCC R0(2) R0(1) INPUT B INPUT A QA QD
R9(1) NC R9(2) QC QB NC GND
OUTPUTS

74293
4-BIT BINARY COUNTERS

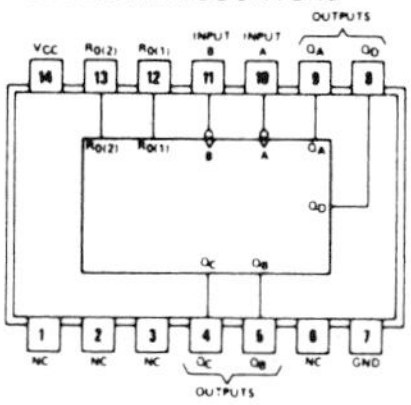
OUTPUTS
VCC R0(2) R0(1) INPUT B INPUT A QA QD
NC NC NC QC QB NC GND
OUTPUTS

74LS295B
4-BIT BIDIRECTIONAL UNIVERSAL SHIFT REGISTERS

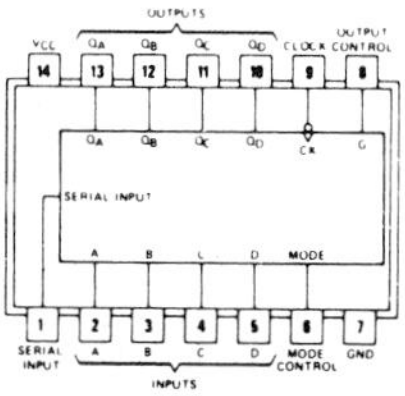
OUTPUTS
VCC QA QB QC QD CLOCK OUTPUT CONTROL
SERIAL INPUT
SERIAL INPUT A B C D MODE CONTROL GND
INPUTS

74298
QUAD 2-INPUT MULTIPLEXERS WITH STORAGE

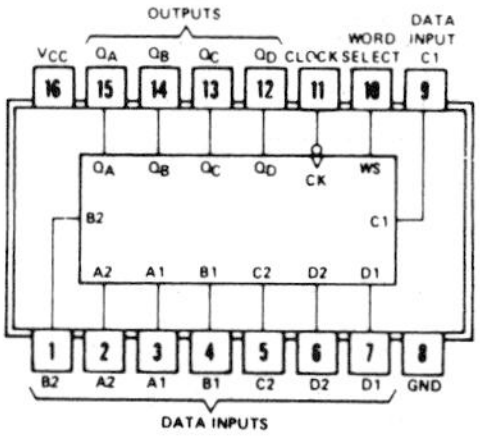
OUTPUTS
VCC QA QB QC QD CLOCK WORD SELECT DATA INPUT C1
B2 A2 A1 B1 C2 D2 D1 GND
DATA INPUTS

74LS299
8-BIT BIDIRECTIONAL UNIVERSAL SHIFT/STORAGE REGISTERS

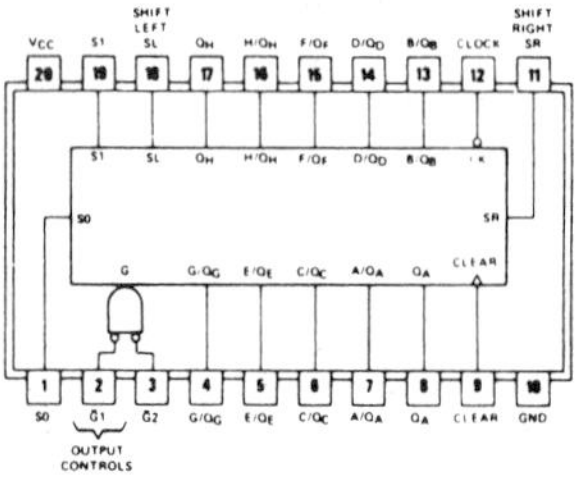
VCC
S1
SHIFT LEFT SL
Q_H
H/Q_H
F/Q_F
D/Q_D
B/Q_B
CLOCK
SHIFT RIGHT SR
S0
$\bar{G}1$
$\bar{G}2$
G/Q_G
E/Q_E
C/Q_C
A/Q_A
Q_A
CLEAR
GND
OUTPUT CONTROLS

74LS300A
256-BIT READ/WRITE MEMORIES

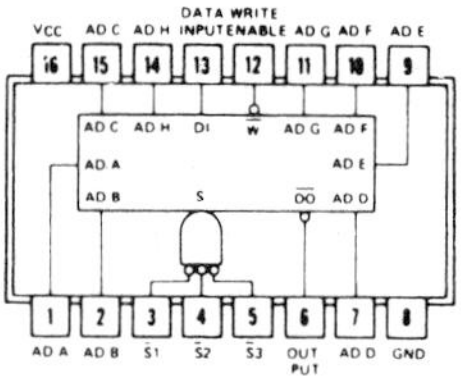
VCC
AD C
AD H
DATA INPUT
WRITE ENABLE
AD G
AD F
AD E
AD A
AD B
$\bar{S}1$
$\bar{S}2$
$\bar{S}3$
OUT PUT
AD D
GND

74S301
256-BIT RANDOM ACCESS MEMORIES

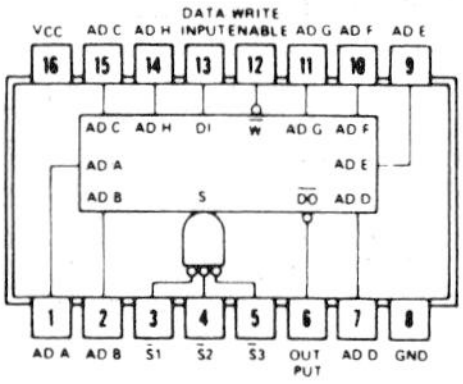
VCC
AD C
AD H
DATA INPUT
WRITE ENABLE
AD G
AD F
AD E
AD A
AD B
$\bar{S}1$
$\bar{S}2$
$\bar{S}3$
OUT PUT
AD D
GND

74LS302
256-BIT READ/WRITE MEMORIES

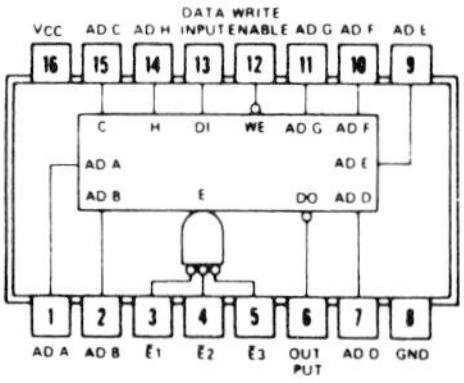
VCC
AD C
AD H
DATA INPUT
WRITE ENABLE
AD G
AD F
AD E
AD A
AD B
$\bar{E}1$
$\bar{E}2$
$\bar{E}3$
OUT PUT
AD D
GND

74LS314 74LS315
1024-BIT RANDOM-ACCESS MEMORIES

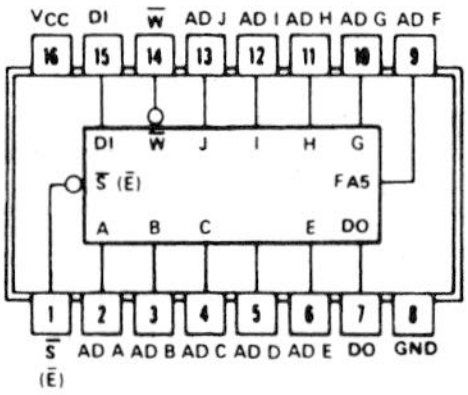
VCC
DI
$\bar{W}$
AD J
AD I
AD H
AD G
AD F
$\bar{S}$ ($\bar{E}$)
AD A
AD B
AD C
AD D
AD E
DO
GND

74LS323
8-BIT BIDIRECTIONAL UNIVERSAL SHIFT/STORAGE REGISTERS

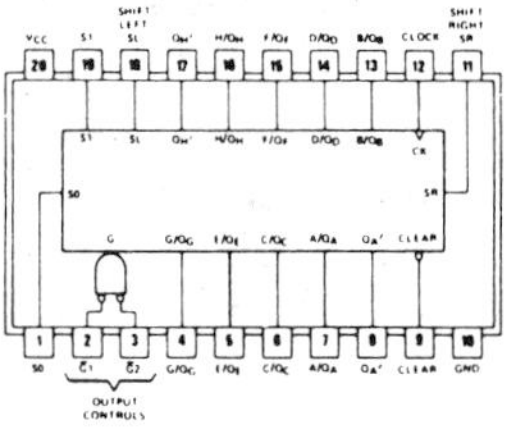

74LS324
VOLTAGE-CONTROLLED OSCILLATORS

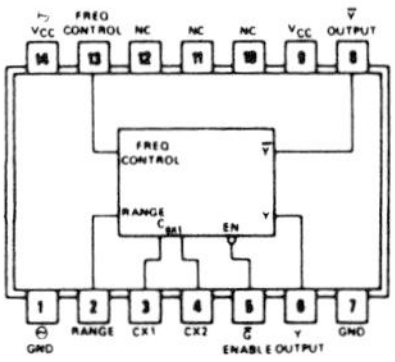
FREQ CONTROL
NC
OUTPUT
RANGE
CX1
CX2
ENABLE
OUTPUT
GND

74LS325
DUAL VOLTAGE-CONTROLLED OSCILLATORS

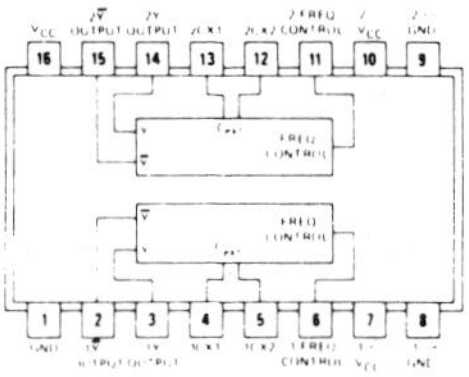

74LS326
DUAL VOLTAGE-CONTROLLED OSCILLATORS

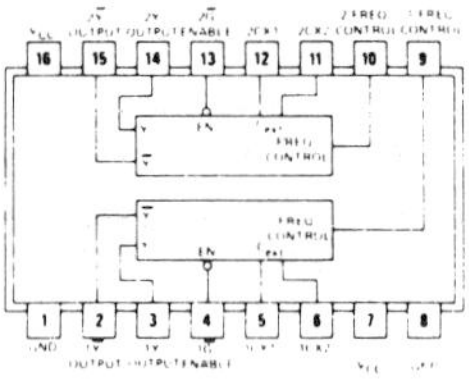

74LS327
DUAL VOLTAGE-CONTROLLED OSCILLATORS

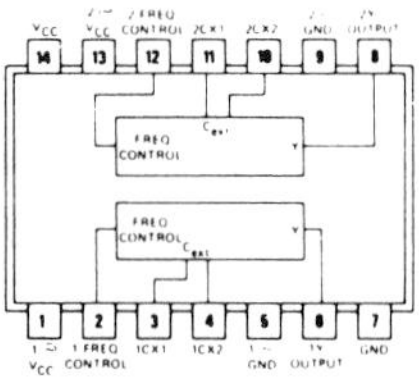

74LS348
8-LINE-TO-3-LINE PRIORITY ENCODERS

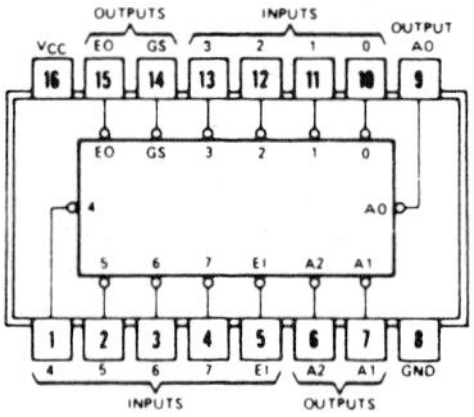
OUTPUTS
INPUTS
OUTPUT
VCC
EO
GS
A0
E1
A2
A1
GND

74351
DUAL 8-LINE-TO-1-LINE DATA SELECTOR/MULTIPLEXER

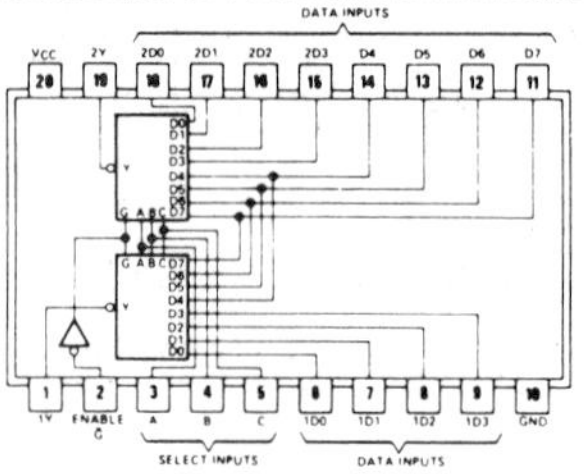
DATA INPUTS
VCC
2Y
1Y
ENABLE $\bar{G}$
SELECT INPUTS
GND

74LS352
DUAL 4-LINE-TO-LINE DATA SELECTORS/MULTIPLEXERS

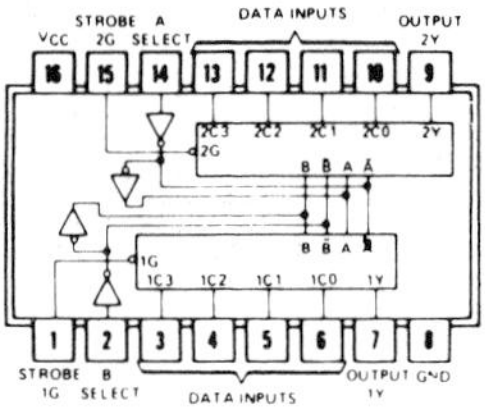
STROBE 2G
A SELECT
DATA INPUTS
OUTPUT 2Y
VCC
STROBE 1G
B SELECT
OUTPUT 1Y
GND

74LS353
DUAL 4-LINE-TO-1-LINE DATA SELECTORS/MULTIPLEXERS

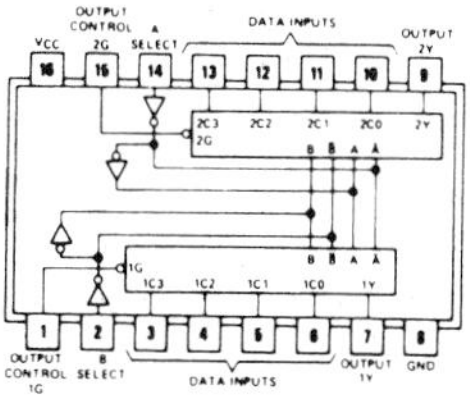
VCC
OUTPUT CONTROL 2G
A SELECT
DATA INPUTS
OUTPUT 2Y
OUTPUT CONTROL 1G
B SELECT
DATA INPUTS
OUTPUT 1Y
GND

74LS362
FOUR-PHASE CLOCK GENERATOR/DRIVER FOR TMS 9900 MICROPROCESSOR

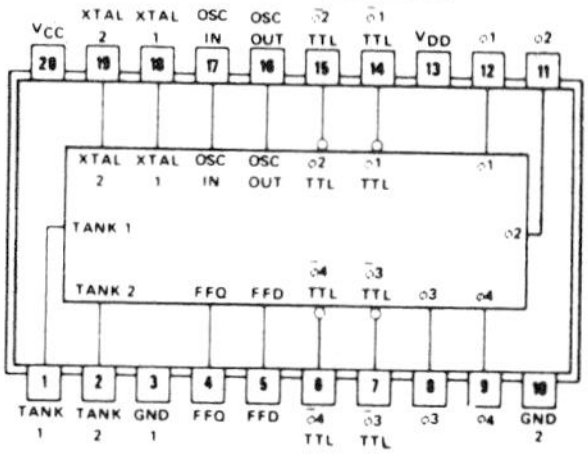
VCC
XTAL 2
XTAL 1
OSC IN
OSC OUT
VDD
TANK 1
TANK 2
FFQ
FFD
TANK 1
TANK 2
GND 1
FFQ
FFD
GND 2

74LS363
OCTAL D-TYPE LATCHES

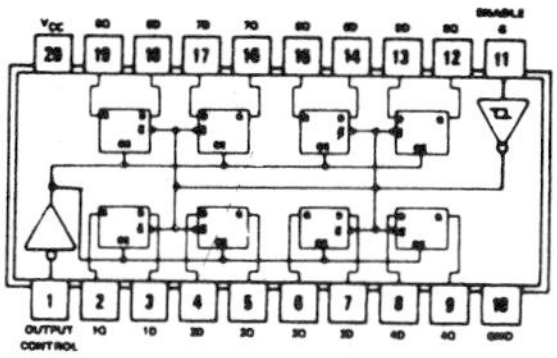
OUTPUT CONTROL
GND

74LS364
OCTAL D-TYPE FLIP-FLOPS

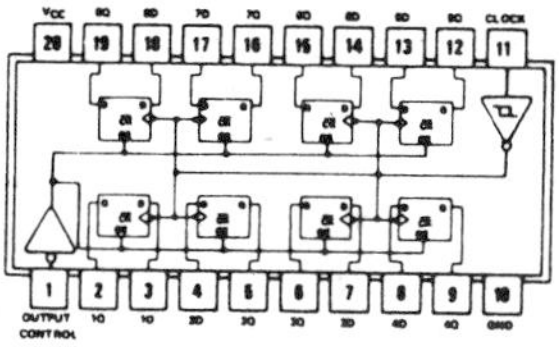
CLOCK
OUTPUT CONTROL
GND

74365A
HEX BUS DRIVERS

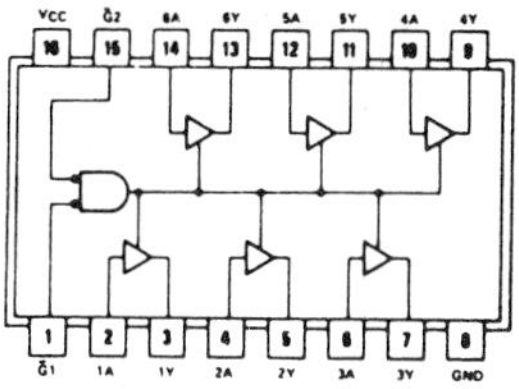
VCC
Ḡ2
6A
6Y
5A
5Y
4A
4Y
Ḡ1
1A
1Y
2A
2Y
3A
3Y
GND

74366A
HEX BUS DRIVERS

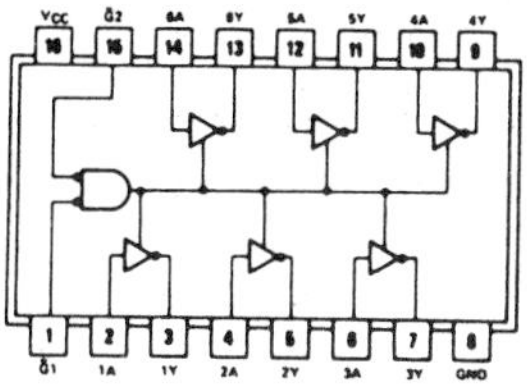
VCC
Ḡ2
6A
6Y
5A
5Y
4A
4Y
Ḡ1
1A
1Y
2A
2Y
3A
3Y
GND

74367A
HEX BUS DRIVERS

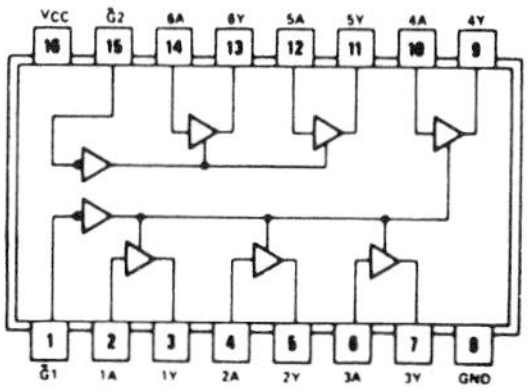
VCC
Ḡ2
6A
6Y
5A
5Y
4A
4Y
Ḡ1
1A
1Y
2A
2Y
3A
3Y
GND

74368A
HEX BUS DRIVERS

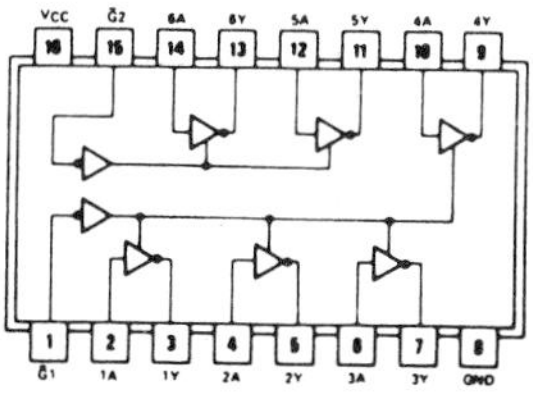
VCC
Ḡ2
6A
6Y
5A
5Y
4A
4Y
Ḡ1
1A
1Y
2A
2Y
3A
3Y
GND

74S370
2048-BIT READ-ONLY MEMORIES

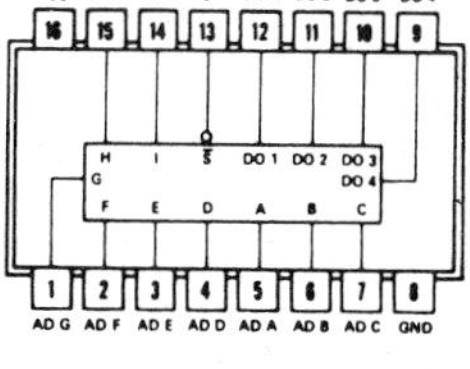
VCC
AD H
AD I
S̄
DO 1
DO 2
DO 3
DO 4
AD G
AD F
AD E
AD D
AD A
AD B
AD C
GND

74S371
2048-BIT READ-ONLY MEMORIES

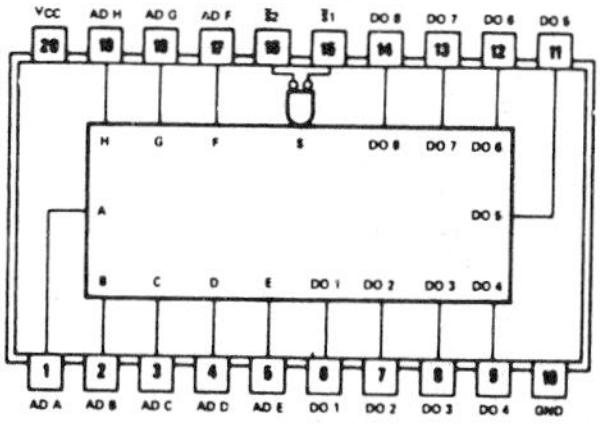
VCC
AD H
AD G
AD F
DO 8
DO 7
DO 6
DO 5
AD A
AD B
AD C
AD D
AD E
DO 1
DO 2
DO 3
DO 4
GND

74LS373
OCTAL D-TYPE LATCHES

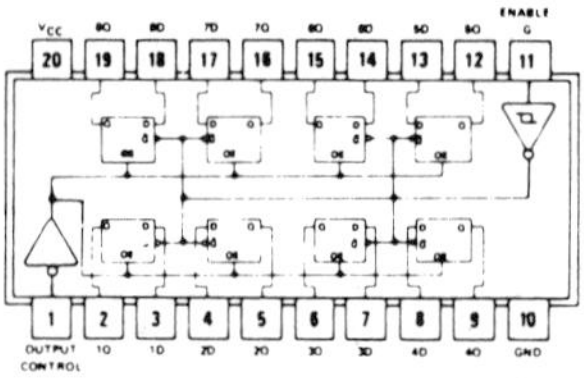

74LS374
OCTAL D-TYPE FLIP-FLOPS

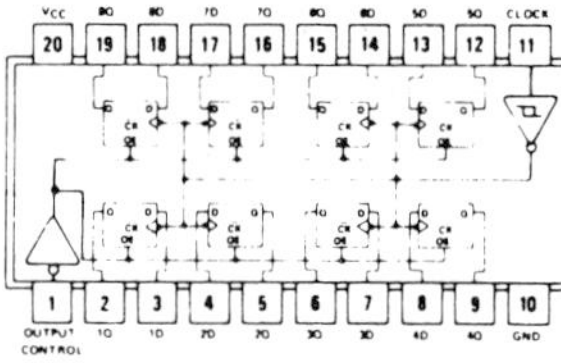

74LS375
4-BIT BISTABLE LATCHES

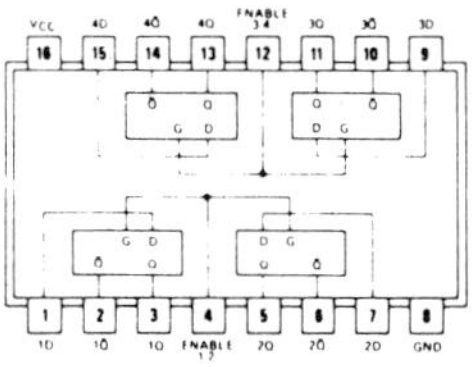

74376
QUAD J-K̄ FLIP-FLOPS

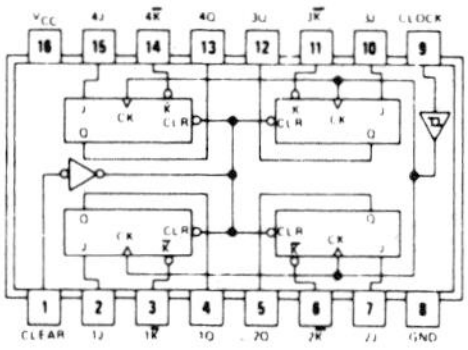

74LS377
OCTAL D-TYPE FLIP-FLOPS

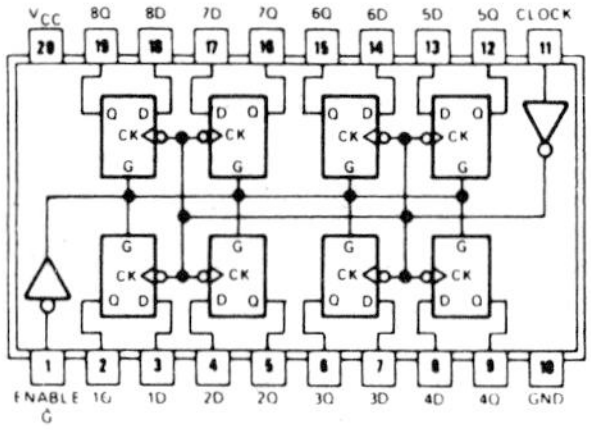
VCC 8Q 8D 7D 7Q 6Q 6D 5D 5Q CLOCK
ENABLE G 1Q 1D 2D 2Q 3Q 3D 4D 4Q GND

74LS378
HEX D-TYPE FLIP-FLOPS

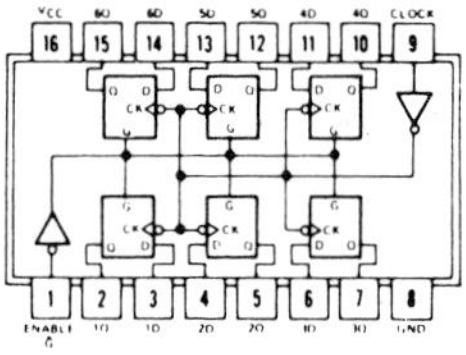

74LS379
QUAD D-TYPE FLIP-FLOPS

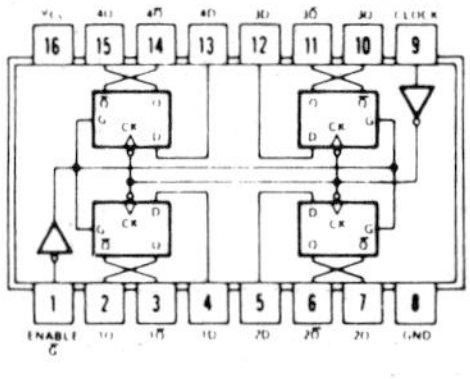

74S381
ARITHMETIC LOGIC UNITS/FUNCTION GENERATORS

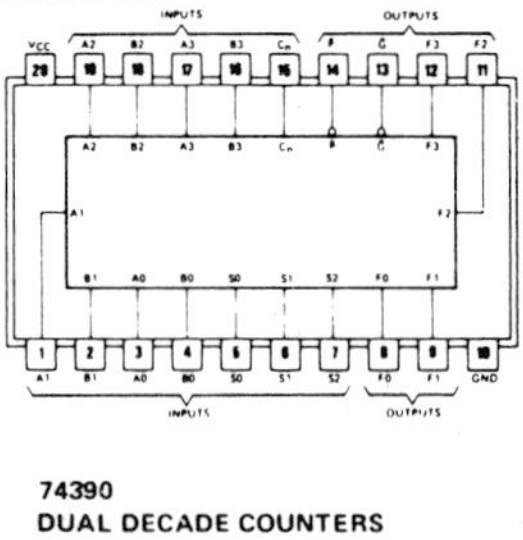

74LS386
QUAD 2-INPUT EXCLUSIVE-OR GATES

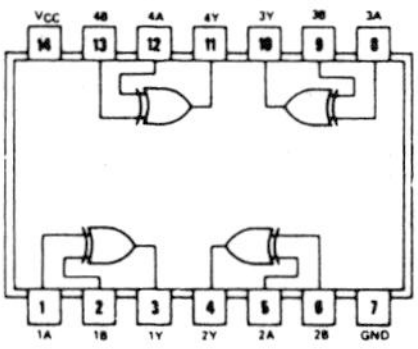
VCC 4B 4A 4Y 3Y 3B 3A
1A 1B 1Y 2Y 2A 2B GND

74S387
1024-BIT PROGRAMMABLE READ-ONLY MEMORIES

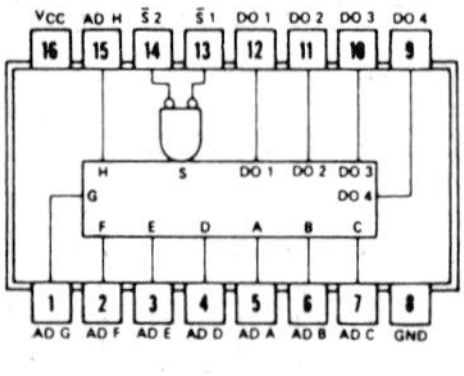
VCC AD H S̄2 S̄1 DO 1 DO 2 DO 3 DO 4
AD G AD F AD E AD D AD A AD B AD C GND

74390
DUAL DECADE COUNTERS

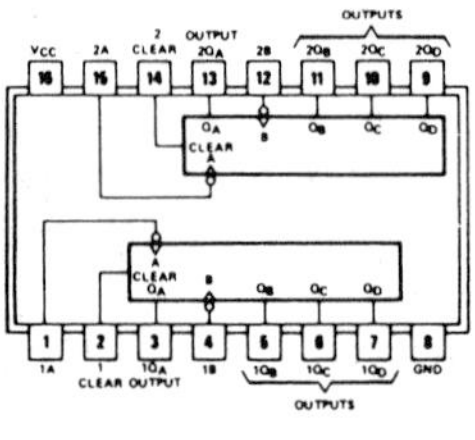

74393
DUAL 4-BIT BINARY COUNTERS

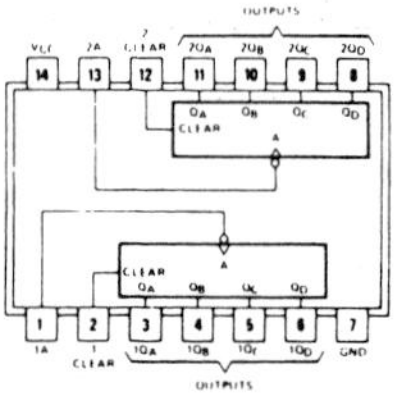

74LS395A
4-BIT UNIVERSAL SHIFT REGISTERS

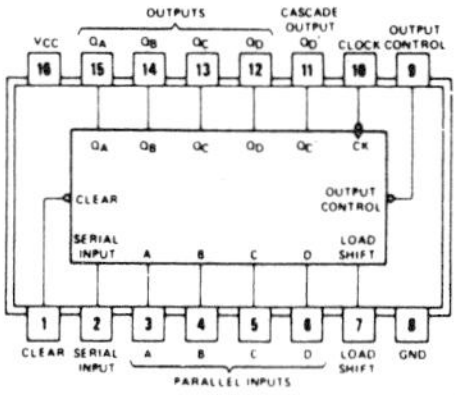
OUTPUTS
CASCADE OUTPUT
OUTPUT CONTROL
VCC
CLOCK
CLEAR
SERIAL INPUT
LOAD SHIFT
GND
PARALLEL INPUTS

74LS398
QUAD 2-INPUT MULTIPLEXERS WITH STORAGE

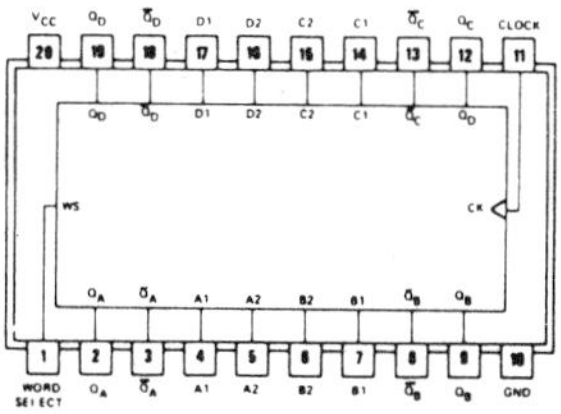
VCC
CLOCK
WS
CK
WORD SELECT
GND

74LS399
QUAD 2-INPUT MULTIPLEXERS WITH STORAGE

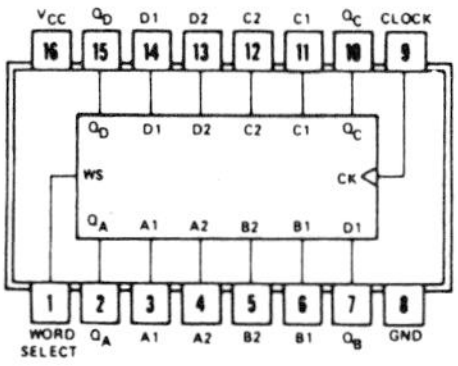
VCC
D1
D2
C2
C1
CLOCK
WS
CK
A1
A2
B2
B1
WORD SELECT
GND

74S412
MULTI-MODE BUFFERED 8-BIT LATCHES

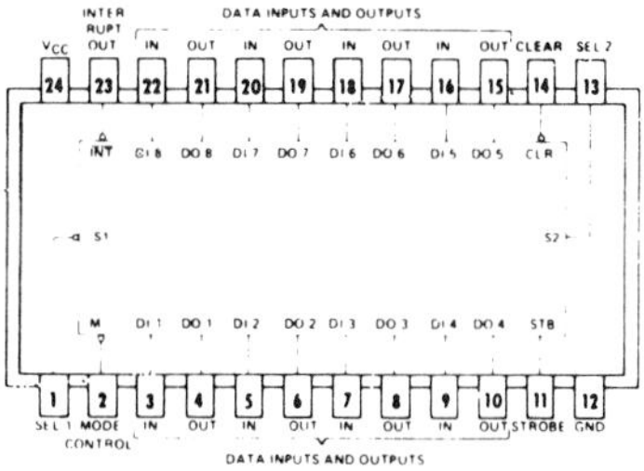
DATA INPUTS AND OUTPUTS
CLEAR
MODE CONTROL
STROBE
GND

74LS424
TWO-PHASE CLOCK GENERATOR/ DRIVER FOR 8080A

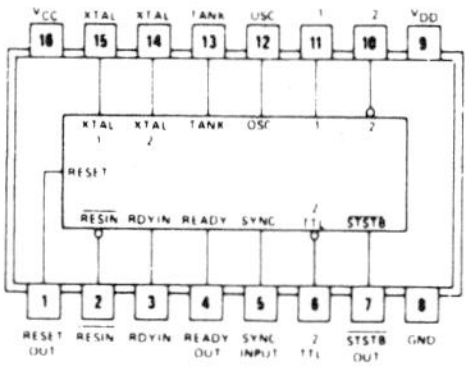

74425
QUAD GATES

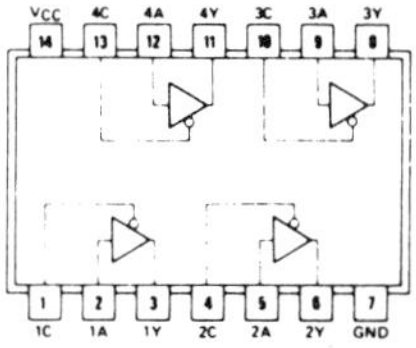
VCC
4C
4A
4Y
3C
3A
3Y
1C
1A
1Y
2C
2A
2Y
GND

74426
QUAD GATES

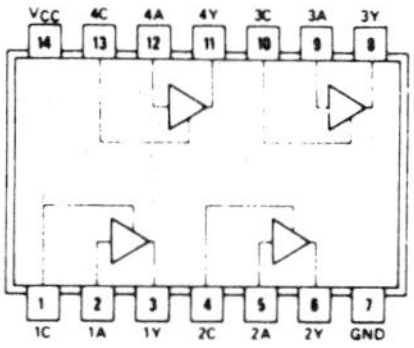
VCC
4C
4A
4Y
3C
3A
3Y
1C
1A
1Y
2C
2A
2Y
GND

74S428 74S438
SYSTEM CONTROLLER FOR 8080A

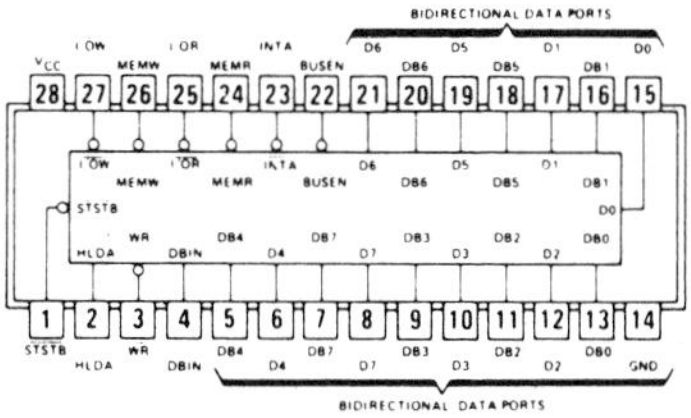
BIDIRECTIONAL DATA PORTS
MEMW
MEMR
INTA
BUSEN
STSTB
WR
HLDA
DBIN
GND

74S470 74S471
PROGRAMMABLE READ-ONLY MEMORIES

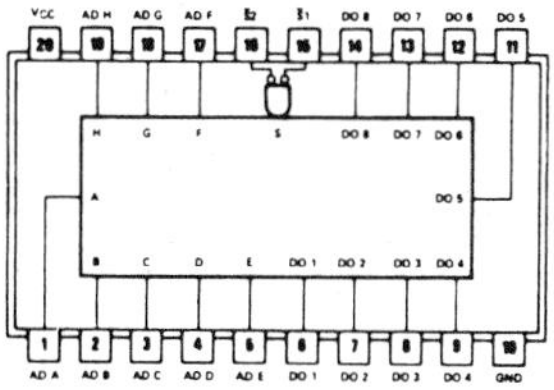

74S472 74S473
PROGRAMMABLE READ-ONLY MEMORIES

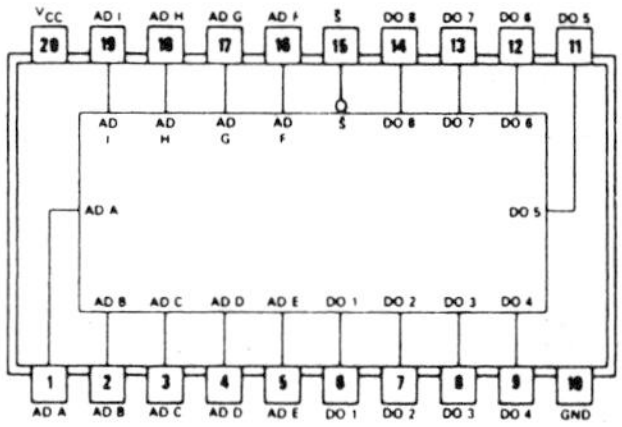
VCC
AD A
AD B
AD C
AD D
AD E
DO 1
DO 2
DO 3
DO 4
GND

74S474 74S475
PROGRAMMABLE READ-ONLY MEMORIES

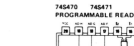

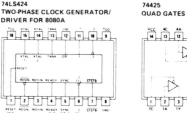

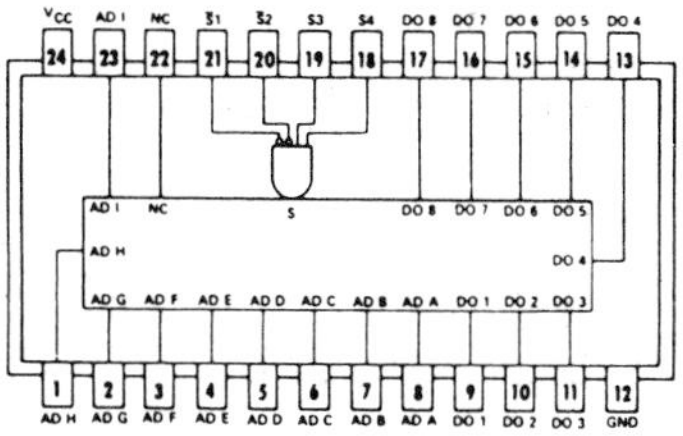
VCC
NC
AD H
AD G
AD F
AD E
AD D
AD C
AD B
AD A
DO 1
DO 2
DO 3
GND

74S481
4-BIT SLICE PROCESSOR ELEMENTS

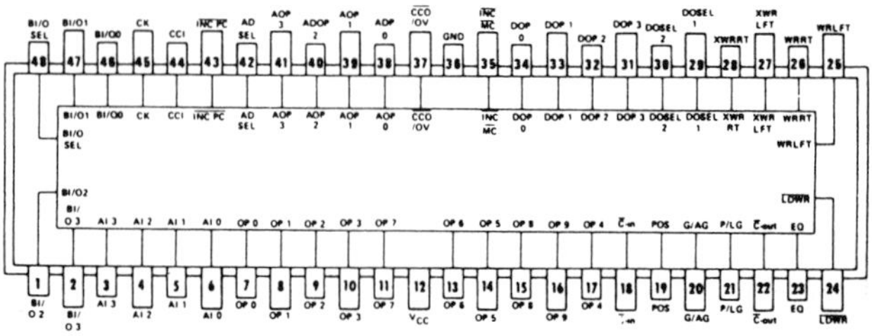
BI/O SEL 48
BI/O1 47
BI/O0 46
CK 45
CCI 44
INC PC 43
AD SEL 42
AOP 3 41
ADOP 2 40
AOP 1 39
ADP 0 38
CCO /OV 37
GND 36
INC MC 35
DOP 0 34
DOP 1 33
DOP 2 32
DOP 3 31
DOSEL 2 30
DOSEL 1 29
XWRRT 28
XWR LFT 27
WRRT 26
WRLFT 25
1 BI/O 2
2 BI/O 3
3 AI 3
4 AI 2
5 AI 1
6 AI 0
7 OP 0
8 OP 1
9 OP 2
10 OP 3
11 OP 7
12 VCC
13 OP 6
14 OP 5
15 OP 8
16 OP 9
17 OP 4
18 C-in
19 POS
20 G/AG
21 P/LG
22 C-out
23 EQ
24 LDWR

74S482
4-BIT-SLICE EXPANDABLE CONTROL ELEMENTS

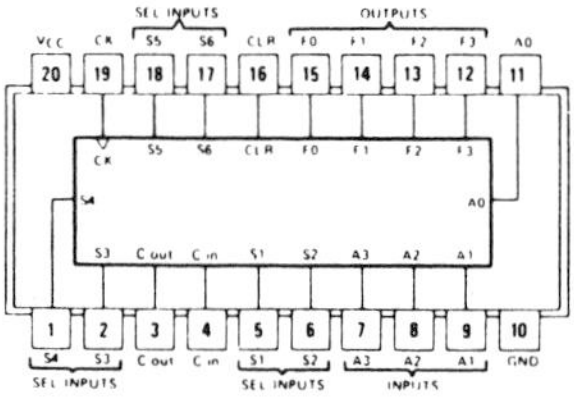
SEL INPUTS
OUTPUTS
VCC 20
CK 19
S5 18
S6 17
CLR 16
F0 15
F1 14
F2 13
F3 12
A0 11
1 S4
2 S3
3 C out
4 C in
5 S1
6 S2
7 A3
8 A2
9 A1
10 GND
SEL INPUTS
SEL INPUTS
INPUTS

74490
DUAL DECADE COUNTERS

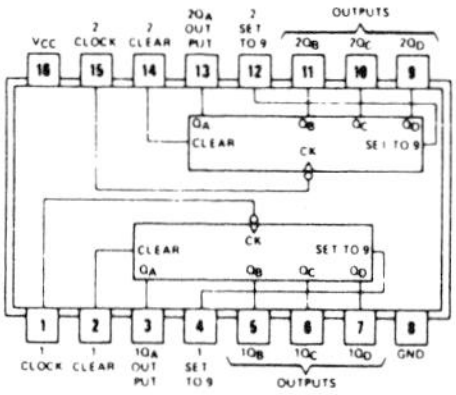
VCC 16
2 CLOCK 15
2 CLEAR 14
2QA OUTPUT 13
2 SET TO 9 12
OUTPUTS
2QB 11
2QC 10
2QD 9
1 1 CLOCK
2 1 CLEAR
3 1QA OUTPUT
4 1 SET TO 9
5 1QB
6 1QC
7 1QD
8 GND
OUTPUTS

74LS670
4-BY-4 REGISTER FILES

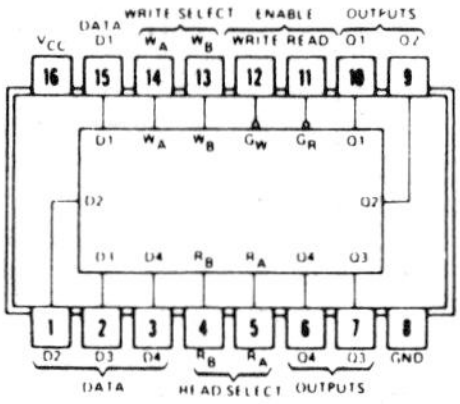
DATA
WRITE SELECT
ENABLE
OUTPUTS
VCC 16
D1 15
WA 14
WB 13
WRITE 12
READ 11
Q1 10
Q2 9
1 D2
2 D3
3 D4
4 RB
5 RA
6 Q4
7 Q3
8 GND
DATA
READ SELECT
OUTPUTS

Table 5.5. CMOS PINOUTS

CD4000
Dual 3-Input NOR Gate Plus Inverter

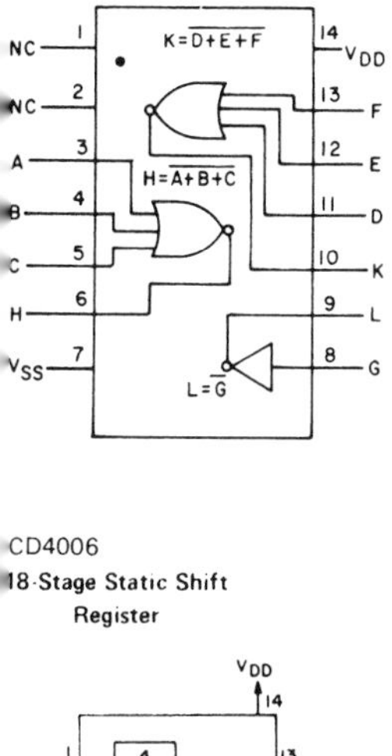
K = $\overline{D+E+F}$
H = $\overline{A+B+C}$
L = $\overline{G}$
NC 1
NC 2
A 3
B 4
C 5
H 6
VSS 7
14 VDD
13 F
12 E
11 D
10 K
9 L
8 G

CD4001
Quad 2-Input NOR Gate

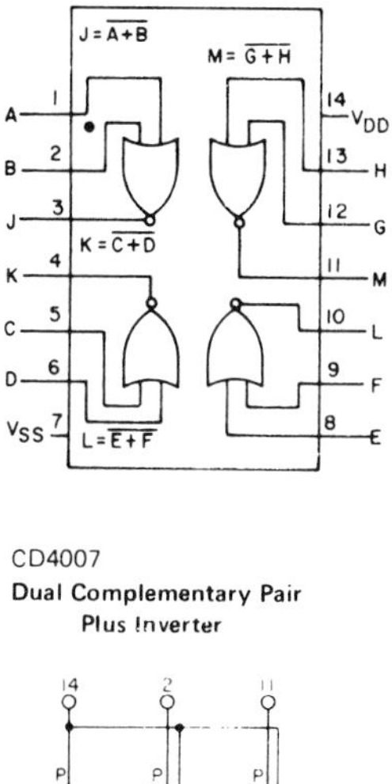
J = $\overline{A+B}$
M = $\overline{G+H}$
K = $\overline{C+D}$
L = $\overline{E+F}$
A 1
B 2
J 3
K 4
C 5
D 6
VSS 7
14 VDD
13 H
12 G
11 M
10 L
9 F
8 E

CD4002
Quad 4-Input NOR Gate

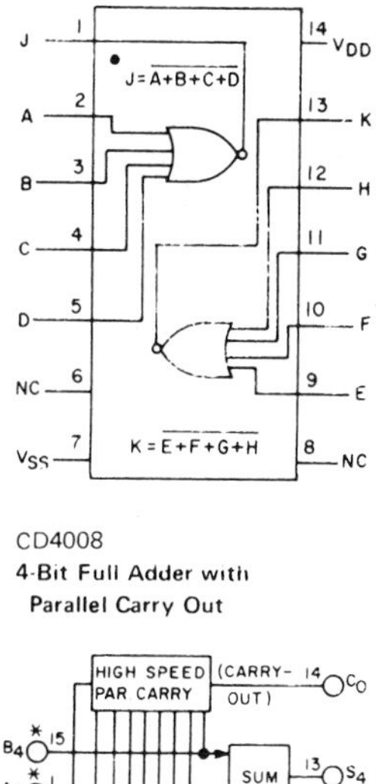
J = $\overline{A+B+C+D}$
K = $\overline{E+F+G+H}$
J 1
A 2
B 3
C 4
D 5
NC 6
VSS 7
14 VDD
13 K
12 H
11 G
10 F
9 E
8 NC

CD4006
18-Stage Static Shift Register

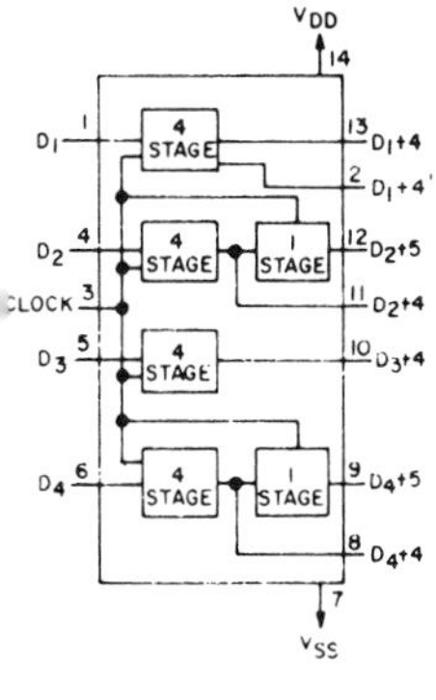
VDD 14
D1 1
D2 4
CLOCK 3
D3 5
D4 6
4 STAGE
1 STAGE
13 D1+4
2 D1+4'
12 D2+5
11 D2+4
10 D3+4
9 D4+5
8 D4+4
7 VSS

CD4007
Dual Complementary Pair Plus Inverter

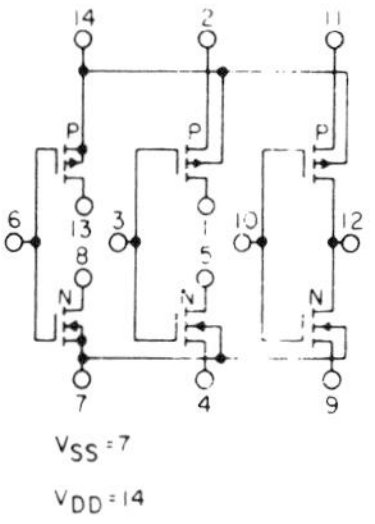
14
2
11
P
N
6
13
8
3
1
5
10
12
7
4
9
VSS = 7
VDD = 14

CD4008
4-Bit Full Adder with Parallel Carry Out

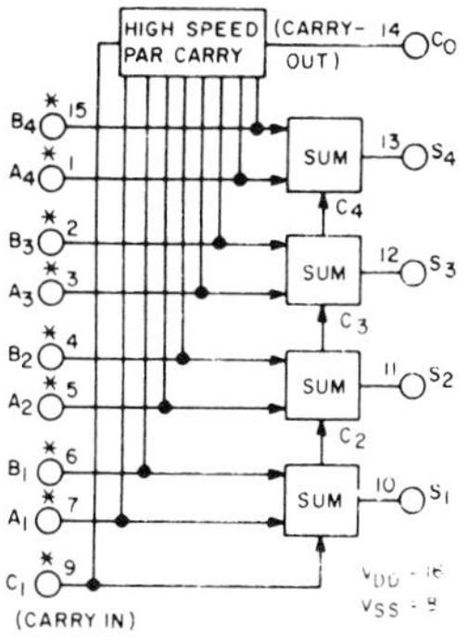
HIGH SPEED PAR. CARRY
(CARRY-OUT)
14 CO
B4 15
A4 1
B3 2
A3 3
B2 4
A2 5
B1 6
A1 7
C1 9
(CARRY IN)
SUM
13 S4
12 S3
11 S2
10 S1
C4
C3
C2

CD4009
Hex Buffer/Converter Inverting Type

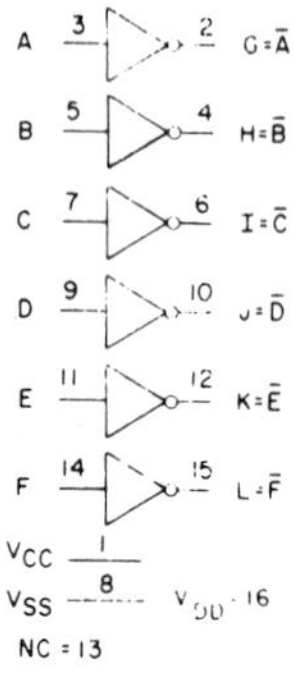
A 3 2 G=Ā
B 5 4 H=B̄
C 7 6 I=C̄
D 9 10 J=D̄
E 11 12 K=Ē
F 14 15 L=F̄
VCC 1
VSS 8 VDD 16
NC = 13

CD4010
Hex Buffer/Converter Non-Inverting Type

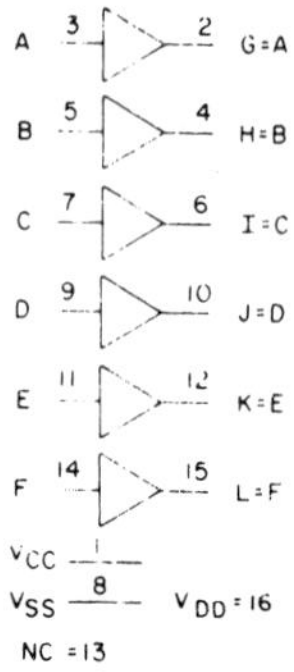
A 3 2 G=A
B 5 4 H=B
C 7 6 I=C
D 9 10 J=D
E 11 12 K=E
F 14 15 L=F
VCC 1
VSS 8 VDD = 16
NC = 13

CD4011
Quad 2-Input NAND Gate

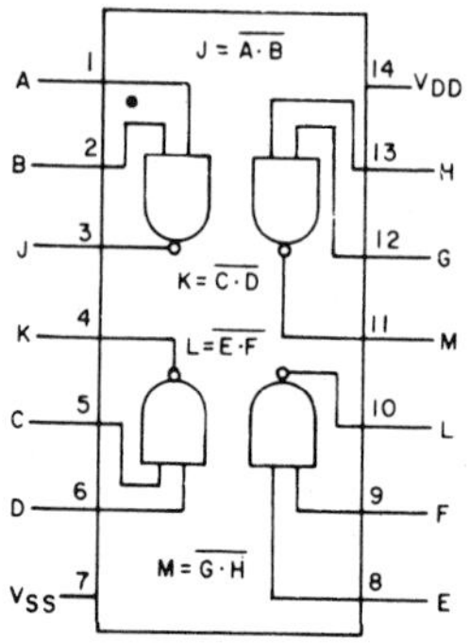
J = $\overline{A \cdot B}$
K = $\overline{C \cdot D}$
L = $\overline{E \cdot F}$
M = $\overline{G \cdot H}$
A 1
B 2
J 3
K 4
C 5
D 6
VSS 7
14 VDD
13 H
12 G
11 M
10 L
9 F
8 E

CD4012
Dual 4-Input NAND Gate

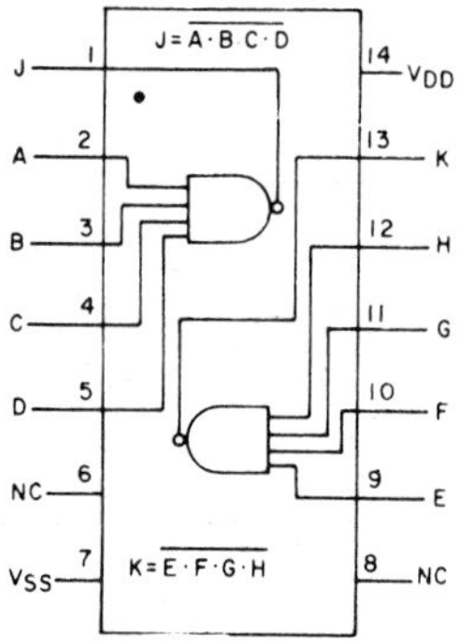
J = $\overline{A \cdot B \cdot C \cdot D}$
K = $\overline{E \cdot F \cdot G \cdot H}$
J 1
A 2
B 3
C 4
D 5
NC 6
VSS 7
14 VDD
13 K
12 H
11 G
10 F
9 E
8 NC

CD4013
Dual "D" Flip-Flop with Set/Reset Capability

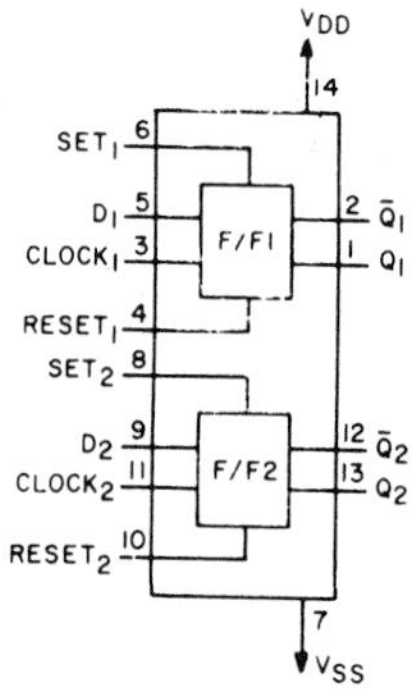
VDD
14
SET1 6
D1 5
CLOCK1 3
RESET1 4
F/F1
2 Q̄1
1 Q1
SET2 8
D2 9
CLOCK2 11
RESET2 10
F/F2
12 Q̄2
13 Q2
7
VSS

CD4014
8-Stage Synchronous Shift Register with Parallel or Serial Input/Serial Output

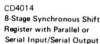

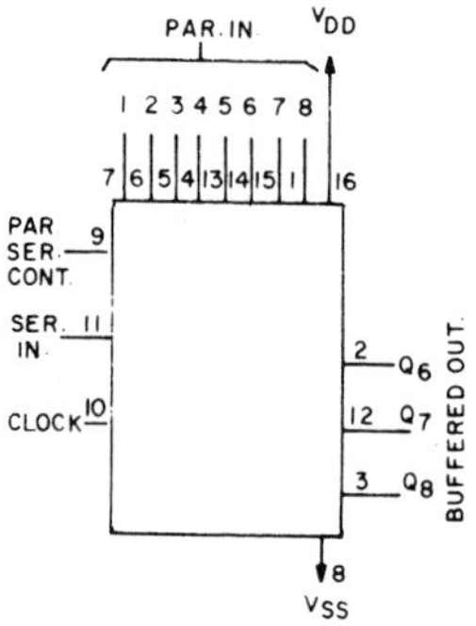
PAR. IN
VDD
1 2 3 4 5 6 7 8
7 6 5 4 13 14 15 1
16
PAR SER. CONT. 9
SER. IN 11
CLOCK 10
2 Q6
12 Q7
3 Q8
BUFFERED OUT.
8
VSS

CD4015
Dual 4-Stage Static Shift Register with Serial Input/Parallel Output

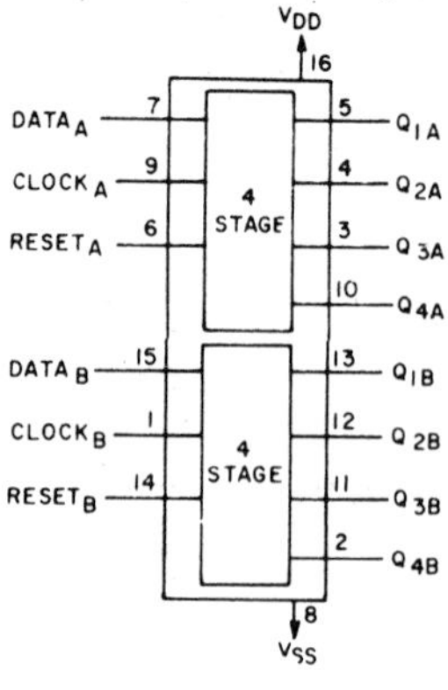
VDD
16
DATA A 7
CLOCK A 9
RESET A 6
4 STAGE
5 Q1A
4 Q2A
3 Q3A
10 Q4A
DATA B 15
CLOCK B 1
RESET B 14
4 STAGE
13 Q1B
12 Q2B
11 Q3B
2 Q4B
8
VSS

CD4016
Quad Bilateral Switch

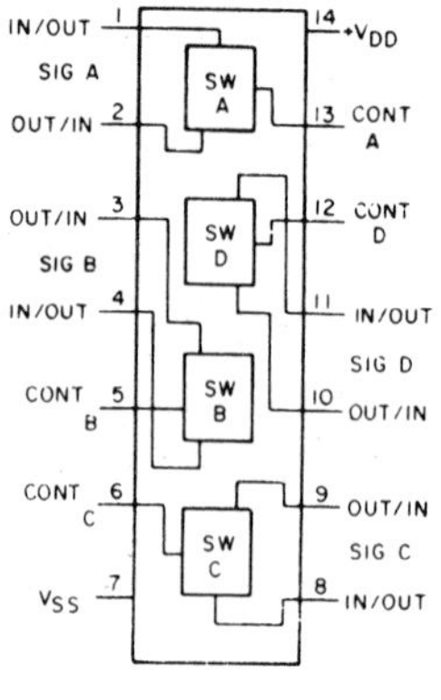
IN/OUT 1
SIG A
OUT/IN 2
OUT/IN 3
SIG B
IN/OUT 4
CONT B 5
CONT C 6
VSS 7
SW A
SW D
SW B
SW C
14 VDD
13 CONT A
12 CONT D
11 IN/OUT
SIG D
10 OUT/IN
9 OUT/IN
SIG C
8 IN/OUT

CD4017
Decade Counter/Divider with 10 Decoded Decimal Outputs

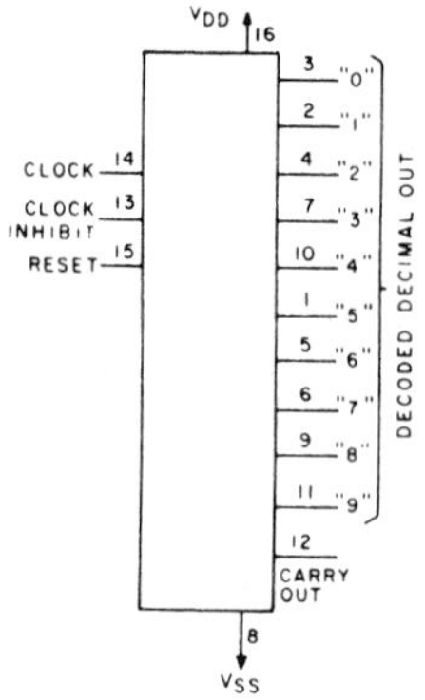
VDD 16
CLOCK 14
CLOCK INHIBIT 13
RESET 15
3 "0"
2 "1"
4 "2"
7 "3"
10 "4"
1 "5"
5 "6"
6 "7"
9 "8"
11 "9"
DECODED DECIMAL OUT
12 CARRY OUT
8
VSS

CD4018

Presettable Divide-by-"N" Counter Fixed or Programmable

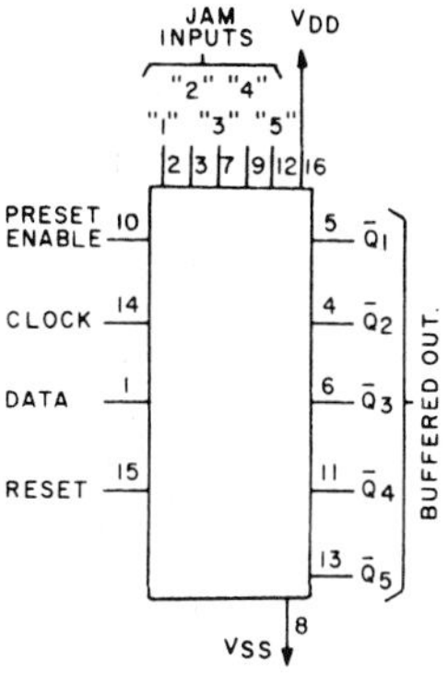
JAM INPUTS
V_{DD}
"2" "4"
"1" "3" "5"
2 3 7 9 12 16
PRESET ENABLE 10
CLOCK 14
DATA 1
RESET 15
5 $\bar{Q}_1$
4 $\bar{Q}_2$
6 $\bar{Q}_3$
11 $\bar{Q}_4$
13 $\bar{Q}_5$
BUFFERED OUT.
8
VSS

CD4019

Quad AND/OR Select Gate

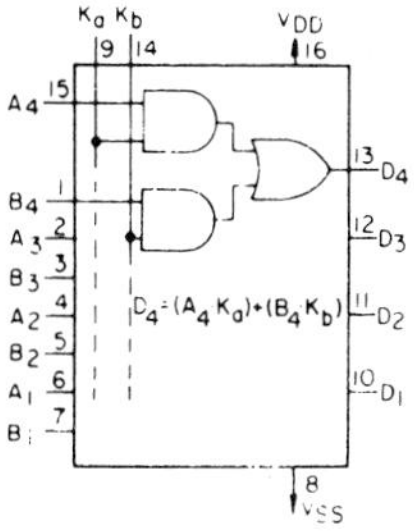
K_a K_b
9 14
V_{DD} 16
A_4 15
B_4 1
A_3 2
B_3 3
A_2 4
B_2 5
A_1 6
B_1 7
13 D_4
12 D_3
11 D_2
10 D_1
$D_4 = (A_4 K_a) + (B_4 K_b)$
8
V_{SS}

CD4020

14-Stage Binary Ripple Counter

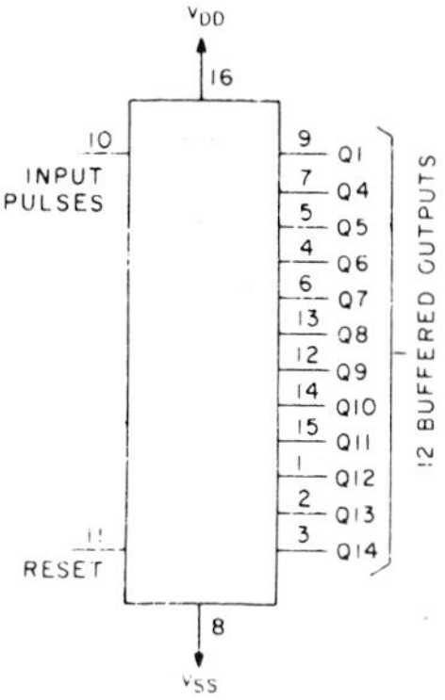
V_{DD}
16
10 INPUT PULSES
11 RESET
9 Q1
7 Q4
5 Q5
4 Q6
6 Q7
13 Q8
12 Q9
14 Q10
15 Q11
1 Q12
2 Q13
3 Q14
12 BUFFERED OUTPUTS
8
V_{SS}

CD4021

8-Stage Static Shift Register Asynchronous Parallel or Synchronous Serial Input/ Serial Output

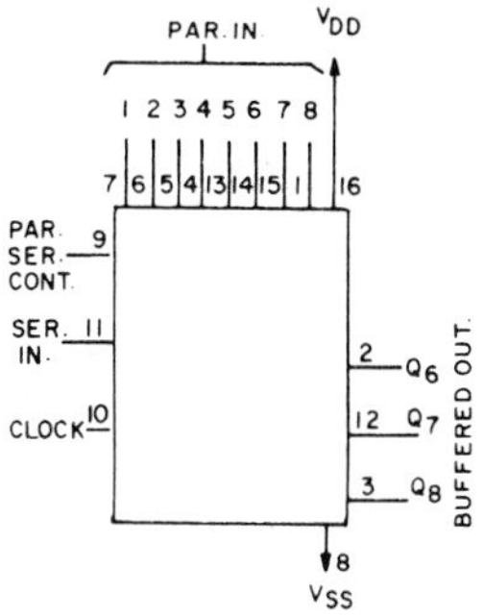
PAR. IN
V_{DD}
1 2 3 4 5 6 7 8
7 6 5 4 13 14 15 1 16
PAR. SER. CONT. 9
SER. IN. 11
CLOCK 10
2 Q_6
12 Q_7
3 Q_8
BUFFERED OUT.
8
V_{SS}

CD4022

Divide-by-8 Counter/Divider with 8 Decoded Decimal Outputs

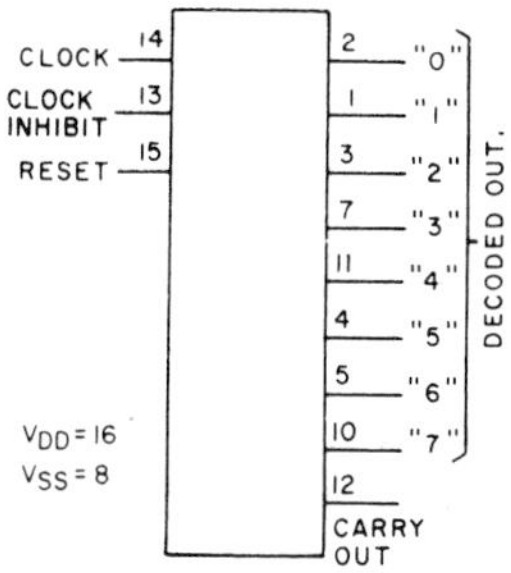
CLOCK 14
CLOCK INHIBIT 13
RESET 15
2 "0"
1 "1"
3 "2"
7 "3"
11 "4"
4 "5"
5 "6"
10 "7"
DECODED OUT.
12 CARRY OUT
$V_{DD} = 16$
$V_{SS} = 8$

CD4023

Triple 3-Input NAND Gate

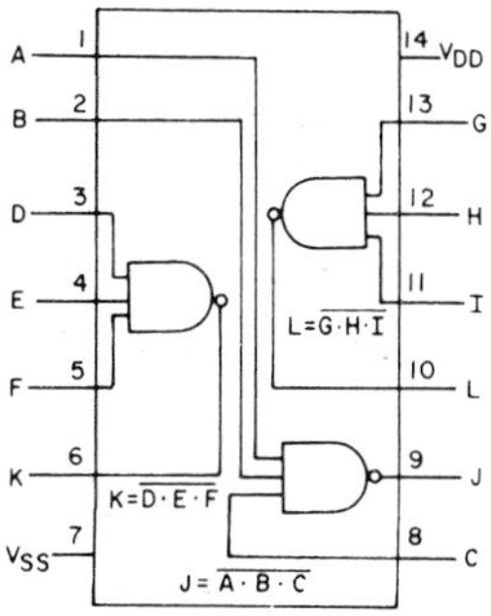
A 1
B 2
D 3
E 4
F 5
K 6
V_{SS} 7
14 V_{DD}
13 G
12 H
11 I
10 L
9 J
8 C
$L = \overline{G \cdot H \cdot I}$
$K = \overline{D \cdot E \cdot F}$
$J = \overline{A \cdot B \cdot C}$

7-Stage Ripple-Carry Binary Counter/Divider

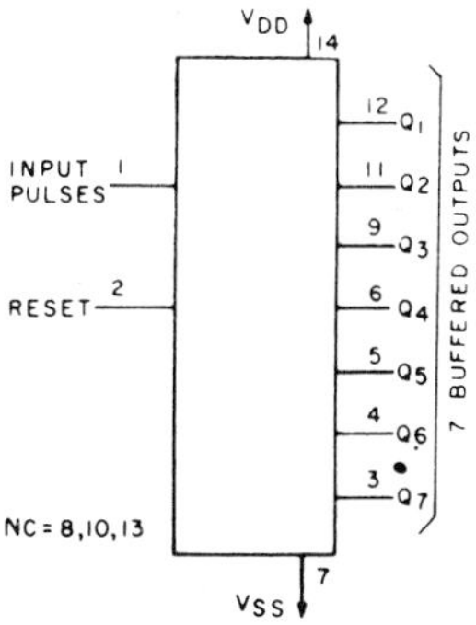
V_{DD} 14
INPUT PULSES 1
RESET 2
12 Q_1
11 Q2
9 Q_3
6 Q_4
5 Q_5
4 Q_6
3 Q_7
7 BUFFERED OUTPUTS
NC = 8,10,13
7
VSS

CD4025

Triple 3-Input NOR Gate

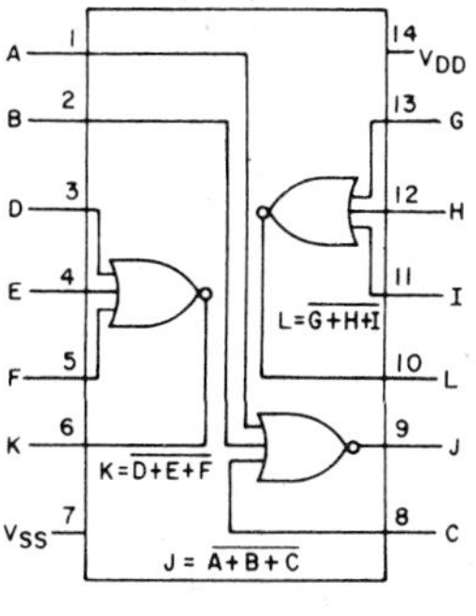
A 1
B 2
D 3
E 4
F 5
K 6
V_{SS} 7
14 V_{DD}
13 G
12 H
11 I
10 L
9 J
8 C
$L = \overline{G+H+I}$
$K = \overline{D+E+F}$
$J = \overline{A+B+C}$

CD4026

Decade Counter/Divider with 7-Segment Display Outputs and Display Enable

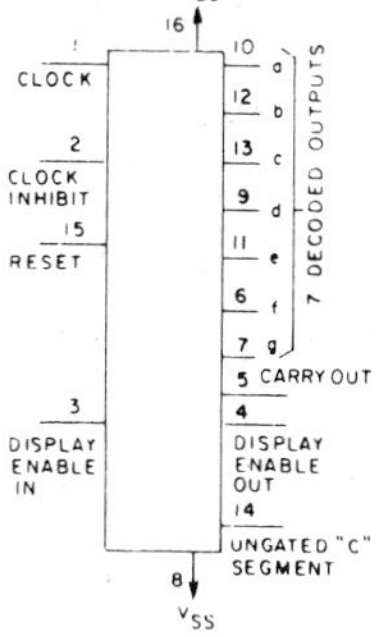
V_{DD}
16
1 CLOCK
2 CLOCK INHIBIT
15 RESET
3 DISPLAY ENABLE IN
10 a
12 b
13 c
9 d
11 e
6 f
7 g
7 DECODED OUTPUTS
5 CARRY OUT
4 DISPLAY ENABLE OUT
14 UNGATED "C" SEGMENT
8
V_{SS}

CD4027

Dual J-K Master-Slave Flip-Flop with Set-Reset Capability

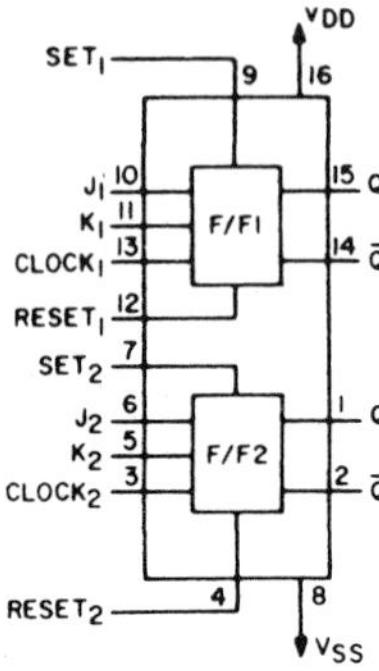
VDD
SET1
9
16
J1
K1
CLOCK1
10
11
13
F/F1
15
Q1
14
Q̄1
RESET1
12
SET2
7
J2
K2
CLOCK2
6
5
3
F/F2
1
Q2
2
Q̄2
RESET2
4
8
VSS

CD4028

BCD-to-Decimal Decoder

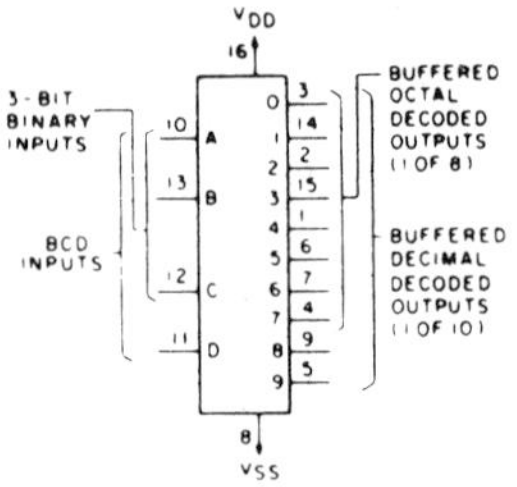
VDD
16
3-BIT BINARY INPUTS
BCD INPUTS
10
A
13
B
12
C
11
D
BUFFERED OCTAL DECODED OUTPUTS (1 OF 8)
BUFFERED DECIMAL DECODED OUTPUTS (1 OF 10)
8
VSS

CD4029

Presettable Up/Down Counter, Binary or BCD-Decade

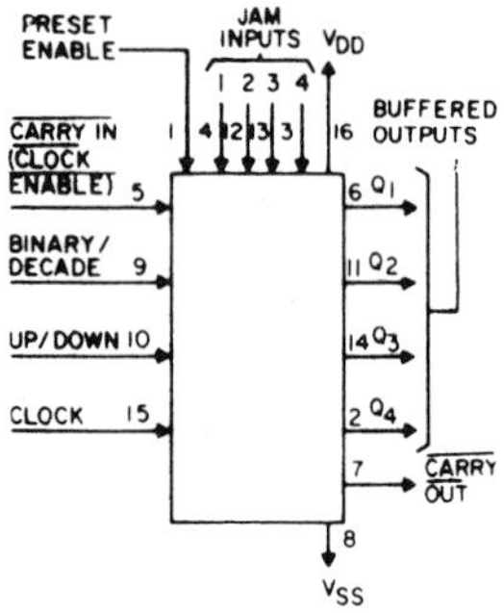
PRESET ENABLE
JAM INPUTS
1 2 3 4
VDD
CARRY IN (CLOCK ENABLE) 5
BINARY/ DECADE 9
UP/DOWN 10
CLOCK 15
BUFFERED OUTPUTS
6 Q1
11 Q2
14 Q3
2 Q4
7
CARRY OUT
8
VSS

CD4030

Quad Exclusive-OR Gate

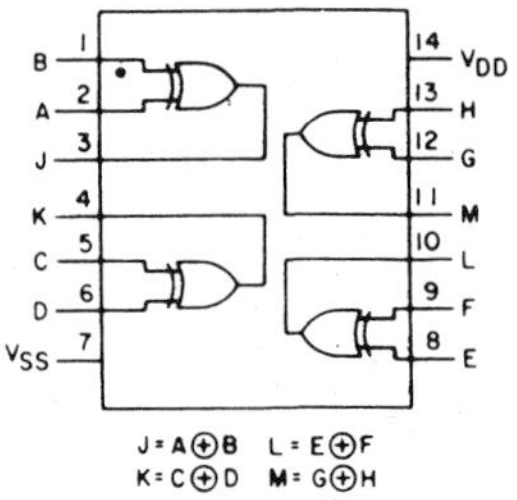
B
A
J
K
C
D
VSS
VDD
H
G
M
L
F
E
J = A⊕B L = E⊕F
K = C⊕D M = G⊕H

CD4031

64-Stage Static Shift Register

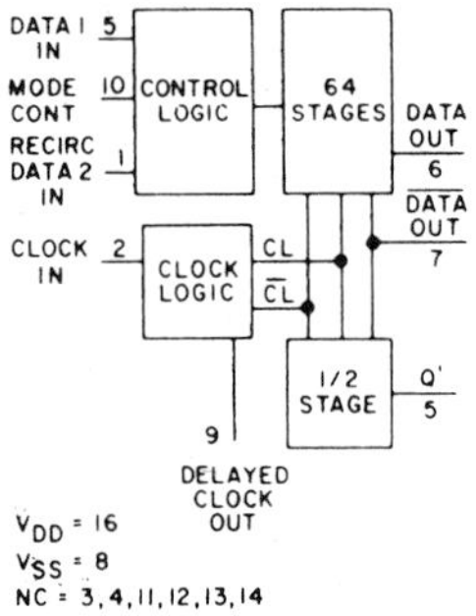
DATA 1 IN 5
MODE CONT 10
RECIRC DATA 2 IN 1
CONTROL LOGIC
64 STAGES
DATA OUT 6
DATA OUT 7
CLOCK IN 2
CLOCK LOGIC
CL
CL̄
1/2 STAGE
Q' 5
9
DELAYED CLOCK OUT
VDD = 16
VSS = 8
NC = 3,4,11,12,13,14

CD4032

Triple Serial Adder Positive Logic

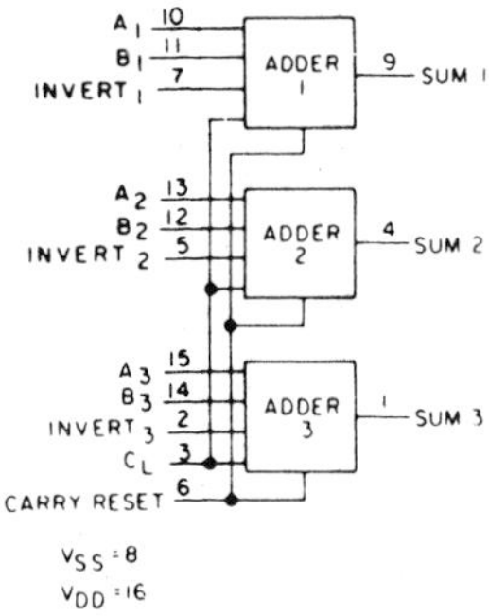
A1 10
B1 11
INVERT1 7
ADDER 1
9 SUM 1
A2 13
B2 12
INVERT2 5
ADDER 2
4 SUM 2
A3 15
B3 14
INVERT3 2
CL 3
ADDER 3
1 SUM 3
CARRY RESET 6
VSS = 8
VDD = 16

CD4033

Decade Counter/Divider with 7-Segment Display Outputs and Ripple Blanking

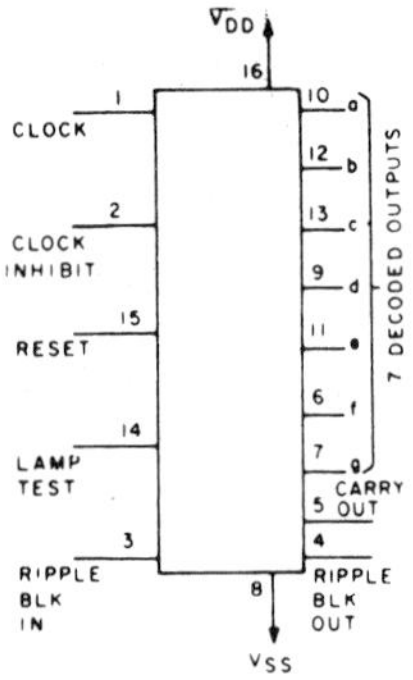
VDD
16
CLOCK 1
CLOCK INHIBIT 2
RESET 15
LAMP TEST 14
RIPPLE BLK IN 3
10 a
12 b
13 c
9 d
11 e
6 f
7 g
7 DECODED OUTPUTS
5 CARRY OUT
4 RIPPLE BLK OUT
8
VSS

CD4034

8-Stage Static Bidirectional Parallel/Serial Input/Output Bus Register

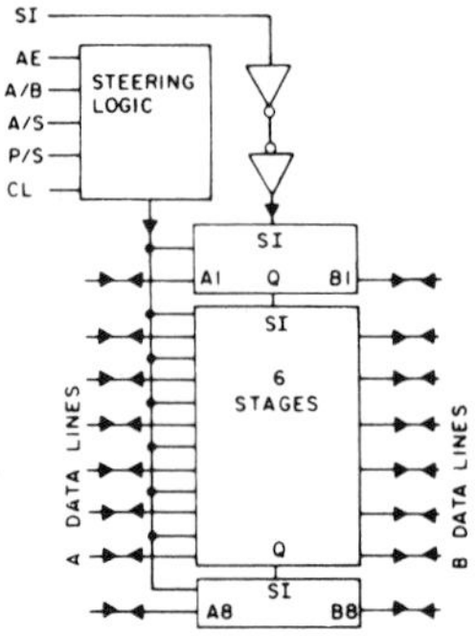
SI
AE
A/B
A/S
P/S
CL
STEERING LOGIC
SI
A1 Q B1
SI
6 STAGES
Q
SI
A8 B8
A DATA LINES
B DATA LINES

CD4035

4-Stage Parallel In/Parallel Out Shift Register with J-K Serial Inputs and True/Complement Outputs

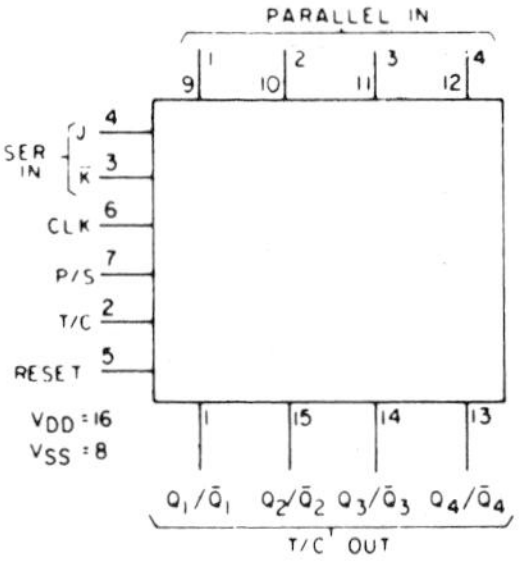
PARALLEL IN
9 1
10 2
11 3
12 4
SER IN J 4
K 3
CLK 6
P/S 7
T/C 2
RESET 5
VDD = 16
VSS = 8
1 15 14 13
Q1/Q̄1 Q2/Q̄2 Q3/Q̄3 Q4/Q̄4
T/C OUT

CD4037
Triple AND/OR Bi-Phase Pair

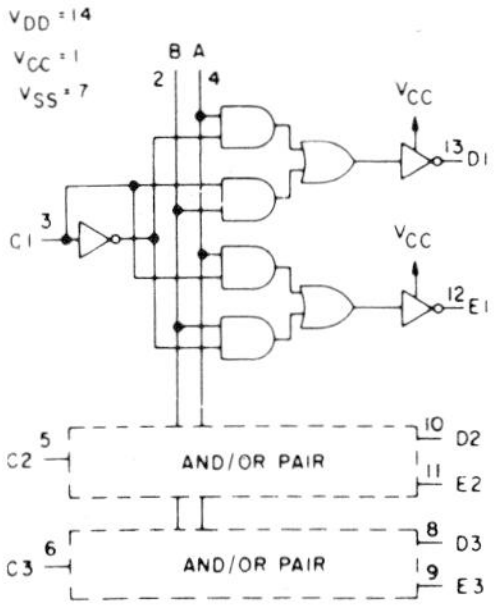
VDD = 14
VCC = 1
VSS = 7
B A
VCC
D1
E1
C1
C2
AND/OR PAIR
D2
E2
C3
AND/OR PAIR
D3
E3

CD4038
Triple Serial Adder Negative Logic

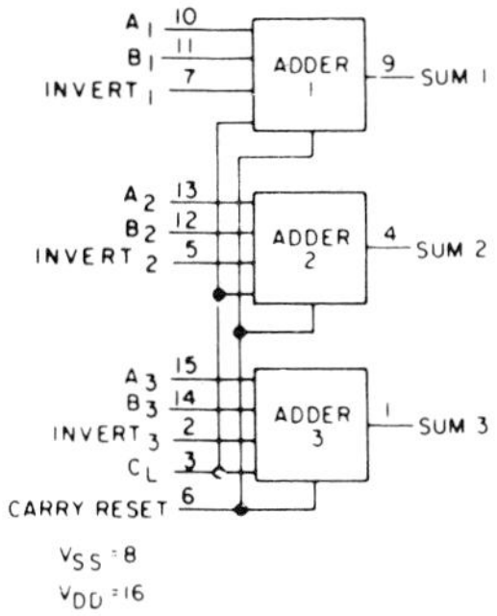
ADDER 1
SUM 1
ADDER 2
SUM 2
ADDER 3
SUM 3
INVERT
CARRY RESET
VSS = 8
VDD = 16

CD4040
12-Stage Ripple-Carry Binary Counter/Divider

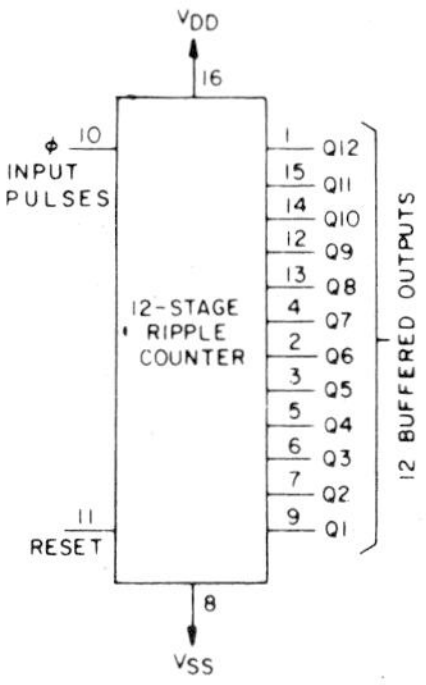
VDD
INPUT PULSES
12-STAGE RIPPLE COUNTER
RESET
VSS
12 BUFFERED OUTPUTS

CD4041
Quad True/Complement Buffer

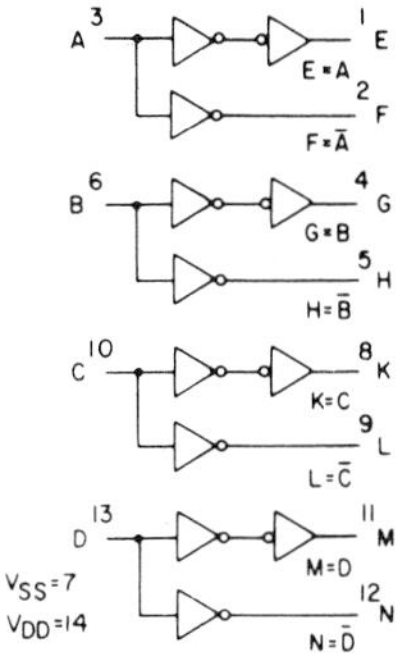
E = A
F = Ā
G = B
H = B̄
K = C
L = C̄
M = D
N = D̄
VSS = 7
VDD = 14

CD4042
Quad Clocked "D" Latch

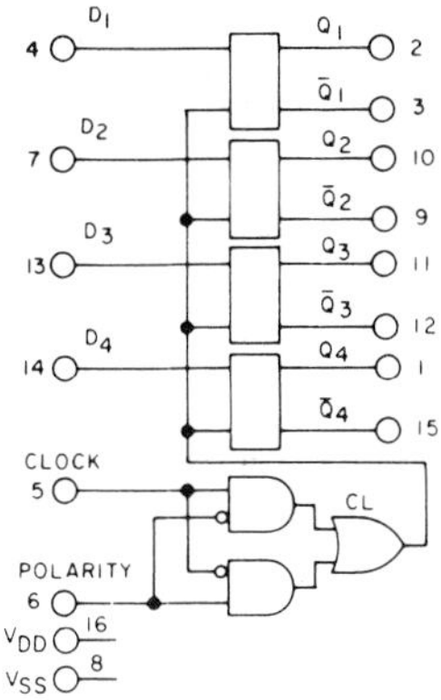
CLOCK
POLARITY
CL
VDD
VSS

CD4043
Quad 3-State NOR R/S Latch

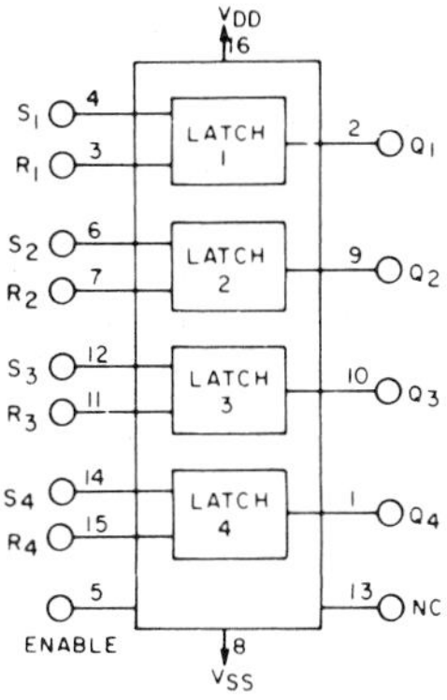
VDD
LATCH 1
LATCH 2
LATCH 3
LATCH 4
ENABLE
NC
VSS

CD4044
Quad 3-State NAND R/S Latch

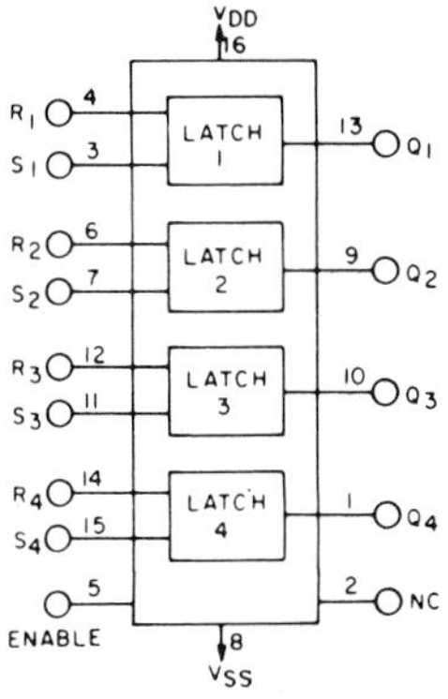
VDD
LATCH 1
LATCH 2
LATCH 3
LATCH 4
ENABLE
NC
VSS

CD4045
21-Stage Counter

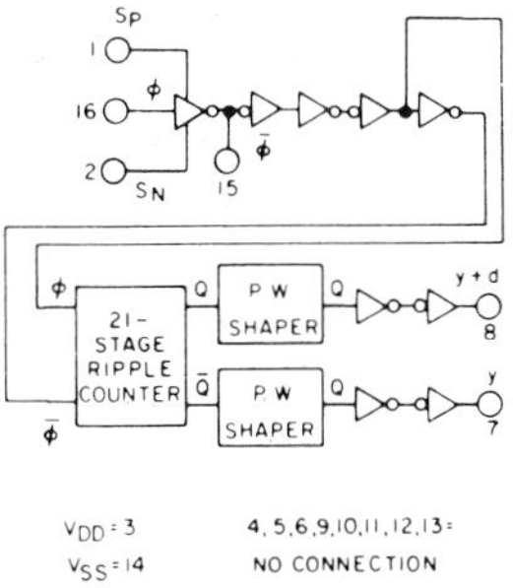
21-STAGE RIPPLE COUNTER
P.W. SHAPER
P.W. SHAPER
VDD = 3
VSS = 14
4, 5, 6, 9, 10, 11, 12, 13 = NO CONNECTION

CD4046
Micropower Phase-Locked Loop

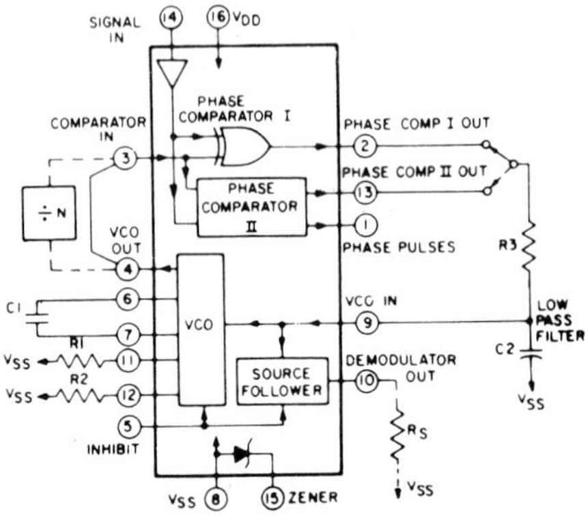
SIGNAL IN
14
16 VDD
PHASE COMPARATOR I
COMPARATOR IN
3
PHASE COMP I OUT
2
PHASE COMP II OUT
13
PHASE COMPARATOR II
1
PHASE PULSES
÷ N
VCO OUT
4
R3
C1
6
VCO
VCO IN
9
LOW PASS FILTER
7
R1
VSS
11
C2
R2
VSS
12
SOURCE FOLLOWER
DEMODULATOR OUT
10
VSS
5
INHIBIT
RS
VSS
VSS 8
15 ZENER

CD4047
Low-Power Monostable/Astable Multivibrator

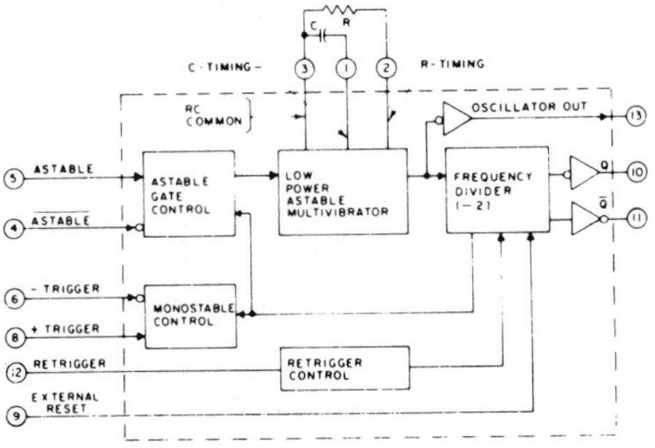
C R
C - TIMING -
3
1
2
R - TIMING
RC COMMON
OSCILLATOR OUT
13
5 ASTABLE
ASTABLE GATE CONTROL
LOW POWER ASTABLE MULTIVIBRATOR
FREQUENCY DIVIDER (÷ 2)
Q 10
4 ASTABLE
Q̄ 11
6 - TRIGGER
MONOSTABLE CONTROL
8 + TRIGGER
12 RETRIGGER
RETRIGGER CONTROL
9 EXTERNAL RESET

CD4048
Multi-Function Expandable 8-Input Gate

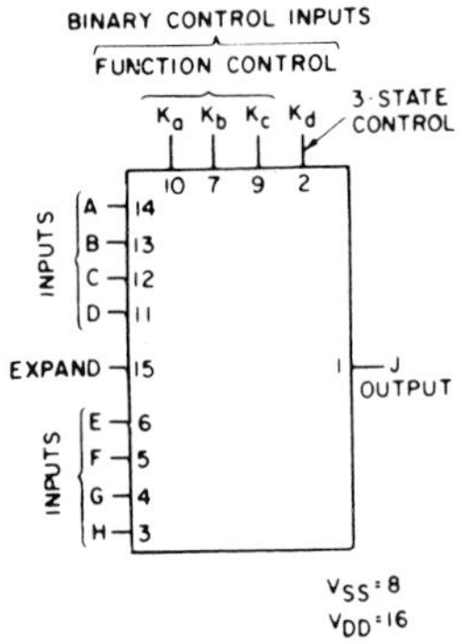
BINARY CONTROL INPUTS
FUNCTION CONTROL
Ka Kb Kc Kd
3-STATE CONTROL
10 7 9 2
INPUTS
A 14
B 13
C 12
D 11
EXPAND 15
1 J OUTPUT
INPUTS
E 6
F 5
G 4
H 3
VSS = 8
VDD = 16

CD4049
Hex Buffer/Converter Inverting Type

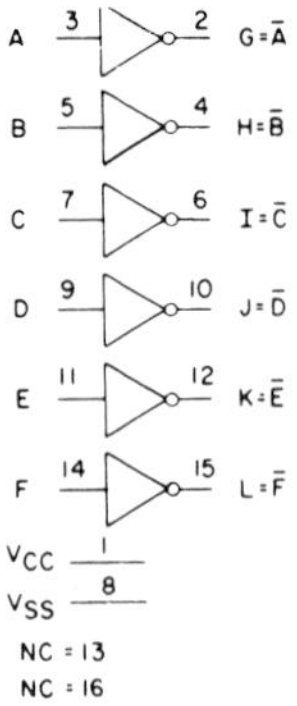
A 3 2 G = Ā
B 5 4 H = B̄
C 7 6 I = C̄
D 9 10 J = D̄
E 11 12 K = Ē
F 14 15 L = F̄
VCC 1
VSS 8
NC = 13
NC = 16

CD4050
Hex Buffer/Converter Non-Inverting Type

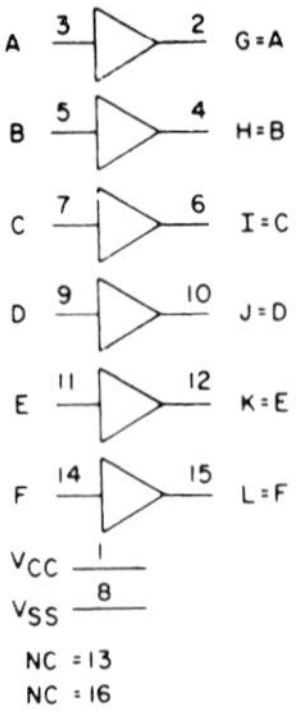
A 3 2 G = A
B 5 4 H = B
C 7 6 I = C
D 9 10 J = D
E 11 12 K = E
F 14 15 L = F
VCC 1
VSS 8
NC = 13
NC = 16

CD4051
Single 8-Channel Analog Multiplexer/Demultiplexer

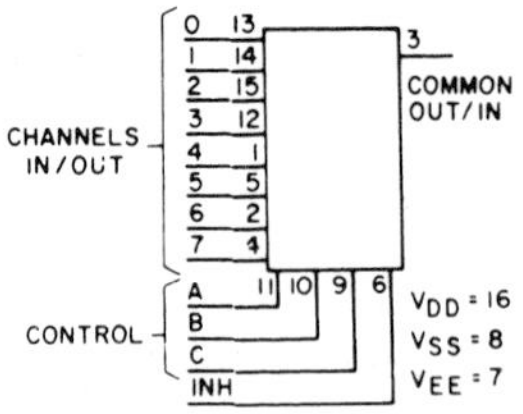
0 13
1 14
2 15
3 12
4 1
5 5
6 2
7 4
3
COMMON OUT/IN
CHANNELS IN/OUT
A 11
B 10
C 9
INH 6
CONTROL
VDD = 16
VSS = 8
VEE = 7

CD4052

Differential 4-Channel Analog Multiplexer/Demultiplexer

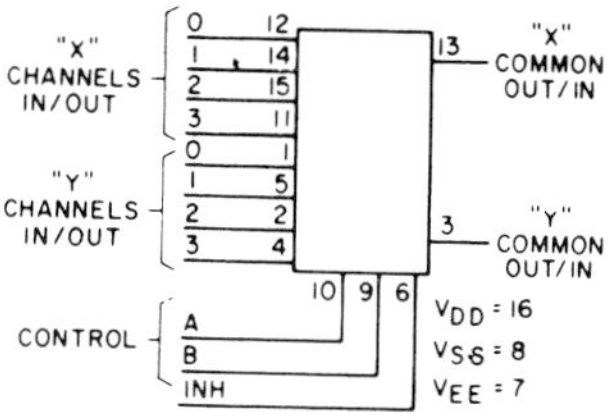
"X" CHANNELS IN/OUT
"Y" CHANNELS IN/OUT
CONTROL
A
B
INH
"X" COMMON OUT/IN
"Y" COMMON OUT/IN
VDD = 16
VSS = 8
VEE = 7

CD4053

Triple 2-Channel Multiplexer/Demultiplexer

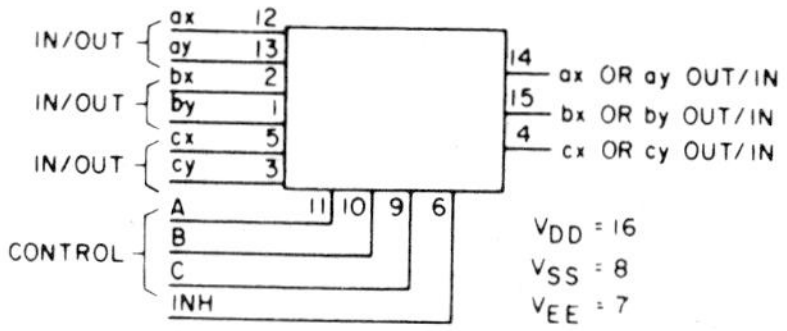
IN/OUT
IN/OUT
IN/OUT
CONTROL
ax
ay
bx
by
cx
cy
A
B
C
INH
ax OR ay OUT/IN
bx OR by OUT/IN
cx OR cy OUT/IN
VDD = 16
VSS = 8
VEE = 7

CD4054

4-Segment Liquid-Crystal Display Driver

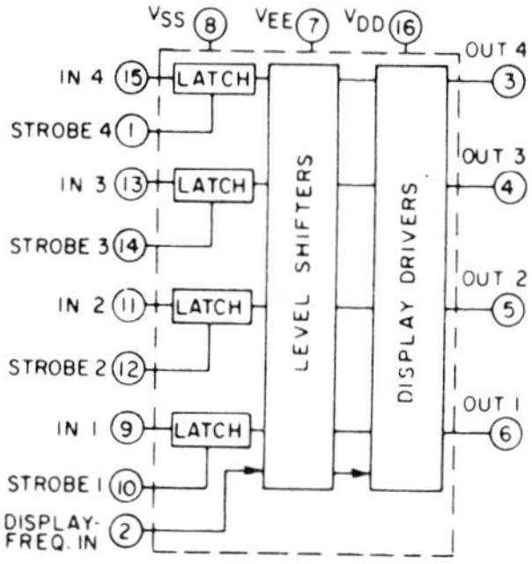
VSS
VEE
VDD
IN 4
STROBE 4
IN 3
STROBE 3
IN 2
STROBE 2
IN 1
STROBE 1
DISPLAY-FREQ. IN
LATCH
LEVEL SHIFTERS
DISPLAY DRIVERS
OUT 4
OUT 3
OUT 2
OUT 1

CD4055

BCD-to-7-Segment Decoder/Driver with "Display-Frequency" Output Liquid-Crystal Display Driver

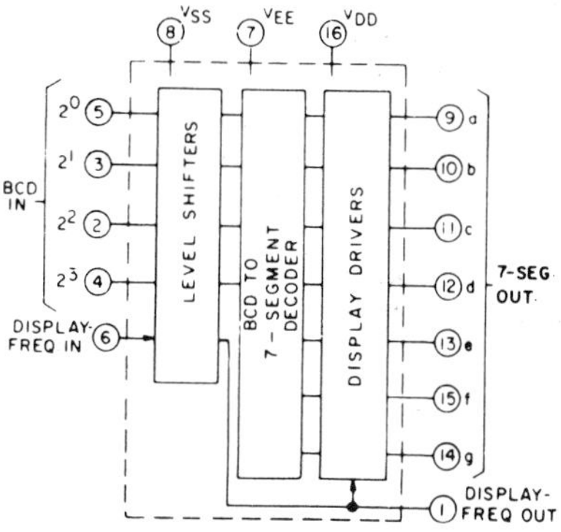
VSS
VEE
VDD
BCD IN
DISPLAY-FREQ IN
LEVEL SHIFTERS
BCD TO 7-SEGMENT DECODER
DISPLAY DRIVERS
7-SEG OUT
DISPLAY-FREQ OUT

CD4056

BCD-to-7-Segment Decoder/Driver with Strobed-Latch Function Liquid-Crystal Display Driver

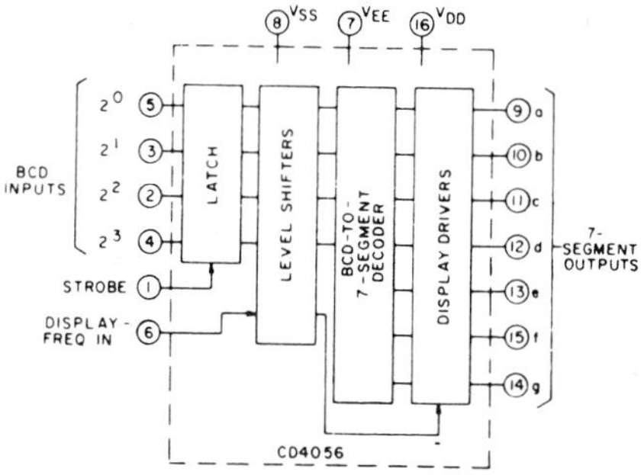
VSS
VEE
VDD
BCD INPUTS
STROBE
DISPLAY-FREQ IN
LATCH
LEVEL SHIFTERS
BCD-TO-7-SEGMENT DECODER
DISPLAY DRIVERS
7-SEGMENT OUTPUTS
CD4056

CD4057

4-Bit Arithmetic Logic Unit

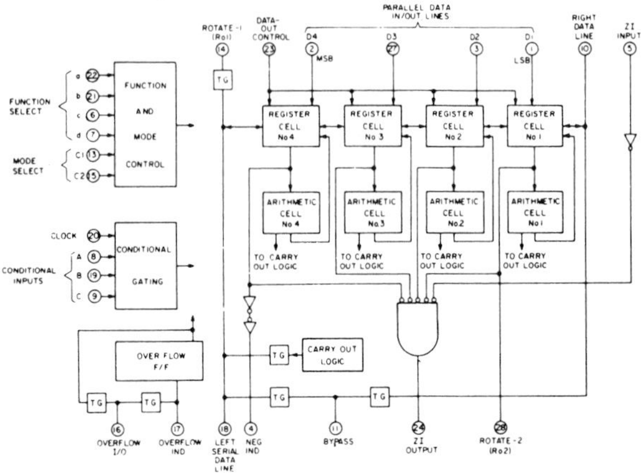

PARALLEL DATA IN/OUT LINES
ROTATE-1 (Ro1)
DATA-OUT CONTROL
D4
D3
D2
D1
MSB
LSB
RIGHT DATA LINE
ZI INPUT
FUNCTION SELECT
MODE SELECT
FUNCTION AND MODE CONTROL
REGISTER CELL No 4
REGISTER CELL No 3
REGISTER CELL No 2
REGISTER CELL No 1
ARITHMETIC CELL No 4
ARITHMETIC CELL No 3
ARITHMETIC CELL No 2
ARITHMETIC CELL No 1
TO CARRY OUT LOGIC
CLOCK
CONDITIONAL INPUTS
CONDITIONAL GATING
OVER FLOW F/F
CARRY OUT LOGIC
TG
OVERFLOW I/O
OVERFLOW IND
LEFT SERIAL DATA LINE
NEG IND
BYPASS
ZI OUTPUT
ROTATE-2 (Ro2)

CD4059

Programmable Divide-by-"N" Counter

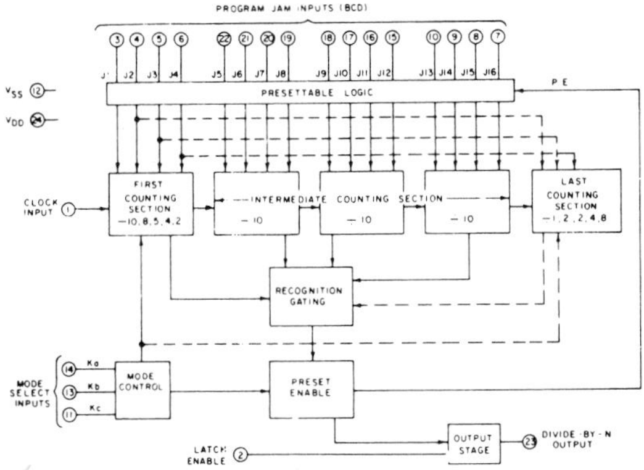

PROGRAM JAM INPUTS (BCD)
J1
J2
J3
J4
J5
J6
J7
J8
J9
J10
J11
J12
J13
J14
J15
J16
PRESETTABLE LOGIC
P.E.
VSS
VDD
CLOCK INPUT
FIRST COUNTING SECTION ÷10,8,5,4,2
INTERMEDIATE COUNTING SECTION
÷10
LAST COUNTING SECTION ÷1,2,2,4,8
RECOGNITION GATING
MODE SELECT INPUTS
Ka
Kb
Kc
MODE CONTROL
PRESET ENABLE
LATCH ENABLE
OUTPUT STAGE
DIVIDE-BY-N OUTPUT

CD4060

14-Stage Ripple-Carry Binary Counter/Divider and Oscillator

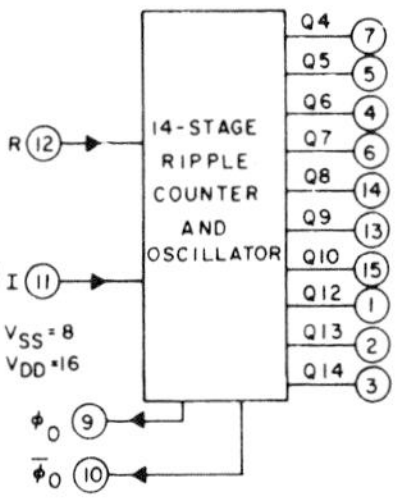

14-STAGE RIPPLE COUNTER AND OSCILLATOR
R
φI
VSS = 8
VDD = 16
φO
φ̄O
Q4
Q5
Q6
Q7
Q8
Q9
Q10
Q12
Q13
Q14

CD4062

200 Stage Dynamic Shift Register

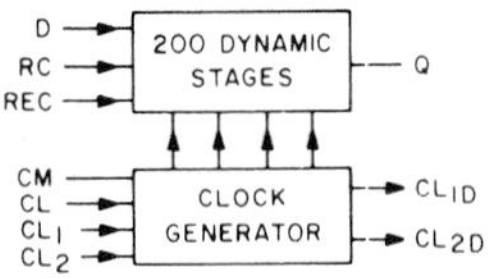

D
RC
REC
200 DYNAMIC STAGES
Q
CM
CL
CL1
CL2
CLOCK GENERATOR
CL1D
CL2D

CD4063

4-Bit Magnitude Comparator

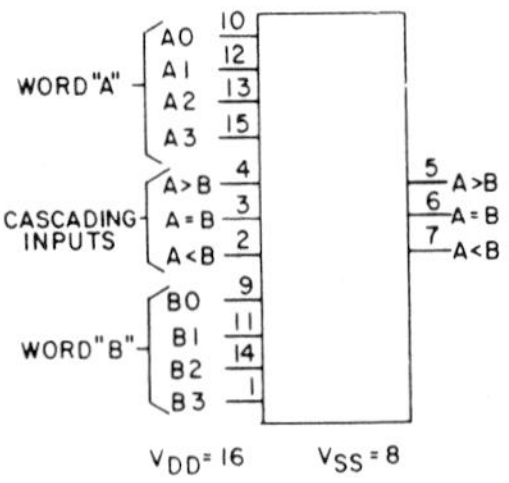

WORD "A"
A0
A1
A2
A3
CASCADING INPUTS
A>B
A=B
A<B
WORD "B"
B0
B1
B2
B3
VDD = 16
VSS = 8

CD4066

Quad Bilateral Switch

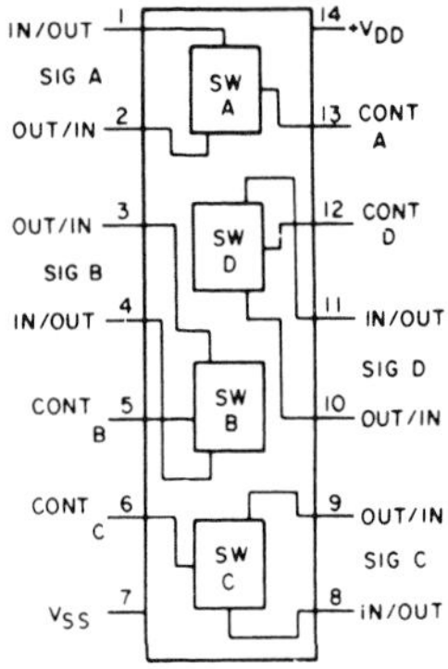

IN/OUT
SIG A
OUT/IN
OUT/IN
SIG B
IN/OUT
CONT B
CONT C
VSS
+VDD
CONT A
CONT D
IN/OUT
SIG D
OUT/IN
OUT/IN
SIG C
IN/OUT
SW A
SW D
SW B
SW C

CD4067
16-Channel Multiplexer/Demultiplexer

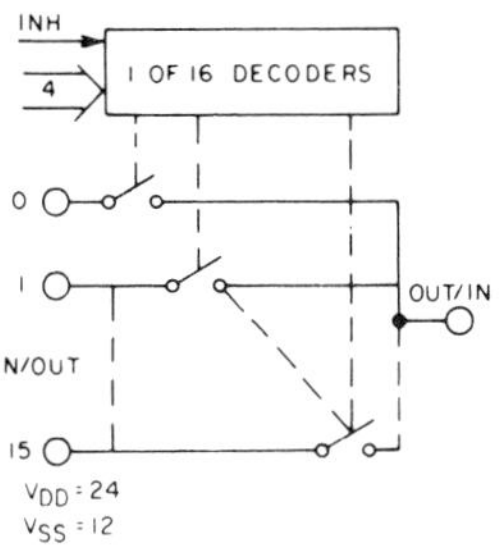
INH
4
1 OF 16 DECODERS
0
1
N/OUT
15
OUT/IN
V_{DD} = 24
V_{SS} = 12

CD4068
8-Input NAND/AND Gate

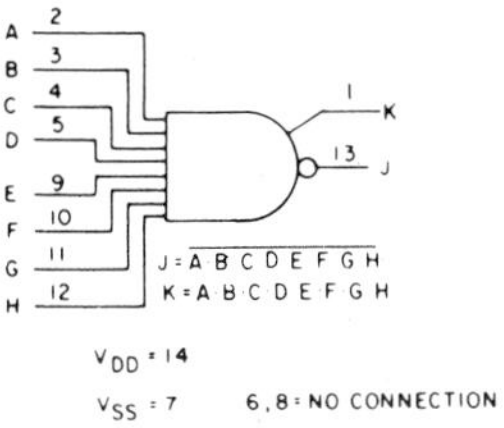
A 2
B 3
C 4
D 5
E 9
F 10
G 11
H 12
1 K
13 J
$J = \overline{A \cdot B \cdot C \cdot D \cdot E \cdot F \cdot G \cdot H}$
$K = A \cdot B \cdot C \cdot D \cdot E \cdot F \cdot G \cdot H$
V_{DD} = 14
V_{SS} = 7
6, 8 = NO CONNECTION

CD4069
Hex Inverter

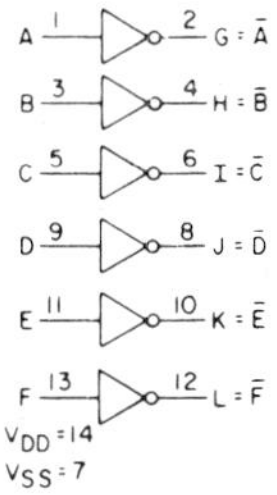
A 1 2 $G = \bar{A}$
B 3 4 $H = \bar{B}$
C 5 6 $I = \bar{C}$
D 9 8 $J = \bar{D}$
E 11 10 $K = \bar{E}$
F 13 12 $L = \bar{F}$
V_{DD} = 14
V_{SS} = 7

CD4070
Quad Exclusive-OR Gate

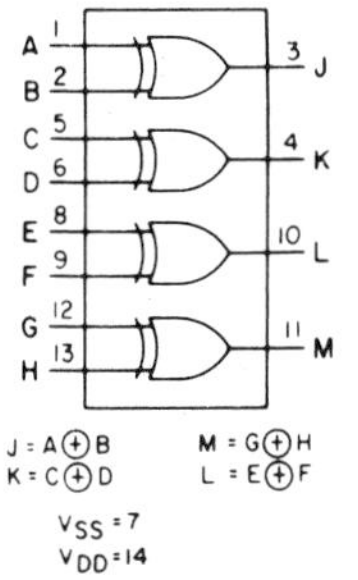
A 1
B 2
C 5
D 6
E 8
F 9
G 12
H 13
3 J
4 K
10 L
11 M
$J = A \oplus B$
$K = C \oplus D$
$M = G \oplus H$
$L = E \oplus F$
V_{SS} = 7
V_{DD} = 14

CD4071
Quad 2-Input OR Gate

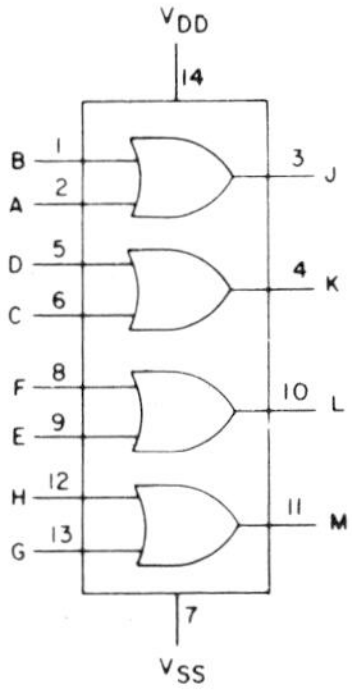
V_{DD}
14
B 1
A 2
D 5
C 6
F 8
E 9
H 12
G 13
3 J
4 K
10 L
11 M
7
V_{SS}

CD4072
Dual 4-Input OR Gate

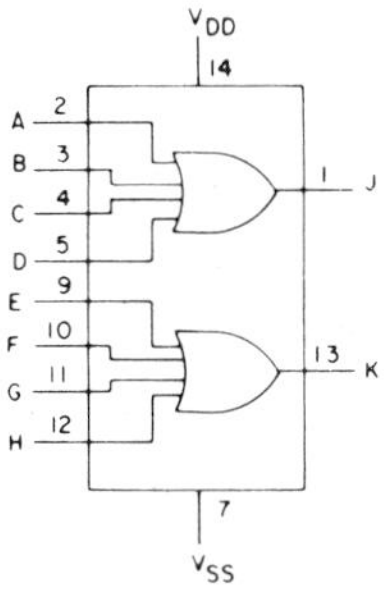
V_{DD}
14
A 2
B 3
C 4
D 5
E 9
F 10
G 11
H 12
1 J
13 K
7
V_{SS}

CD4073
Triple 3-Input AND Gate

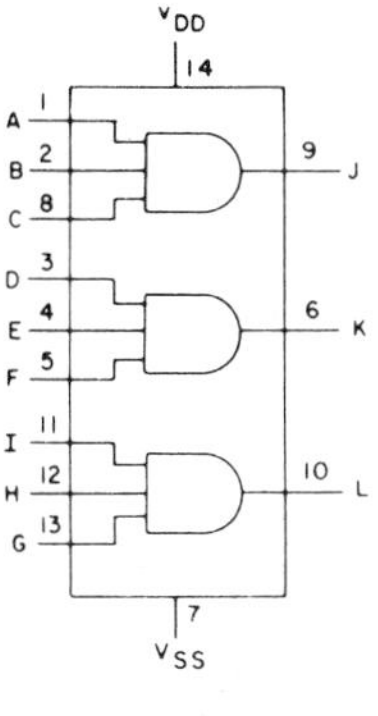
V_{DD}
14
A 1
B 2
C 8
D 3
E 4
F 5
I 11
H 12
G 13
9 J
6 K
10 L
7
V_{SS}

CD4075
Triple 3-Input OR Gate

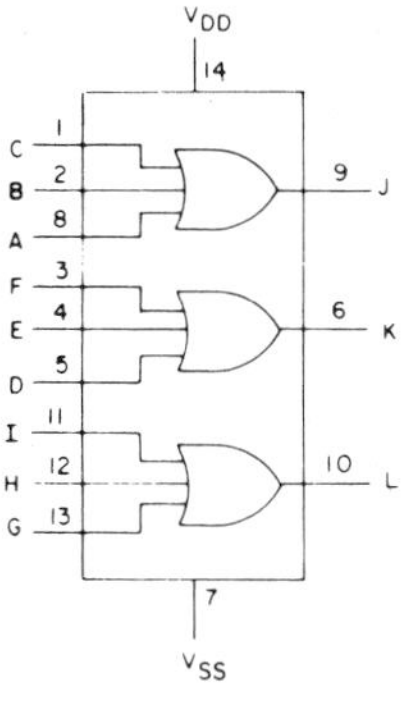
V_{DD}
14
C 1
B 2
A 8
F 3
E 4
D 5
I 11
H 12
G 13
9 J
6 K
10 L
7
V_{SS}

CD4076
4-Bit D-Type Register

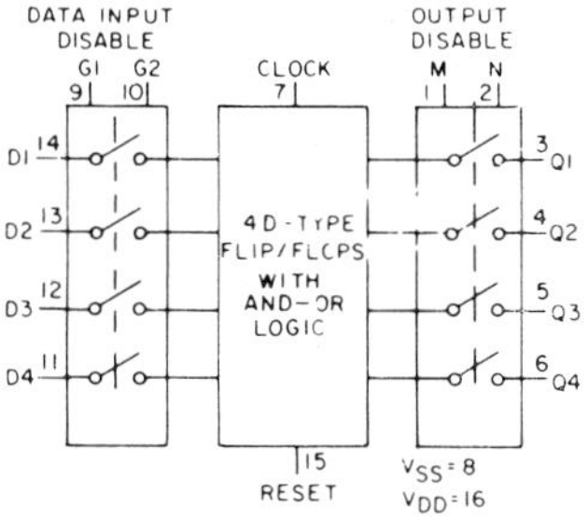
DATA INPUT DISABLE
G1 G2
9 10
CLOCK
7
OUTPUT DISABLE
M N
1 2
D1 14
D2 13
D3 12
D4 11
4 D-TYPE FLIP/FLOPS WITH AND-OR LOGIC
3 Q1
4 Q2
5 Q3
6 Q4
15
RESET
V_{SS} = 8
V_{DD} = 16

CD4077
Quad Exclusive-NOR Gate

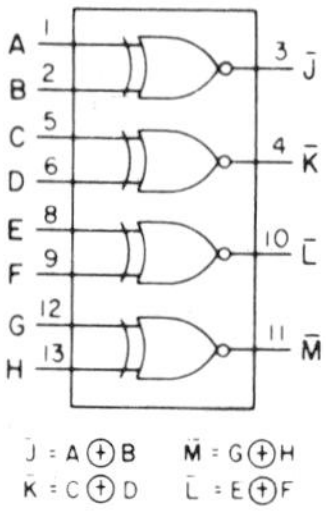
A 1
B 2
C 5
D 6
E 8
F 9
G 12
H 13
3 J̄
4 K̄
10 L̄
11 M̄
J̄ = A ⊕ B
M̄ = G ⊕ H
K̄ = C ⊕ D
L̄ = E ⊕ F

CD4078
8-Input NOR/OR Gate

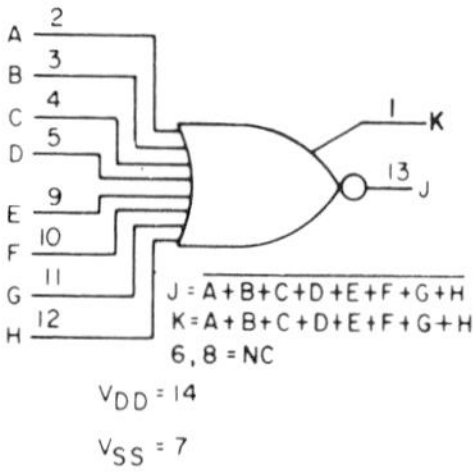
A 2
B 3
C 4
D 5
E 9
F 10
G 11
H 12
1 K
13 J
J = $\overline{A+B+C+D+E+F+G+H}$
K = A+B+C+D+E+F+G+H
6, 8 = NC
V_{DD} = 14
V_{SS} = 7

CD4081
Quad 2-Input AND Gate

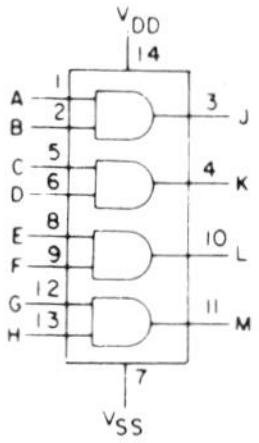
V_{DD}
14
A 1
B 2
C 5
D 6
E 8
F 9
G 12
H 13
3 J
4 K
10 L
11 M
7
V_{SS}

CD4082
Dual 4-Input AND Gate

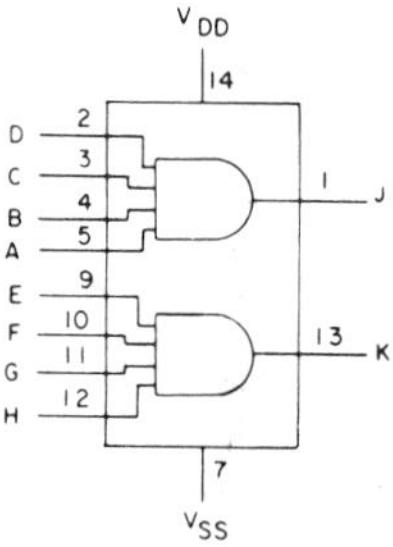
V_{DD}
14
D 2
C 3
B 4
A 5
E 9
F 10
G 11
H 12
1 J
13 K
7
V_{SS}

CD4085
Dual 2-Wide, 2-Input AND-OR-INVERT (AOI) Gate

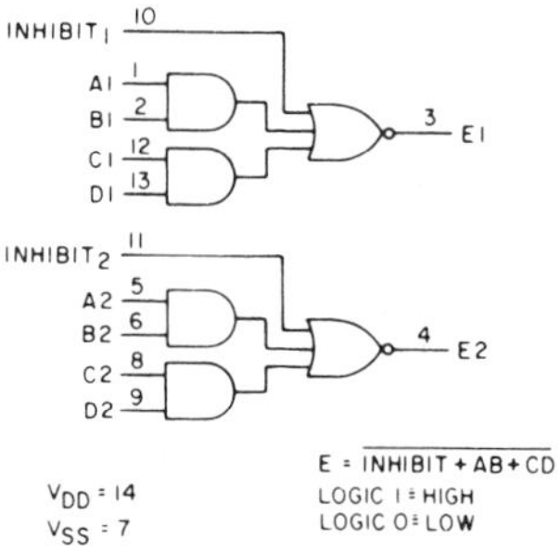
INHIBIT1 10
A1 1
B1 2
C1 12
D1 13
3 E1
INHIBIT2 11
A2 5
B2 6
C2 8
D2 9
4 E2
E = $\overline{INHIBIT + AB + CD}$
LOGIC 1 = HIGH
LOGIC 0 = LOW
V_{DD} = 14
V_{SS} = 7

CD4086
Expandable 4-Wide, 2-Input AND-OR-INVERT (AOI) Gate

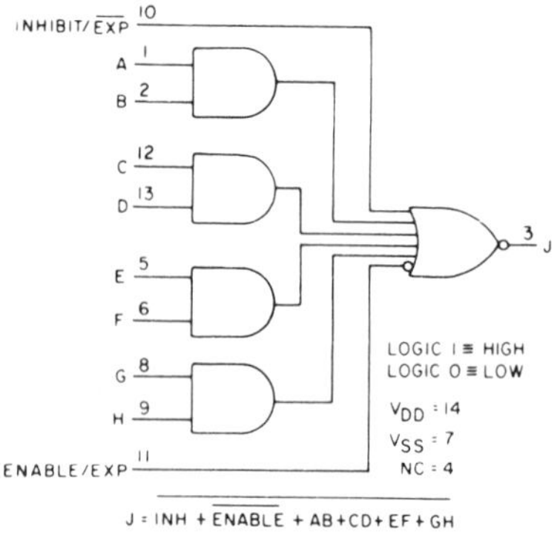
INHIBIT/$\overline{EXP}$ 10
A 1
B 2
C 12
D 13
E 5
F 6
G 8
H 9
3 J
ENABLE/EXP 11
LOGIC 1 ≡ HIGH
LOGIC 0 ≡ LOW
V_{DD} = 14
V_{SS} = 7
NC = 4
J = $\overline{INH + \overline{ENABLE} + AB+CD+EF+GH}$

CD4089
Binary Rate Multiplier

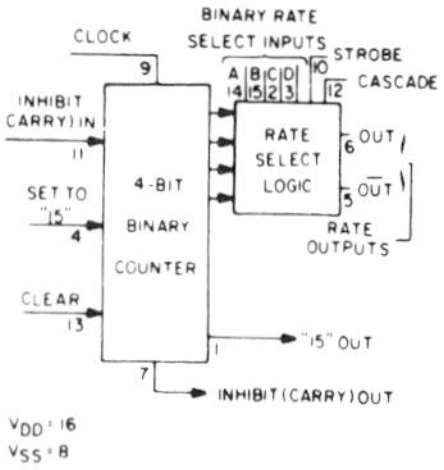
BINARY RATE SELECT INPUTS
CLOCK 9
A B C D
14 15 2 3
10 STROBE
12 CASCADE
INHIBIT (CARRY) IN 11
SET TO "15" 4
CLEAR 13
4-BIT BINARY COUNTER
RATE SELECT LOGIC
6 OUT
5 $\overline{OUT}$
RATE OUTPUTS
1 "15" OUT
7 INHIBIT (CARRY) OUT
V_{DD} = 16
V_{SS} = 8

CD4093

Quad 2-Input NAND Schmitt Trigger

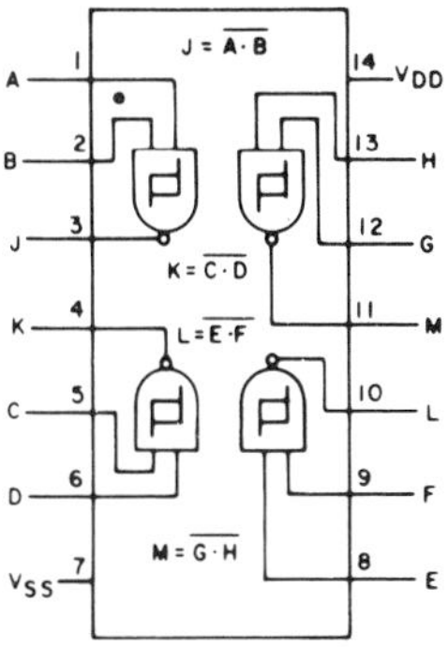
J = $\overline{A \cdot B}$
K = $\overline{C \cdot D}$
L = $\overline{E \cdot F}$
M = $\overline{G \cdot H}$
A 1
B 2
J 3
K 4
C 5
D 6
V_{SS} 7
14 V_{DD}
13 H
12 G
11 M
10 L
9 F
8 E

CD4094

8-Stage Shift-and-Store Bus Register

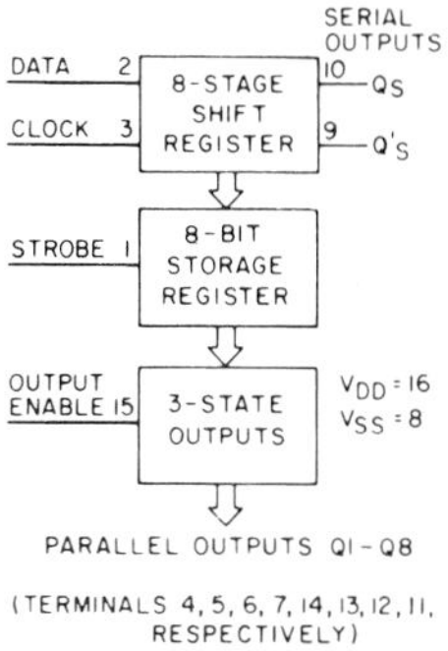
SERIAL OUTPUTS
DATA 2
CLOCK 3
8-STAGE SHIFT REGISTER
10 Q_S
9 Q'_S
STROBE 1
8-BIT STORAGE REGISTER
OUTPUT ENABLE 15
3-STATE OUTPUTS
V_{DD} = 16
V_{SS} = 8
PARALLEL OUTPUTS Q1-Q8
(TERMINALS 4, 5, 6, 7, 14, 13, 12, 11, RESPECTIVELY)

CD4095

Gated J-K Master-Slave Flip-Flop, Non-Inverting Inputs

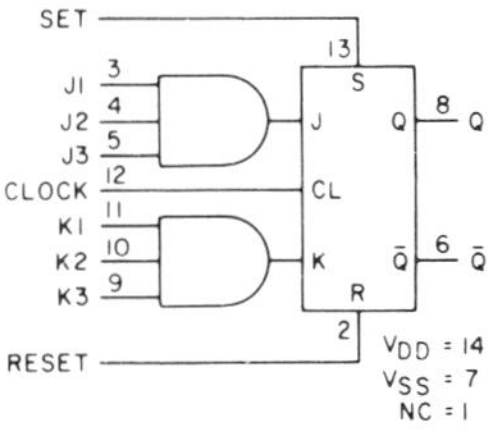
SET 13
J1 3
J2 4
J3 5
CLOCK 12
K1 11
K2 10
K3 9
RESET 2
S
J
CL
K
R
Q 8 Q
$\bar{Q}$ 6 $\bar{Q}$
V_{DD} = 14
V_{SS} = 7
NC = 1

CD4096

Gated J-K Master-Slave Flip-Flop, Inverting and Non-Inverting Inputs

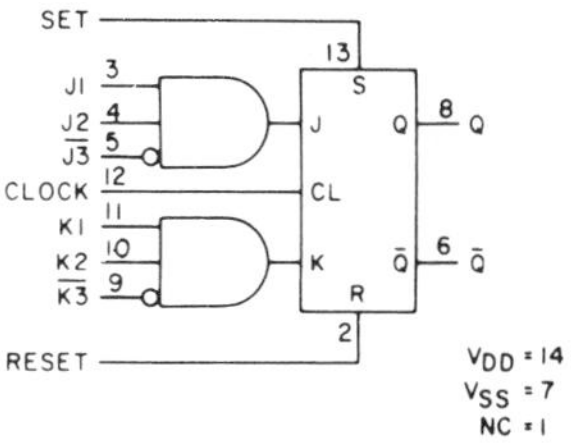
SET 13
J1 3
J2 4
$\overline{J3}$ 5
CLOCK 12
K1 11
K2 10
$\overline{K3}$ 9
RESET 2
S
J
CL
K
R
Q 8 Q
$\bar{Q}$ 6 $\bar{Q}$
V_{DD} = 14
V_{SS} = 7
NC = 1

CD22100

4-by-4 Crosspoint Switch with Control Memory

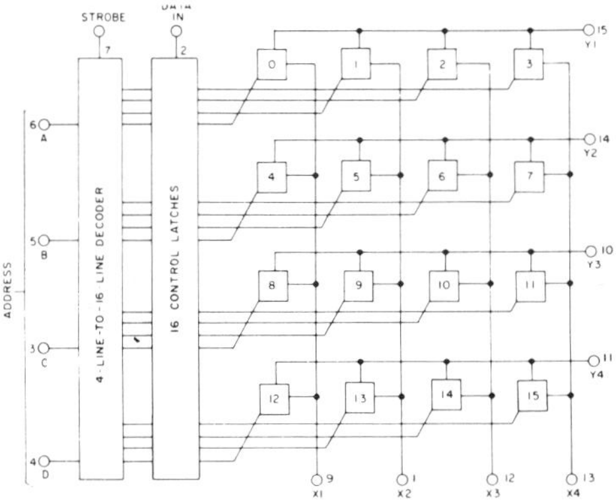
STROBE 7
DATA IN 2
ADDRESS
6 A
5 B
3 C
4 D
4-LINE-TO-16 LINE DECODER
16 CONTROL LATCHES
0 1 2 3
4 5 6 7
8 9 10 11
12 13 14 15
15 Y1
14 Y2
10 Y3
11 Y4
9 X1
1 X2
12 X3
13 X4

CD22101 CD22102
4-by-4-by-2 Crosspoint Switch with Control Memory

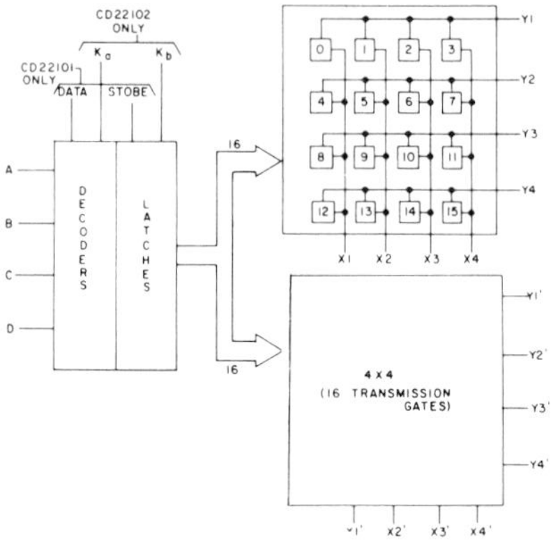

CD40100
32-Stage Static Left/Right Shift Register

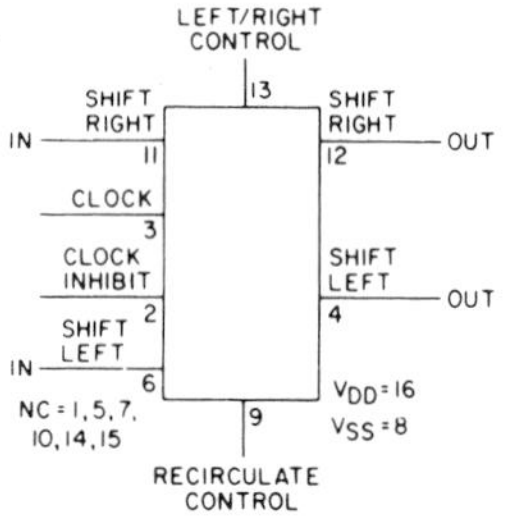
LEFT/RIGHT CONTROL
13
SHIFT RIGHT
IN
11
SHIFT RIGHT
12
OUT
CLOCK
3
CLOCK INHIBIT
2
SHIFT LEFT
4
OUT
SHIFT LEFT
IN
6
NC = 1, 5, 7, 10, 14, 15
9
RECIRCULATE CONTROL
VDD = 16
VSS = 8

CD40101
9-Bit Parity Generator/Checker

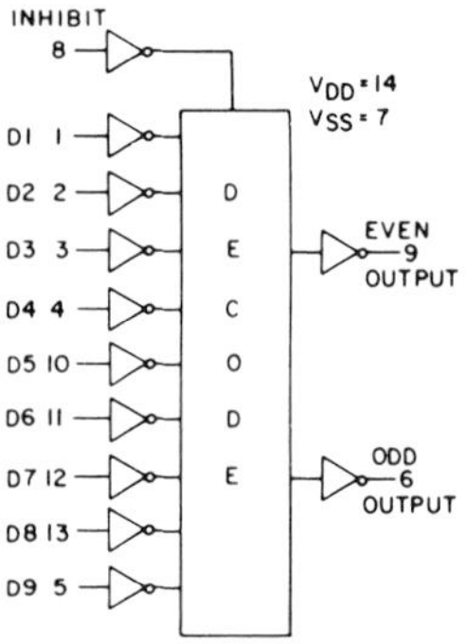
INHIBIT
8
VDD = 14
VSS = 7
D1 1
D2 2
D3 3
D4 4
D5 10
D6 11
D7 12
D8 13
D9 5
DECODE
EVEN 9 OUTPUT
ODD 6 OUTPUT

CD40102 2-Decade BCD
CD40103 8-Bit Binary
8-Stage Presettable Synchronous Down Counter

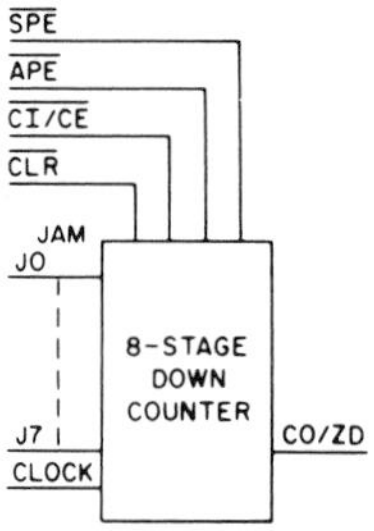
SPE
APE
CI/CE
CLR
JAM
J0
J7
CLOCK
8-STAGE DOWN COUNTER
CO/ZD

CD40105
FIFO Register
4-Bits Wide by 16-Bits Long

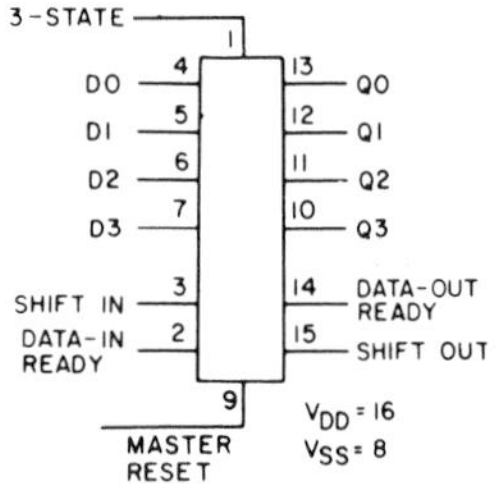
3-STATE
1
D0 4
D1 5
D2 6
D3 7
13 Q0
12 Q1
11 Q2
10 Q3
SHIFT IN 3
DATA-IN READY 2
14 DATA-OUT READY
15 SHIFT OUT
9
MASTER RESET
VDD = 16
VSS = 8

CD40106
Hex Schmitt Trigger

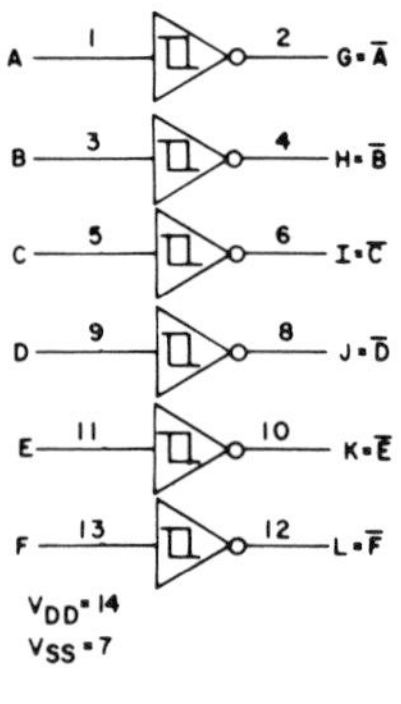
A 1 2 G = Ā
B 3 4 H = B̄
C 5 6 I = C̄
D 9 8 J = D̄
E 11 10 K = Ē
F 13 12 L = F̄
VDD = 14
VSS = 7

CD40107
Dual 2-Input NAND Buffer/Driver

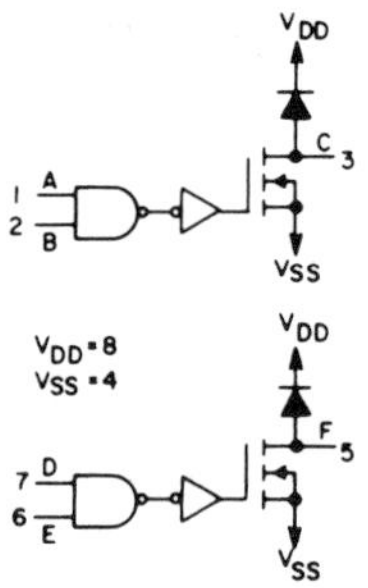
VDD
C
3
1 A
2 B
VSS
VDD
VDD = 8
VSS = 4
F
5
7 D
6 E
VSS

CD40108
4-by-4 Multiport Register

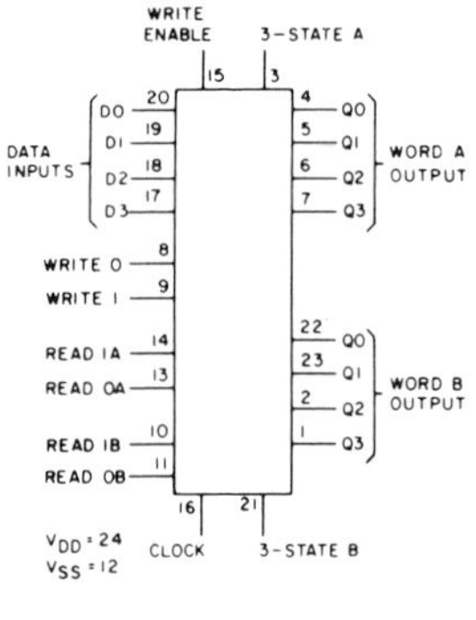
WRITE ENABLE
3-STATE A
15
3
DATA INPUTS
D0 20
D1 19
D2 18
D3 17
4 Q0
5 Q1
6 Q2
7 Q3
WORD A OUTPUT
WRITE 0 8
WRITE 1 9
READ 1A 14
READ 0A 13
READ 1B 10
READ 0B 11
22 Q0
23 Q1
2 Q2
1 Q3
WORD B OUTPUT
16
21
CLOCK
3-STATE B
VDD = 24
VSS = 12

CD40109

Quad Low-to-High Voltage Level Shifter

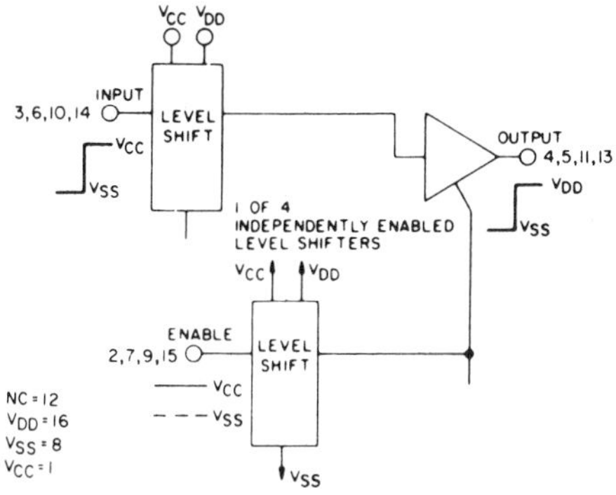

V_{CC} V_{DD}
INPUT
3,6,10,14
LEVEL SHIFT
V_{CC}
V_{SS}
OUTPUT
4,5,11,13
V_{DD}
V_{SS}
1 OF 4 INDEPENDENTLY ENABLED LEVEL SHIFTERS
ENABLE
2,7,9,15
NC = 12
V_{DD} = 16
V_{SS} = 8
V_{CC} = 1

CD40110

Decade Up-Down Counter/ Decoder/Latch/Driver

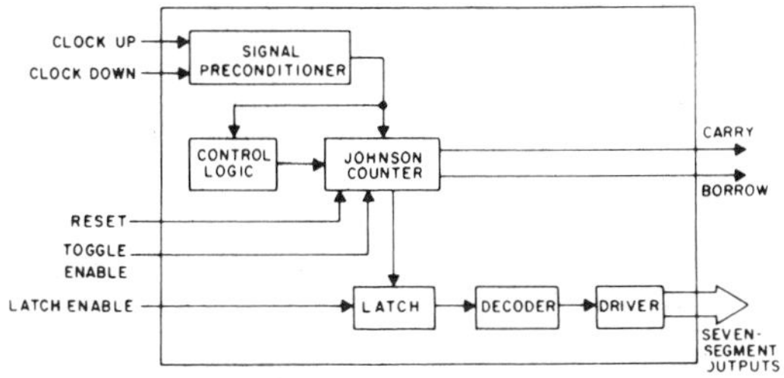

CLOCK UP
CLOCK DOWN
SIGNAL PRECONDITIONER
CONTROL LOGIC
JOHNSON COUNTER
CARRY
BORROW
RESET
TOGGLE ENABLE
LATCH ENABLE
LATCH
DECODER
DRIVER
SEVEN-SEGMENT OUTPUTS

CD40114

64-Bit Random-Access Memory

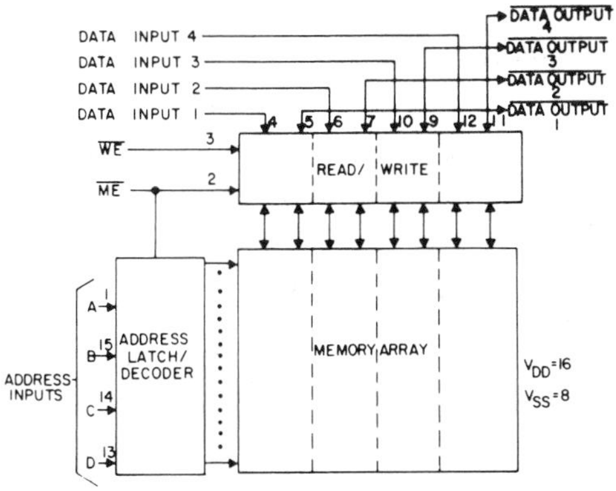

DATA INPUT 4
DATA INPUT 3
DATA INPUT 2
DATA INPUT 1
DATA OUTPUT 4
DATA OUTPUT 3
DATA OUTPUT 2
DATA OUTPUT 1
WE
ME
READ/ WRITE
ADDRESS LATCH/ DECODER
MEMORY ARRAY
ADDRESS INPUTS
V_{DD}=16
V_{SS}=8

CD40115

8-Bit Bidirectional CMOS/TTL Level Converter

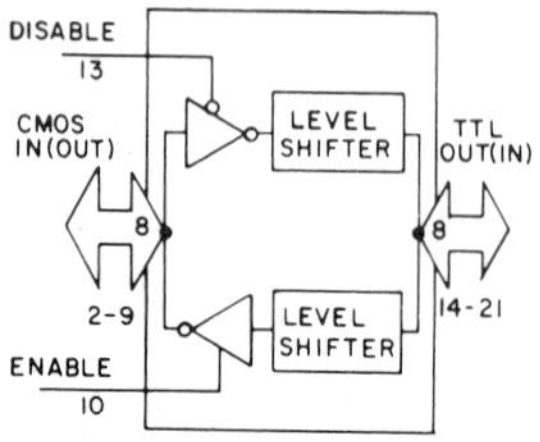
DISABLE
13
CMOS
IN(OUT)
LEVEL
SHIFTER
TTL
OUT(IN)
8
8
2-9
14-21
LEVEL
SHIFTER
ENABLE
10

$V_{DD} = 1$

$V_{CC} = 22$

$V_{SS} = 11$

CD40147

10-Line-to-4-Line BCD Priority Encoder

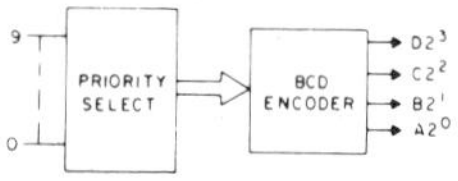
9
0
PRIORITY
SELECT
BCD
ENCODER
D2³
C2²
B2¹
A2⁰

CD40160 Decade with Asynchronous Clear
CD40161 Binary with Asynchronous Clear
CD40162 Decade with Synchronous Clear
CD40163 Binary with Synchronous Clear

Synchronous 4-Bit Counter

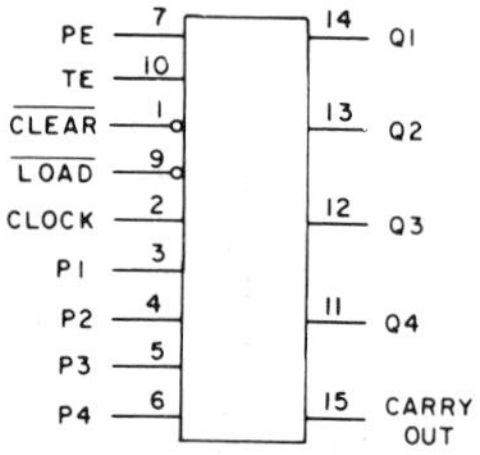
PE 7
TE 10
CLEAR 1
LOAD 9
CLOCK 2
P1 3
P2 4
P3 5
P4 6
14 Q1
13 Q2
12 Q3
11 Q4
15 CARRY OUT

$V_{DD} = 16$

$V_{SS} = 8$

CD40174

Hex "D" Type Flip-Flop

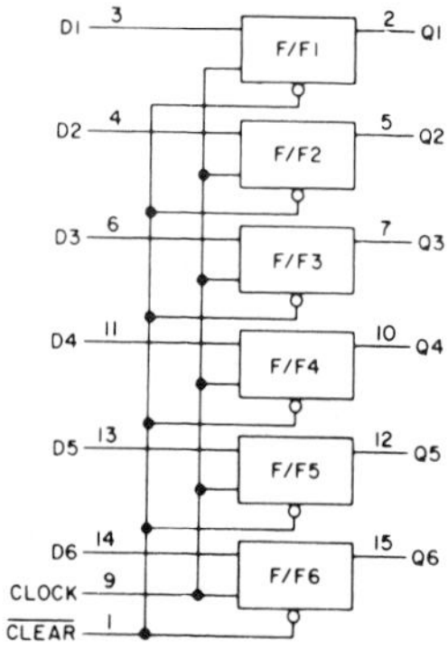
D1 3
D2 4
D3 6
D4 11
D5 13
D6 14
CLOCK 9
CLEAR 1
F/F1
F/F2
F/F3
F/F4
F/F5
F/F6
2 Q1
5 Q2
7 Q3
10 Q4
12 Q5
15 Q6

$V_{SS} = 8$

$V_{DD} = 16$

CD4097

Differential 8-Channel Multiplexer/Demultiplexer

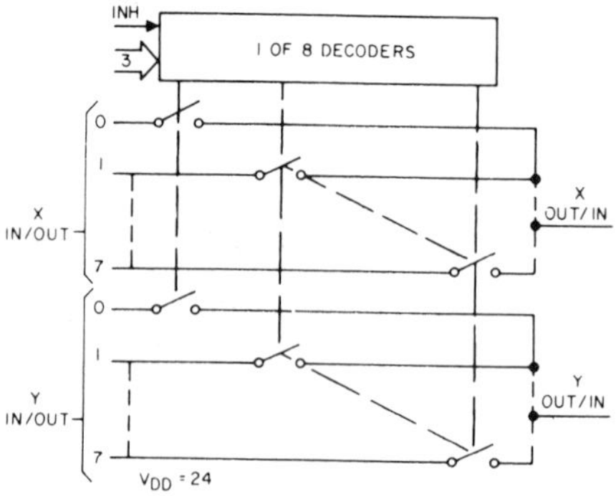
INH
3
1 OF 8 DECODERS
0
1
7
X
IN/OUT
X
OUT/IN
0
1
7
Y
IN/OUT
Y
OUT/IN

$V_{DD} = 24$

$V_{SS} = 12$

CD4098

Dual Monostable Multivibrator

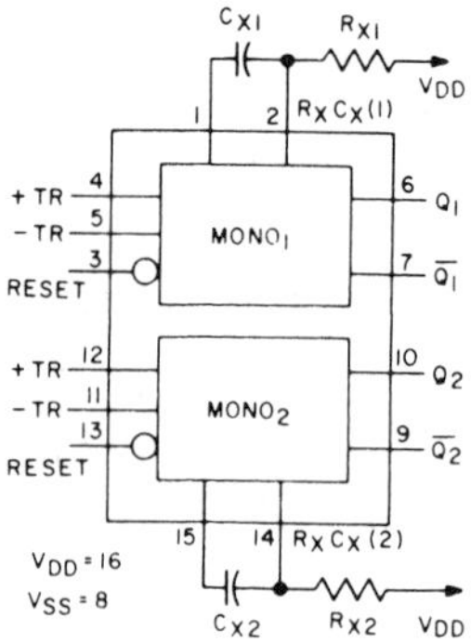
CX1
RX1
VDD
1
2
RX CX (1)
+TR 4
−TR 5
3
RESET
MONO1
6 Q1
7 Q1
+TR 12
−TR 11
13
RESET
MONO2
10 Q2
9 Q2
15
14
RX CX (2)
VDD = 16
VSS = 8
CX2
RX2
VDD

CD4099
8-Bit Addressable Latch

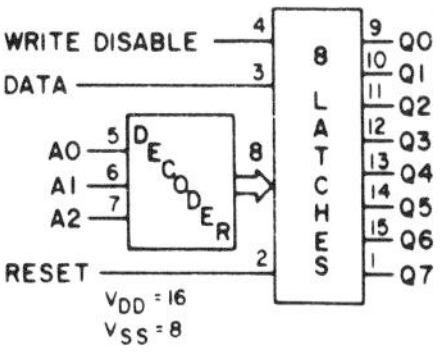
WRITE DISABLE 4
DATA 3
A0 5
A1 6
A2 7
DECODER
8
8 LATCHES
RESET 2
9 Q0
10 Q1
11 Q2
12 Q3
13 Q4
14 Q5
15 Q6
1 Q7
VDD = 16
VSS = 8

CD4502
Strobed Hex Inverter/Buffer

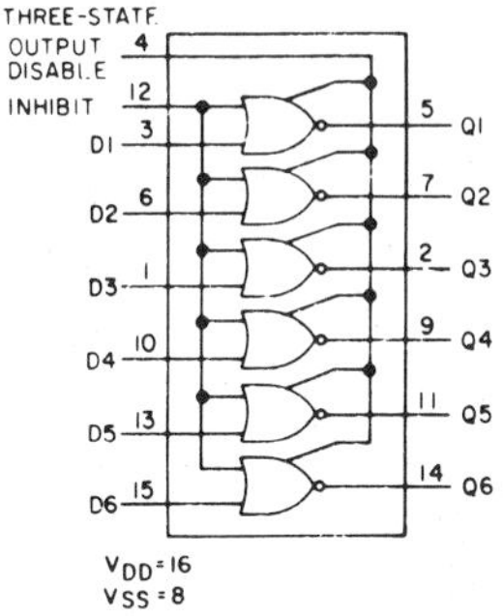
THREE-STATE OUTPUT DISABLE 4
INHIBIT 12
D1 3
D2 6
D3 1
D4 10
D5 13
D6 15
5 Q1
7 Q2
2 Q3
9 Q4
11 Q5
14 Q6
VDD = 16
VSS = 8

CD4508
Dual 4-Bit Latch

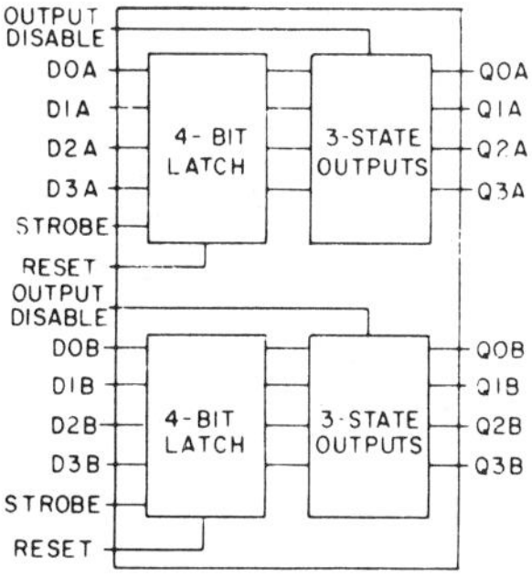
OUTPUT DISABLE
D0A
D1A
D2A
D3A
STROBE
RESET
4-BIT LATCH
3-STATE OUTPUTS
Q0A
Q1A
Q2A
Q3A
OUTPUT DISABLE
D0B
D1B
D2B
D3B
STROBE
RESET
4-BIT LATCH
3-STATE OUTPUTS
Q0B
Q1B
Q2B
Q3B

CD4510
BCD Presettable Up/Down Counter

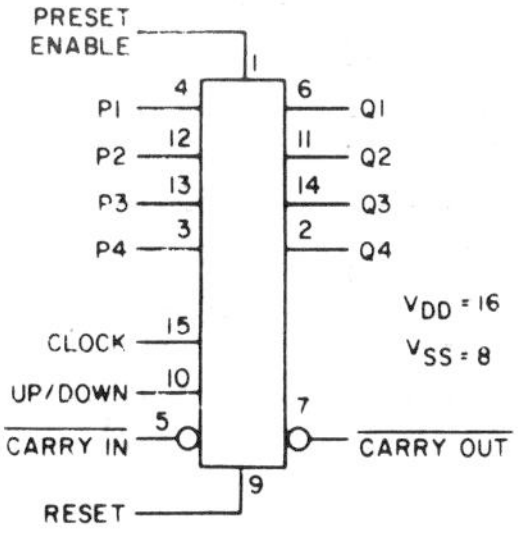
PRESET ENABLE 1
P1 4
P2 12
P3 13
P4 3
6 Q1
11 Q2
14 Q3
2 Q4
VDD = 16
VSS = 8
CLOCK 15
UP/DOWN 10
CARRY IN 5
7 CARRY OUT
RESET 9

CD4511
BCD-to-7-Segment Latch Decoder Driver

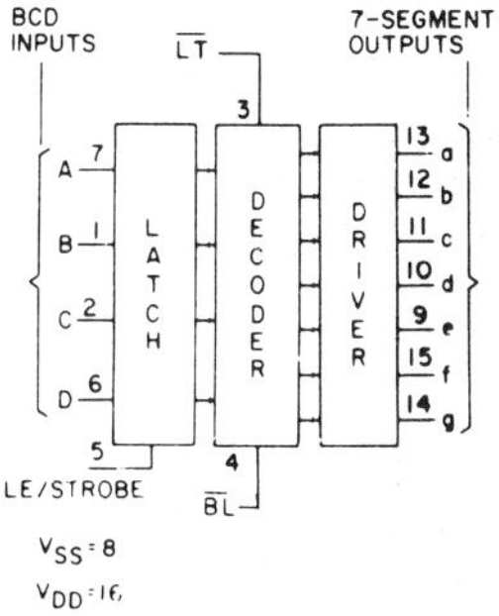
BCD INPUTS
LT
7-SEGMENT OUTPUTS
3
A 7
B 1
C 2
D 6
LATCH
DECODER
DRIVER
13 a
12 b
11 c
10 d
9 e
15 f
14 g
5
LE/STROBE
4
BL
VSS = 8
VDD = 16

CD4512
8-Channel Data Selector

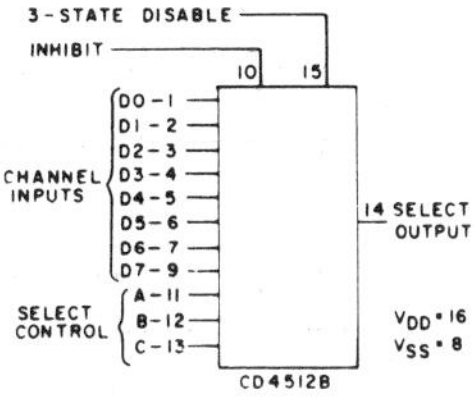
3-STATE DISABLE
INHIBIT
10
15
CHANNEL INPUTS
D0 - 1
D1 - 2
D2 - 3
D3 - 4
D4 - 5
D5 - 6
D6 - 7
D7 - 9
SELECT CONTROL
A - 11
B - 12
C - 13
14 SELECT OUTPUT
VDD = 16
VSS = 8
CD4512B

CD4514
CD4515 Output "Low" on Select
4-Bit Latch/4-to-16 Line Decoder

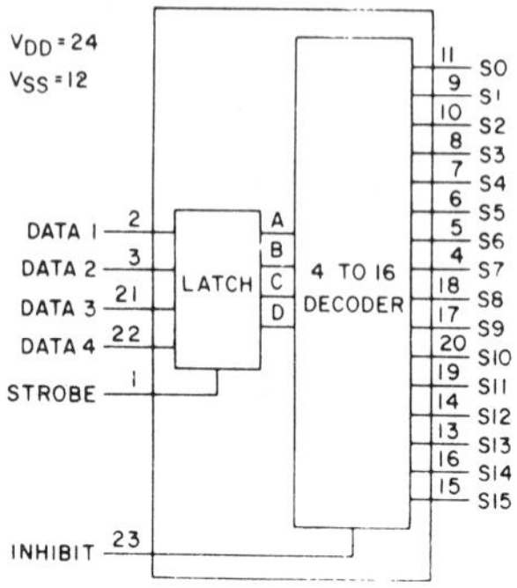

CD4516
Binary Presettable Up/Down Counter

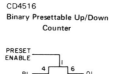

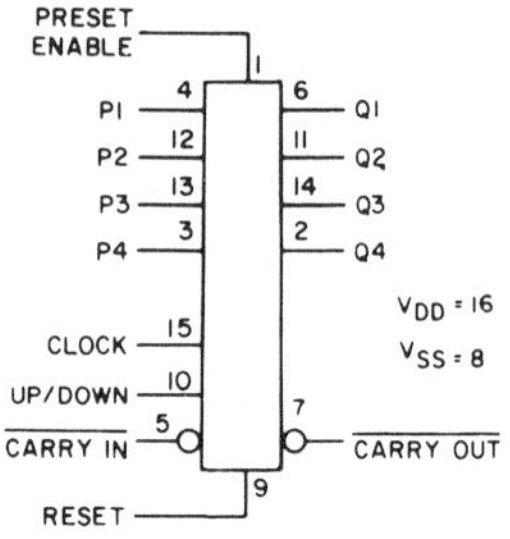

CD4517
Dual 64-Bit Shift Register

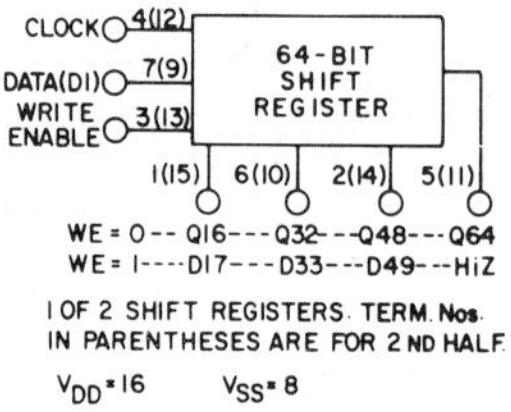

CD4518 BCD
CD4520 Binary
Dual Up Counter

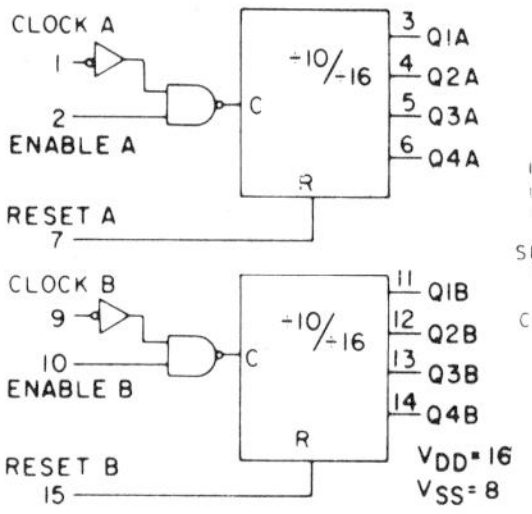

CD4527
BCD Rate Multiplier

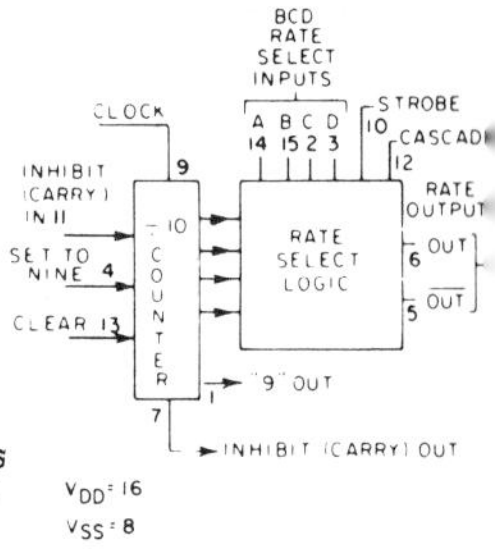

CD4532
8-Bit Priority Encoder

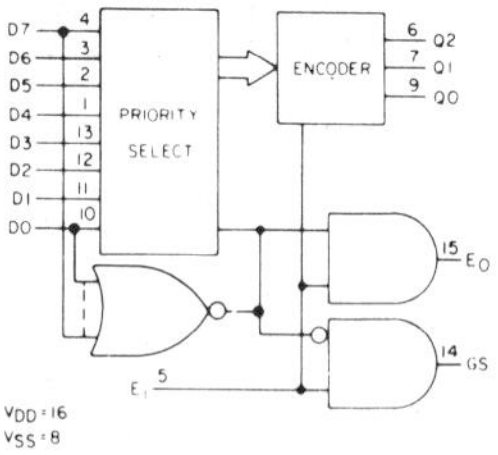

CD4536
Programmable Timer

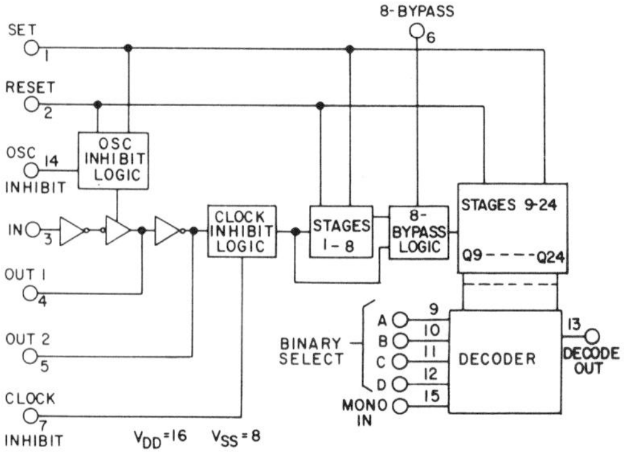

CD4555

Dual Binary-to-1-of-4 Decoder/Demultiplexer Output "High" on Select

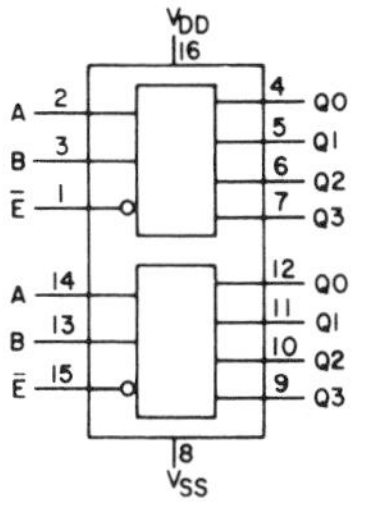
VDD 16
A 2
B 3
Ē 1
4 Q0
5 Q1
6 Q2
7 Q3
A 14
B 13
Ē 15
12 Q0
11 Q1
10 Q2
9 Q3
8 VSS

CD4556

Dual Binary-to-1-of-4 Decoder/Demultiplexer Output "Low" on Select

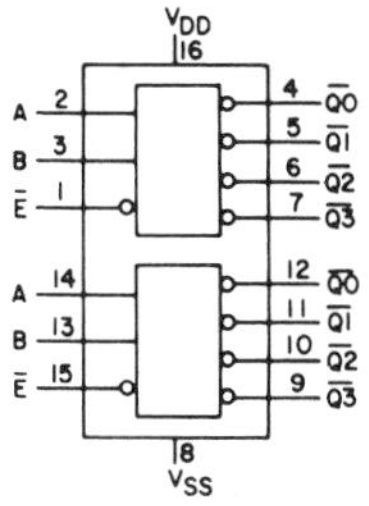
VDD 16
A 2
B 3
Ē 1
4 Q̄0
5 Q̄1
6 Q̄2
7 Q̄3
A 14
B 13
Ē 15
12 Q̄0
11 Q̄1
10 Q̄2
9 Q̄3
8 VSS

CD4585

4-Bit Magnitude Comparator

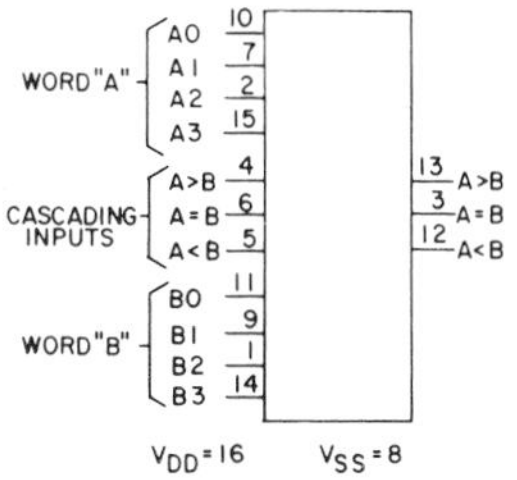
WORD "A"
A0 10
A1 7
A2 2
A3 15
CASCADING INPUTS
A>B 4
A=B 6
A<B 5
WORD "B"
B0 11
B1 9
B2 1
B3 14
13 A>B
3 A=B
12 A<B
VDD=16
VSS=8

CD4724

8-Bit Addressable Latch

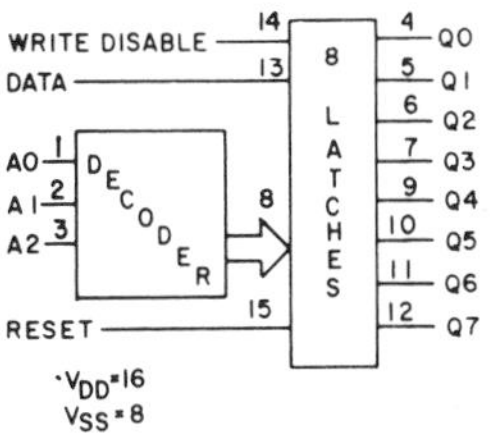
WRITE DISABLE 14
DATA 13
A0 1
A1 2
A2 3
DECODER
8
8 LATCHES
RESET 15
4 Q0
5 Q1
6 Q2
7 Q3
9 Q4
10 Q5
11 Q6
12 Q7
VDD=16
VSS=8

CD40181

4-Bit Arithmetic Logic Unit

Active-Low Data

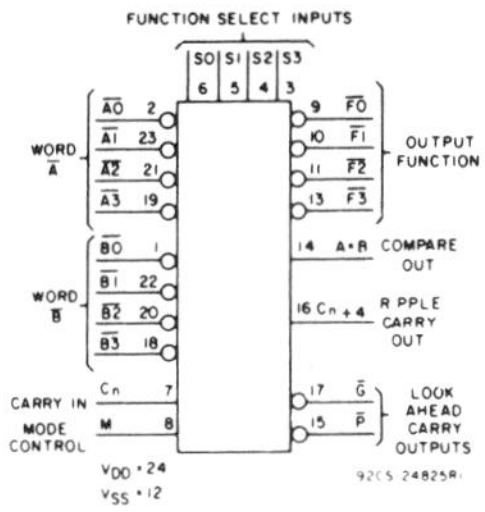
FUNCTION SELECT INPUTS
S0 S1 S2 S3
6 5 4 3
WORD A
WORD B
CARRY IN
MODE CONTROL
OUTPUT FUNCTION
COMPARE OUT
RIPPLE CARRY OUT
LOOK AHEAD CARRY OUTPUTS
VDD=24
VSS=12

Active-High Data

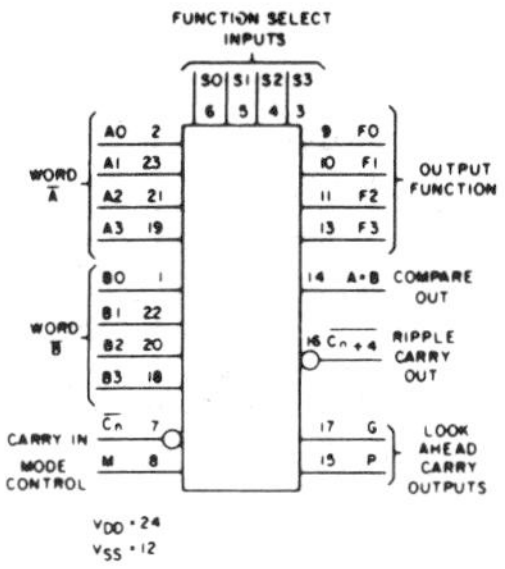
FUNCTION SELECT INPUTS
S0 S1 S2 S3
6 5 4 3
WORD A
WORD B
CARRY IN
MODE CONTROL
OUTPUT FUNCTION
COMPARE OUT
RIPPLE CARRY OUT
LOOK AHEAD CARRY OUTPUTS
VDD=24
VSS=12

CD40182

Look-Ahead Carry Generator

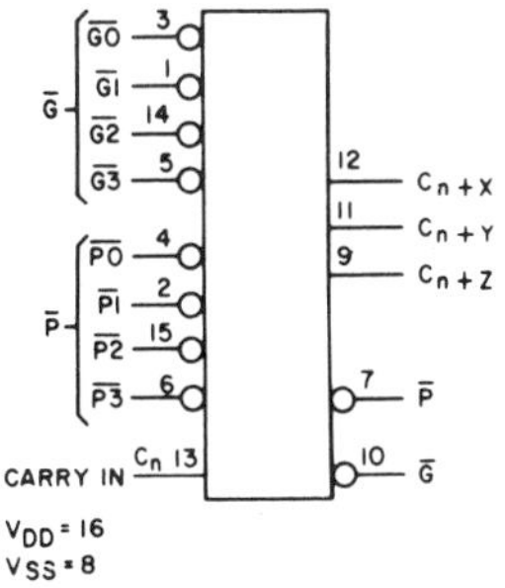
Ḡ
Ḡ0 3
Ḡ1 1
Ḡ2 14
Ḡ3 5
P̄
P̄0 4
P̄1 2
P̄2 15
P̄3 6
CARRY IN Cn 13
12 Cn+X
11 Cn+Y
9 Cn+Z
7 P̄
10 Ḡ
VDD=16
VSS=8

CD40192 BCD

CD40193 Binary

Presettable Up/Down Counter (Dual Clock with Reset)

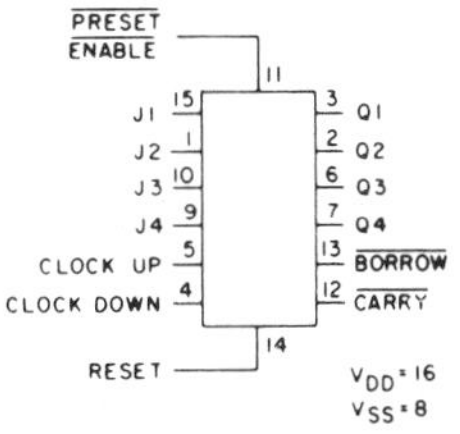
PRESET ENABLE 11
J1 15
J2 1
J3 10
J4 9
CLOCK UP 5
CLOCK DOWN 4
RESET 14
3 Q1
2 Q2
6 Q3
7 Q4
13 BORROW
12 CARRY
VDD=16
VSS=8

CD40208

4-by-4 Multiport Register

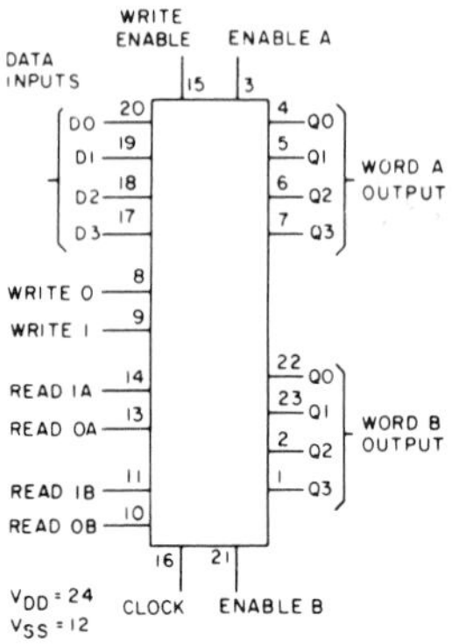
WRITE ENABLE
ENABLE A
DATA INPUTS
D0
D1
D2
D3
20
19
18
17
15
3
4
5
6
7
Q0
Q1
Q2
Q3
WORD A OUTPUT
WRITE 0
WRITE 1
8
9
READ 1A
READ 0A
READ 1B
READ 0B
14
13
11
10
22
23
2
1
Q0
Q1
Q2
Q3
WORD B OUTPUT
16
21
CLOCK
ENABLE B
VDD = 24
VSS = 12

CD40257

Quad 2-Line-to-1-Line Data Selector/Multiplexer

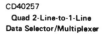

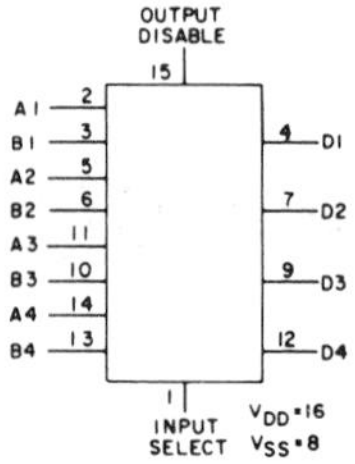
OUTPUT DISABLE
15
A1
B1
A2
B2
A3
B3
A4
B4
2
3
5
6
11
10
14
13
4
7
9
12
D1
D2
D3
D4
1
INPUT SELECT
VDD = 16
VSS = 8

Chapter 6

Transferring Digital Data

Serial and parallel

Some applications of digital circuitry make use of the digital data at the time when it is obtained. A digital voltmeter, for example, displays the obtained data as soon as the data is collected, and refreshes this data at intervals of, typically, half a second. Even in this case, however, there is some storage in the sense that the display retains the reading until a new reading is made. The storage is transitory, however, using latches, and there is no requirement to transfer digital data from one piece of equipment to another. In many other applications, however, and particularly for computing, data has to be transferred over distances that range from a metre or less up to the maximum distance that a radio signal can reach. In this chapter, we shall look at data transfer methods.

The simplest method of transferring digital data is to connect to a microprocessor bus, usually by way of buffer or driver circuits. The word 'buffer' is used here in the electronics sense of a circuit that acts as an impedance transformer, reducing loading on the source. The computing term means a piece of memory used to gather up data until it is needed. The transfer of data from one board to another in a digital system makes use of the microprocessor bus either directly or by way of buffer circuits, the *bus drivers*. In such connections, all of the microprocessor signals are transferred, including data lines, address lines and all of the synchronising and timing lines.

The more difficult requirement is the transfer of digital data between different pieces of equipment, of which the most common example in computing is the use of a printer. For industrial and instrumentation purposes, the requirements are much more varied but the basic methods are much the same. The choice is of either parallel or serial transmission and reception, and in many cases only serial transmission is possible. No matter which method is used, some form of synchronisation will be needed because the rate at which data can be received by a device such as a printer is never as fast as the rate at which it can be transmitted. Both serial and parallel data transfer systems must therefore ensure synchronisation of transmission and reception, and the problem is more acute for serial links. The signals that are used for this purpose are called *handshaking* signals.

Parallel transmission means that all the data lines of the microprocessor bus, or a set of data lines, will be used, along with a few synchronising lines. For instrumentation purposes, a more complete set of signals will be needed than is the case for a computer printer. Many of the microprocessors used in industrial equipment are of the 8-bit variety, and for parallel transmission of data all eight data lines will be used. In applications which use 16-bit microprocessors, industrial requirements will usually call for all sixteen data lines to be used in a parallel connection, but computer printers require only eight lines at most, and some use only seven.

The difference is owing to the way that data is used. Transfer of data for instrumentation purposes will generally require as many data lines as the microprocessor itself can use, because the nature of the data will require the use of all the lines. Printable characters in computing applications make use of the ASCII code (see *Figure 5.22*) whose numbers can be expressed by a 7-bit binary code, so that only seven lines of data are ever required for a connection to a printer for text. Some printers make use of (non-standard) codes in the range 128 to 255, and these require eight binary bits and, hence, eight data lines.

The main problems of parallel data transfer are of line length and pulse frequency. These two problems are interconnected because they both arise from the stray capacitances between the leads of the cable. A parallel cable will be driven from a low impedance source (the usual double emitter-follower type of circuit) and will connect in to a comparatively high impedance at the receiver end. The stray capacitance between leads, together with the very fast rise and fall times of the pulses, can therefore induce a 1 signal in a line which should be at level 0. The longer the line, the greater the induced signal, until the voltage becomes great enough to drive the receiver circuit, at which point a false signal will be received.

The practical effect is to restrict parallel printer leads from computers to 1–2 metres. Greater lengths can be obtained by using correctly matched 50 Ω lines, but these are rare in computing applications though fairly common for instrumentation applications. Long parallel links can be used if repeaters are connected at intervals. The repeater consists of

Schmitt trigger circuits which will give a 1 output only for signals which reach a preset minimum level, and a 0 output for all others, combined with a very low output impedance. Repeaters for computer printer leads are comparatively simple (but only recently available), because most of the signals are in one direction only – from computer to printer – but for instrumentation purposes the repeaters will normally be required to cope with signals in either direction.

Table 6.1 shows the signals that are used in a Centronics printer output, in this case from a computer that is IBM compatible. Not all printers make use of all of these signals, nor do all computers, but the Centronics standard is sufficiently flexible to ensure that any computer that provides a parallel printer output to Centronics standard can be matched to any printer with a Centronics input. There are various minor deviations between printers and computers, all of which can be dealt with by omitting one or more links in the cable. Plugs and sockets are standardised only at the printer end of the cable which uses a 36-pin plug of the Amphenol 57-30360 type. The plug and socket that is used at the computer end of the cable is not standardised, and small computers in particular seem to use any plug/socket that comes to hand. The IBM compatible machines use a 25-pin D-connector for both the parallel output and the serial input/output, with a socket used for the serial connection and a plug for the parallel printer connection.

Only about twenty-two pins of the 36-pin connector are likely to be used, and on the 36-pin connector, pins 2 to 9 inclusive are used for the

Table 6.1. THE SIGNALS AT THE CENTRONICS OUTPUT OF A TYPICAL COMPUTER

Signal pin	*Return pin*	*Name*	*Notes*
1	19	$\overline{\text{STROBE}}$	Low to send data
2	20	DØ	
3	21	D1	
4	22	D2	
5	23	D3	
6	24	D4	Data bus
7	25	D5	
8	26	D6	
9	27	D7	
10	28	$\overline{\text{ACKNOWLEDGE}}$	Low when printer ready
11	29	BUSY	High when printer not ready
12	30	PE	High when out of paper
13	–	SELECT	On/off line (out)
14	–	AUTOFEED	
15	32	ERROR	Out of paper, off line, error
16	–	INIT	Reset
17	–	$\overline{\text{SLCT IN}}$	On/off line (in)
18	–	NC	

eight data lines of the data bus, D0 to D7. The Centronics standard provides for signal return lines, each at signal ground level, so that twisted pairs of signal/ground return wires can be used. Return pins 20 to twenty-seven inclusive are used in this way for data, but this provision is not always used, particularly for short cables. Pin 1 (return pin 19) handles a strobe signal which is driven low by the computer in order to send data to the printer. The width of the strobe pulse must be at least 0.5 μs, and this pulse is one of three that forms the main handshaking provisions in this type of interface. The other two are BUSY (pin 11, return on 29) and ACKNOWLEDGE (pin 10, return on 28).

The BUSY signal is a steady-level signal that is set high by the printer to indicate that no more data can be received. The BUSY line is taken high when data starts to be entered, during printing and when the printer is off-line, or disabled because of a fault. Most printers contain buffer memory which can range from one line to several pages of printed characters, and the BUSY signal is taken high when this buffer is full. For such a printer, transmission of signals is intermittent because of the time needed to fill the buffer at the normal rate of parallel transmission, which is as fast as the signals can be clocked (subject to the minimum pulse widths that can be used). The use of a printer buffer allows the computer to be used during a print out, and some computers provide buffering in their own memory to assist this detachment of printing from other operations. The snag is that if you want to stop the printer you cannot do so immediately by stopping data being sent out from the computer because the printer will stop only when the buffer is empty. You can, of course, switch off or reset the printer, but this will empty the buffer and you will need to retransmit this data when the printer is ready for use again.

The ACKNOWLEDGE pulse signal is sent out by the printer to indicate that the printer has received data and is now ready to receive more data; usually when the buffer is empty. The relationship of these signals to each other is fairly flexible, as the timing diagrams of *Figure 6.1* indicate. These have been taken from a computer manual and two printer manuals, and they show noticeable differences in the way that the pulses are related, though all are within the tolerance of the Centronics standards. The important point is the maintenance of the 0.5 μs minimum pulse width and timing intervals, so that the clock rate of output is usually considerably lower than the clock rate for the computer itself.

The data, strobe, BUSY and ACKNOWLEDGE signals are the most important parts of the Centronics interfacing, and the remaining signals and their uses are summarised in *Table 6.2*. There is some minor variation between printers in the use of these signals. Not all printers, for example, make use of the Autofeed signal, but such variations are not generally important unless you are trying to use one printer with a

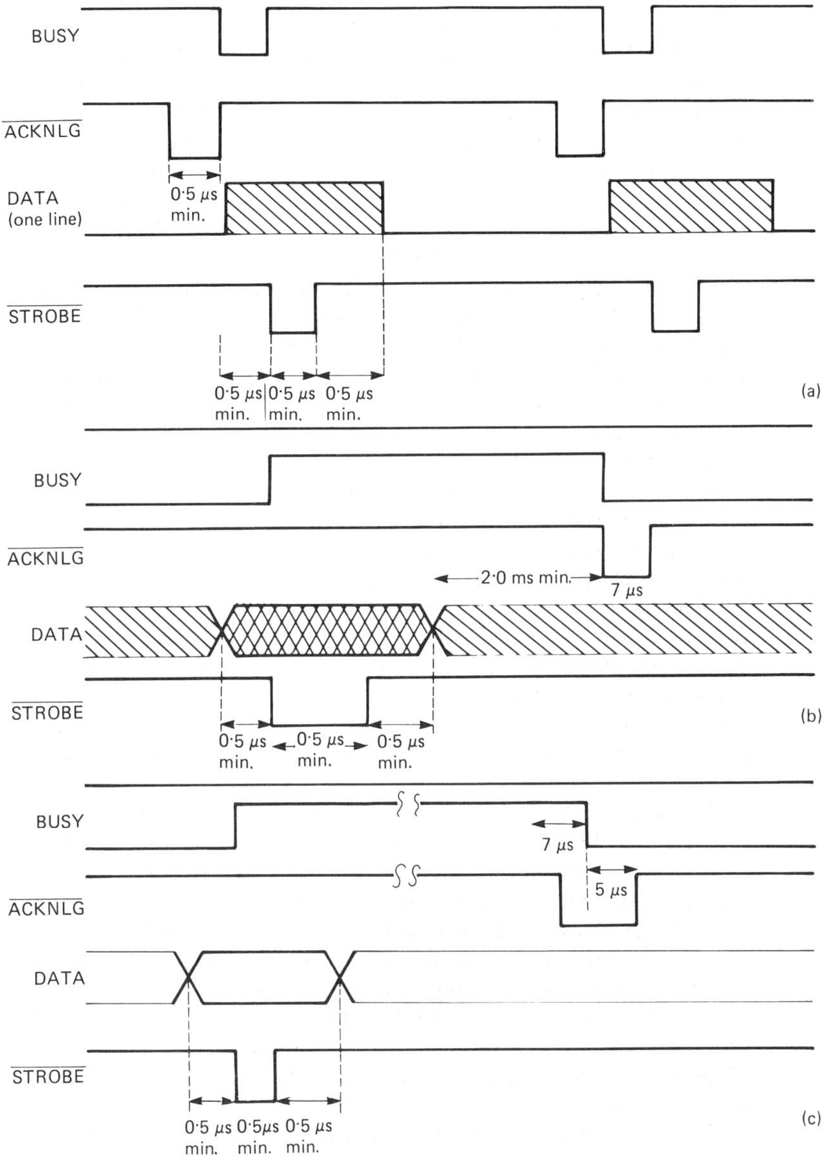
BUSY
ACKNLG
DATA (one line)
0·5 μs min.
STROBE
0·5 μs min.
0·5 μs min.
0·5 μs min.
(a)
BUSY
ACKNLG
2·0 ms min.
7 μs
DATA
STROBE
0·5 μs min.
0·5 μs min.
0·5 μs min.
(b)
BUSY
7 μs
ACKNLG
5 μs
DATA
STROBE
0·5 μs min.
0·5μs min.
0·5 μs min.
(c)

Figure 6.1. Typical timing specifications. (a) Computer, (b) Daisywheel printer, (c) Dot-matrix printer

Table 6.2. THE OTHER SIGNALS OF THE CENTRONICS SET

Signal	*Direction*	*Use*
PE	From printer	High when out of paper
$\overline{\text{AUTOFEEDXT}}$	To printer	Low to feed one line after printing
INIT	To printer	Low to reset printer and clear buffer. Width > 50 μs
$\overline{\text{ERROR}}$	From printer	Low to indicate no paper, off line error
$\overline{\text{SLCT IN}}$	To printer	Low to allow data to be sent

cable that was intended for another. It is very unusual to find major problems of compatibility between computers and printers using the Centronics interface – one example in the past was owing to the computer manufacturer earthing pin 14 so that printers which used the Autofeed signal were forced to take an additional line spacing.

The IEEE-488 bus

The IEEE-488 bus is a parallel data transfer system for connecting complete systems rather than parts of systems and it is widely used in digital electronic instrumentation. The standard dates back to 1974 and much of the detail of the standard is owing to Hewlett-Packard, who hold patents on the handshaking method. Because of this, a licence must be purchased if this handshaking method is used.

The bus is used to connect devices that can carry out actions described as controlling, listening and talking. A device might carry out just one of the functions, any two of these functions, or all three. A controller device will control other devices and is almost always a microcomputer or microprocessor-based controller system. A talker device will place data on to the bus, but does not receive data, and a listener will receive data from the bus but does not place any data on the bus. A counter might, for example, be connected as a talker, placing the data from its count on to the bus but not receiving any data (though it would obey command signals) from the bus. By contrast, a signal generator might be used as a listener, generating signals as commanded by data read from the bus, though not placing any digital signals on to the bus. Many devices will be used as talkers and listeners, receiving signals from the bus (for changing range or function) and placing signals on to the bus to indicate readings. Since the IEEE-488 is primarily intended for instrumentation, the prime example of a talker/listener is a digital multimeter.

The bus, like the Centronics parallel system, consists mainly of data lines with no address information and uses a total of sixteen lines on a

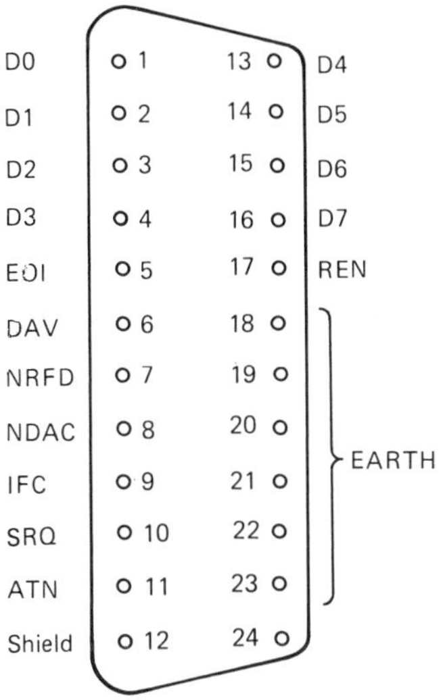
D0
D1
D2
D3
EOI
DAV
NRFD
NDAC
IFC
SRQ
ATN
Shield
1
2
3
4
5
6
7
8
9
10
11
12
13
14
15
16
17
18
19
20
21
22
23
24
D4
D5
D6
D7
REN
EARTH

Figure 6.2. The IEEE-488 connector pinout

24-pin connector. Of these, eight are data lines that are bidirectional, five are bus control lines, and three are handshaking lines. *Figure 6.2* shows the standard pin layout, with data on lines 1 to 4 and 13 to 16 inclusive. The handshaking lines are on pins 6, 7, and 8, with the 'transfer-control' lines on 5, 9 to 11, and 17. The handshaking lines use open-collector outputs, active low, so that these lines can be connected to other output lines (wired-OR connection) without risk to the internal circuitry.

The handshaking lines are DAV (data valid) on pin 6, NRFD (not ready for data) on pin 7 and NDAC (no data accepted) on pin 8. The DAV signal is sent out by a talker device to signal that data has been placed on the data lines and is valid for use. The other two lines are controlled by listeners, with NRFD signifying not ready for accepting the data, and NDAC signifying that data has not been read. When both NRFD and NDAC lines go high, the data is read. The action for a single talker and listener is as shown in *Figure 6.3*. The DAV line from the talker remains high even in the presence of data until the NRFD signal goes high. This is not such a simple action when several listeners are present because the NRFD line is ANDed; it cannot go high until all listening devices are ready. When the NRFD line from the listener(s) goes high, the talker activates the DAV line (low state) so that data can be transferred. The data transfer is complete when the NDAC line rises to the high level and the rate of transfer is controlled by the slowest

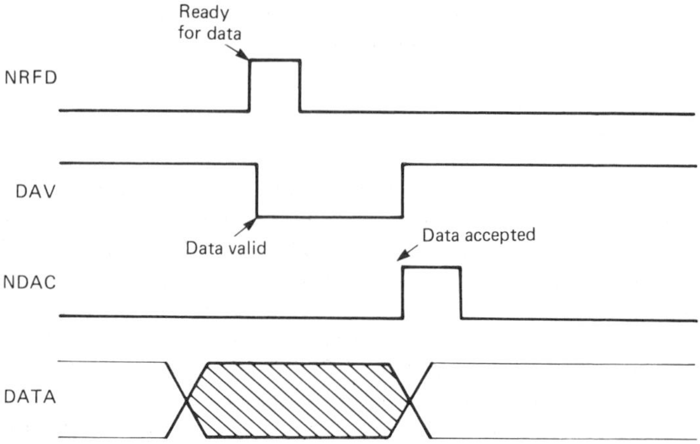
Ready
for data
NRFD
DAV
Data valid
Data accepted
NDAC
DATA

Figure 6.3. The timing of a single talker-listener IEEE-488 exchange. Where more than one listener exists, data is not valid until all listeners are ready

listener. Typical maximum rates range between 50 and 250 kilobytes per second if the listeners are fast-acting devices.

The bus control lines are used to determine how devices interact with the controller. The simplest of these is the interface clear on pin 9, taken low to reset the system in preparation for use. By contrast, the end or identify (EOI) signal on pin 5 is used to indicate that data transfer is complete. The attention (ATN) line, pin 11, decides the use of the eight data lines. When this pin voltage is low, all eight data lines are used for data, which need not necessarily use all eight data lines. When the ATN pin voltage is high, the lower lines of the data bus are used to hold an address number to which a specific device will respond. In the absence of such a specific address, all listening devices can receive signals sent over the data lines.

The other two bus control signals are service request (SRQ) and remote enable (REN). The SRQ, pin 10, is used by a device to indicate to the controller that the device needs attention. This is the equivalent of an interrupt signal to a microprocessor, and is normally used by a talker to indicate that it has data to transfer, or by a listener which needs data. The REN signal allows any device to be operated either from the IEEE-488 bus (remotely) or locally, as from its own front panel or from a test connector.

Though the IEEE-488 bus is used to a considerable extent in the computer control of electronics systems, few small computers have been equipped with the bus, and for the most popular controller computer, the IBM PC (and its many 'clones'), the IEEE-488 bus interface is fitted as an extra by way of a plug-in card. More specialised microprocessor controllers, however, use the IEEE-488 bus as a standard interface in addition to the serial RS-232.

Serial transfer

The serial transfer of data makes use of only one line (plus a ground return) for data, with the data being transmitted one bit at a time. The standard system is known as RS-232, and has been in use for a considerable time. Unfortunately, because the standard is so old and its full implementation is so seldom required nowadays, many manufacturers have made use of serial transfer systems that look like RS-232 and are sometimes referred to as RS-232, but which are not to RS-232 standards. One obvious deviation concerns signal voltages. The original RS-232 system called for voltage levels of around +15 V and −15 V to be used for logic 1 and 0 levels respectively. Many modern systems make use of the TTL levels of +5 V and 0 V in a system that is otherwise RS-232 but which cannot be compatible with a true RS-232 system.

The second complication of RS-232 relates to its two different uses. When RS-232 was originally specified, two types of device were specified as data terminal equipment (DTE) and as data communications equipment (DCE). A DTE device can send out or receive serial signals, and is a terminal in the sense that the signals are not routed elsewhere.

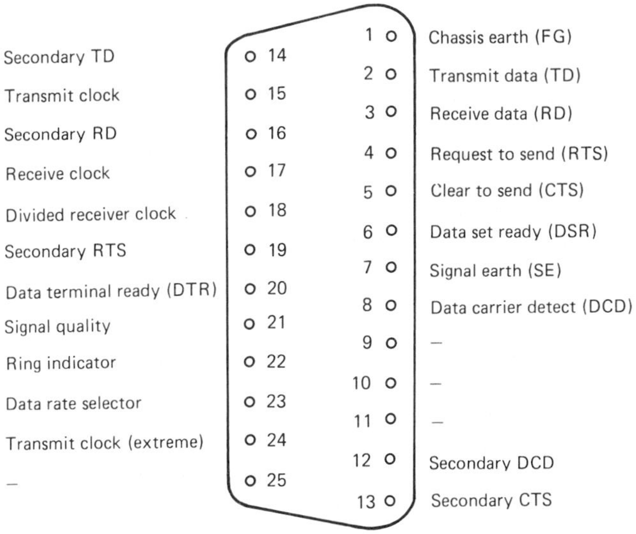

Secondary TD
Transmit clock
Secondary RD
Receive clock
Divided receiver clock
Secondary RTS
Data terminal ready (DTR)
Signal quality
Ring indicator
Data rate selector
Transmit clock (extreme)
—
14
15
16
17
18
19
20
21
22
23
24
25
1
2
3
4
5
6
7
8
9
10
11
12
13
Chassis earth (FG)
Transmit data (TD)
Receive data (RD)
Request to send (RTS)
Clear to send (CTS)
Data set ready (DSR)
Signal earth (SE)
Data carrier detect (DCD)
—
—
—
Secondary DCD
Secondary CTS

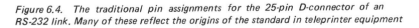

Figure 6.4. The traditional pin assignments for the 25-pin D-connector of an RS-232 link. Many of these reflect the origins of the standard in teleprinter equipment

A DCE device is a half-way house for signals, like a modem which converts serial data signals into tones for communication over telephone lines or converts received tones into digital signals. The original conception of RS-232 was that a DTE device would always be connected to a DCE device, but with the development of microcomputers and their associated printers it is now more common to need to connect two DTE devices to each other. This requires the connections in the cable to be changed, as we shall see. The original specification also stipulated that DTE equipment would use a male connector (plug) and the DCE equipment would use a female connector (socket), but you are likely to find either gender of connector on either type of device nowadays.

The original specification was for a connecting cable of twenty-five leads, as shown in *Figure 6.4*. Many of these reflect the use of old-fashioned telephone equipment and teleprinters, and very few applications of RS-232 now make use of more than eight lines. The standard connector is the D-type 25 pin, but even in this respect standards are widely ignored and some manufacturers use Din, Amphenol or Cannon connectors. Worse still, some equipment makes use of the full 25-pin systems, but uses the 'spare' pins to carry other signals or even d.c. supply lines. The moral is that any link that is alleged to be RS-232 must be regarded with suspicion unless the wiring is known from a wiring diagram or from investigation of the connections.

The majority of RS-232 links can make use of eight pins of the 25-pin connector, and these are pins 1 to 7 and 20. *Table 6.3* shows this arrangement on a 25-pin connector; for a lot of equipment, the distinction between chassis earth (pin 1) and signal earth (pin 7) is not necessary. The main data pins are pin 2, the output pin for data transmitted from the DTE to the DCE, and pin 3, the input pin for data from the DCE to the DTE. The use of separate transmit and receive lines means that the serial channel can be used in duplex, allowing transmission and reception of data simultaneously. A full RS-232 implementation allows for two sets of these transmit and receive lines.

Table 6.3. MOST EQUIPMENT NOWADAYS USES A SUBSET OF THE RS-232 SIGNALS SUCH AS GIVEN HERE, AND EVEN THIS CAN BE REDUCED

Pin	*Signal*	*Action*
1	FG	Chassis earth (frame ground)
2	TD	Serial output from DTE to DCE
3	RD	Serial input from DCE to DTE
4	RTs	DTE ready to send data to DCE
5	CTs	DCE ready to accept data from DTE
6	DSR	DCE connected and ready
7	SG	Signal earth
20	DTR	DTE ready to send data

Table 6.4. THE STANDARD BAUD RATES OF RS-232. THE SLOWER RATES ARE SELDOM USED, APART FROM THE PRESTEL USE OF 75 BAUD FOR TRANSMISSION FROM A TERMINAL AND 1200 FOR RECEIVING. NO OTHER DATA COMMUNICATIONS SERVICES USE THIS SPLIT RATE

50	75	110	150	Slow rates, seldom used
300	600	1200	2400	Used for printers, modems
4800	9600	19200		Fast rate, used for VDU terminals

Now when a DTE is to be connected to a DCE, the pin 2 of the DTE is connected to pin 2 of the DCE and pin 3 of the DTE is connected to pin 3 of the DCE. The other pins of the DTE are also connected to their corresponding numbers on the DCE. When two DTE devices are connected to each other, however, some links must be crossed. The pin 2 on one DTE must be connected to the pin 3 on the other DTE, and similar cross connection may be needed on handshaking lines. This difference in cabling is indicated by the naming of cables as modem (DTE to DCE) or non-modem (DTE to DTE), and failure to get a serial link working is very often due to this very elementary difference.

A serial link can be operated either synchronously (a data bit sent at each clock pulse) or asynchronously (data sent when ready), and since practically all modern applications of RS-232 make use of asynchronous operation, the pin connections for synchronous use can be omitted. For asynchronous use, each transmitted byte has to be preceded by a start bit and ended by one or more stop bits. Ten or eleven bits must therefore be transmitted for each byte of data, and both transmitter and receiver must use the same number of stop bits. In addition, both transmitter and receiver must use the same baud rate, the rate of alternation of signal voltage. *Table 6.4* shows the RS-232 standard baud rates, of which 1200, 2400 and 9600 are the most common. The rates below 300 baud are hardly used other than by the painfully slow Prestel rate of 75 baud (for transmitting), and even 300 baud is becoming unusual.

The serially transmitted data will almost certainly use ASCII code (the older EBDIC code is almost obsolete), which will require only 7 of the 8 data bits that can be sent. Once again, this is not a hard-and-fast rule, because a serial link to a dot-matrix printer may require all eight data bits. If only 7-bit ASCII is needed, then the eighth bit can be used as a parity bit, a check on the integrity of the data. The parity system can be even or odd. In the even parity system, the number of logic 1s in the remainder of the byte is counted, and the parity bit made either 1 or 0 so that the total number of 1s is even. In the odd parity system, the parity bit will be adjusted so as to make the number of 1s an odd number. At the receiver, the parity can be checked and an error flagged if the parity is found to be incorrect. This simple scheme will detect a

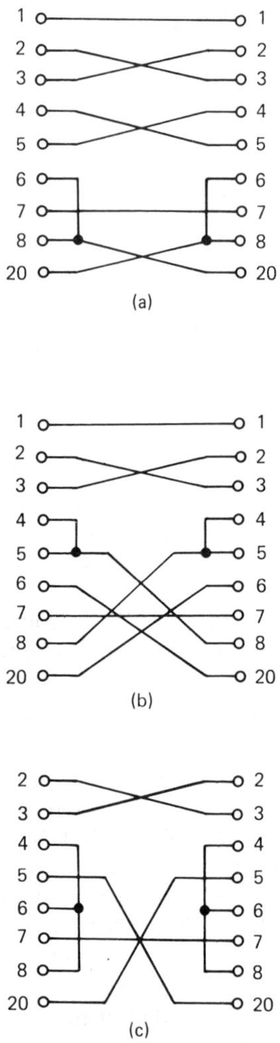
1 1
2 2
3 3
4 4
5 5
6 6
7 7
8 8
20 20
(a)
1 1
2 2
3 3
4 4
5 5
6 6
7 7
8 8
20 20
(b)
2 2
3 3
4 4
5 5
6 6
7 7
8 8
20 20
(c)

Figure 6.5. Three common ways of wiring cables or null-modem links. (c) is the recommended linking for the Amstrad PC

single-bit error in a byte, but cannot detect multiple errors nor correct errors. Methods such as cyclic redundancy checking (CRC) and Hamming or Reed-Solomon codes are needed to perform such correction and are outwith the scope of this book.

Given that the transmitter and the receiver are set up to the correct protocols, meaning that the same baud rate, number of stop bits and use of parity will be identical, we still need hardware methods of hand-

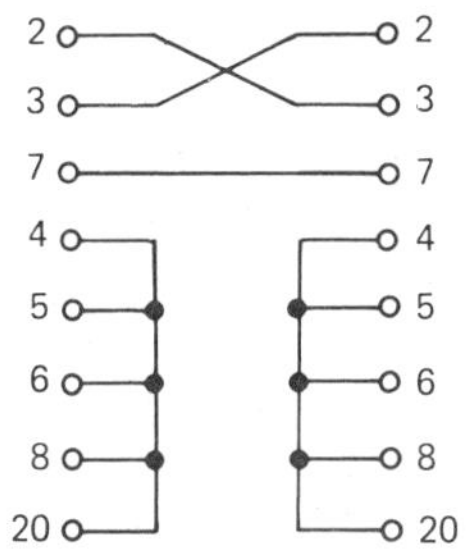
2
3
7
4
5
6
8
20
2
3
7
4
5
6
8
20

Figure 6.6. An 'auto loopback' connection which dispenses with handshaking. This is useful for testing, and can be used with the XON/XOFF form of software handshaking, or with very low speed transmissions

shaking to ensure that signals are transferred only when both transmitter and receiver are ready. The signals that are used for handshaking are referred to as RTS (ready to send), CTS (clear to send), DSR (data set ready) and DTR (data terminal ready). As usual, you are likely to find that not all four will be used, and only RTS and CTS are common to nearly all equipment.

When a DTE is connected to a DCE (computer to modem, for example), the RTS signal is sent out by the DTE to the DCE to indicate that the DTE has data to transmit. The DCE then responds with the CTS signal to indicate that data can be accepted, and data will be sent until the end of the data – note that handshaking is used only at the start and end of each block of data, not between bytes. The handshaking can be correctly implemented for connection of a DTE to DCE simply by using a cable that contains all the necessary strands (a 25-strand cable, for example). For the connection of two DTE devices to each other, a non-modem (or null-modem) cable can be used in which leads are crossed. *Figure 6.5* shows three common connections for such a cable, including the connections for the popular Amstrad PC 1512/PC 1640 machines. An alternative to keeping sets of modem and non-modem cables is the use of 'null-modem' connectors, back-to-back cable connectors which incorporate the reversed leads.

Some more recent developments are useful to note. The RS-423 connection is physically identical to RS-232 but makes use of TTL voltage levels and will tolerate 450 Ω impedance levels. There is a proposed S5/8 standard which makes use of the skeleton of RS-232 in a system that would, if adopted, clear out the present confusion over RS-232, standardise a DiN type of plug and socket for all equipment, and remove the DTE/DCE distinction.

Finally, if you are connecting a serial printer to the serial output of a computer, you can use an 'auto loop back' connection of the form shown in *Figure 6.6*. This makes no use of hardware handshaking and will usually work with no problems at the slower baud rates. If some form of handshaking is still needed, it can be implemented in software by using the XON/XOFF system. This uses the ASCII codes 17 and 19 between computer and printer. Data can be sent to the printer following the ASCII 17 code, and disabled following the ASCII 19. Since these codes are sent over the normal data lines, only the data lines and earth need be connected. The rate of data transfer is slower because of the time that is needed to send the XON/XOFF signals.

Appendix I

Standard metric wire table

Diameter (mm)	*Resistance (ohms per metre)*	*Current rating (mA)*
0.025	35.1	2.3
0.032	21.4	3.7
0.040	13.7	5.8
0.050	8.8	9.1
0.063	5.5	14.5
0.080	3.4	23.4
0.100	2.2	36.5
0.125	1.4	57.1
0.140	1.1	71.6
0.160	0.86	93.5
0.180	0.68	118.3
0.200	0.55	146.1
0.250	0.35	228.3
0.280	0.28	286.3
0.315	0.22	362.4
0.400	0.14	584.3
0.450	0.11	739.5
0.500	0.088	913.0
0.56	0.070	1.14 (amperes)
0.63	0.055	1.45
0 71	0.043	1.84
0.75	0.039	2.05
0.80	0.034	2.34
0.085	0.030	2.64
0.90	0.027	2.96
0 95	0.024	3.30
1.00	0.022	3.65

The values of resistance per metre and of current rating have been rounded off. Only the smaller gauges are tabulated, representing the range of wire gauges which might be used in constructing r.f. and a.f. transformers.

Appendix II

Bibliography

Some of the most useful reference books in electronics are either out of print or difficult to obtain. They have been included in the list below because they can often be found in libraries, or in second-hand shops.

Components: Understanding Electronic Components (Sinclair) Fountain Press, 1972.

Formulae and tables: Reference Data for Radio Engineers (ITT).

Audio & Radio: Radio Designers's Handbook (Langford-Smith) Iliffe, 4th edn, 1967. A wealth of data, though often on valve circuits. Despite the age of the book, now out of print, it is still the most useful source book for audio work.

Radio Amateurs Handbook (ARRL). A mine of information of transmitting and receiving circuits. An excellent British counterpart is available, but the US publication contains more varied circuits, because the US amateur is not so restricted in his operations.

GE Transistor Manual (General Electric of USA). Even the early editions are extremely useful.

Oscilloscopes: The Oscilloscope in Use (Sinclair) Argus Books, 1976.

Manufacturer's Databooks by Texas, RCA, SGS–ATES, Motorola, National Semiconductor and Mullard contain detailed information on semiconductors, with many applications circuits.

Microprocessors: The textbooks by Dr. Adam Osborne are by far the most useful for anyone who already has some knowledge of microprocessors. For beginners, the last chapter of *Beginner's Guide to Digital Electronics* is a useful introduction.

A useful book on home computers and programming is *Microprocessors for Hobbyists* (Coles) Newnes Technical Books, 1979.

Other, up-to-date, titles from Newnes Technical Books include the following:

Newnes Radio and Electronics Engineer's Pocket Book (16th Edition), 1986, by Keith Brindley.

The Practical Electronics Microprocessor Handbook, 1986, by Ray Coles.

16-Bit Microprocessor Handbook, 1986, by Trevor Raven.

Oscilloscopes: how to use them, how they work (2nd Edition), 1986, by Ian Hickman.

Op-Amps: their principles and applications (2nd Edition), 1986, by J. Brian Dance.

The Art of Micro Design, 1984, by A.A. Berk.

Practical Design of Digital Circuits, 1983, by Ian Kampel.

In addition there are *Beginner's Guides* on the following subjects: *Amateur Radio, Hi-Fi, Computers, Digital Electronics, Electronics, Microcomputing, Microprocessors, Radio, Television, Video,* and *Videocassette Recorders.*

For further information on Newnes Technical titles write to the publishers for a catalogue.

Index